Edited by
H. Schmidt-Traub
Preparative Chromatography

Also of Interest

J. Weiss
Handbook of Ion Chromatography
Third, Completely Revised and Enlarged Edition

2 Volumes
2004
ISBN 3-527-28701-9

J. S. Fritz, D. T. Gjerde
Ion Chromatography

Third, Completely Revised and Enlarged Edition
2000
ISBN 3-527-29914-9

D. T. Gjerde, C. P. Hanna, D. Hornby
DNA Chromatography

2001
ISBN 3-527-30244-1

K. S. Sutherland
Solid/Liquid Separation Equipment

2005
ISBN 3-527-29600-X

A. Rushton, A. S. Ward, R. G. Holdich
Solid-Liquid Filtration and Separation Technology
Second, Completely Revised Edition

2000
ISBN 3-527-29604-2

R. Bott, T. Langeloh (Eds.)
Solid/Liquid Separation Lexicon

2002
ISBN 3-527-30522-X

Preparative Chromatography

of Fine Chemicals and
Pharmaceutical Agents

Edited by
Henner Schmidt-Traub

WILEY-VCH

WILEY-VCH Verlag GmbH & Co. KGaA

Editor

Prof. Dr.-Ing. Henner Schmidt-Traub
Universität Dortmund
Fachbereich Bio- und Chemieingenieurwesen
Lehrstuhl für Anlagentechnik
Emil-Figge-Str. 70
44221 Dortmund
Germany

■ All books published by Wiley-VCH are carefully produced. Nevertheless, authors, editor, and publisher do not warrant the information contained in these books, including this book, to be free of errors. Readers are advised to keep in mind that statements, data, illustrations, procedural details or other items may inadvertently be inaccurate.

Library of Congress Card No.: Applied for

British Library Cataloging-in-Publication Data:
A catalogue record for this book is available from the British Library

Bibliographic information published by Die Deutsche Bibliothek
Die Deutsche Bibliothek lists this publication in the Deutsche Nationalbibliografie; detailed bibliographic data is available in the Internet at <http://dnb.ddb.de>.
© 2005 WILEY-VCH Verlag GmbH & Co. KGaA, Weinheim

Printed in the Federal Republic of Germany
Printed on acid-free paper

Composition Mitterweger & Partner Kommunikationsgesellschaft mbH, Plankstadt
Printing Strauss GmbH, Mörlenbach
Bookbinding Litges & Dopf Buchbinderei GmbH, Heppenheim

ISBN-13 978-3-527-30643-5
ISBN-10 3-527-30643-9

Introduction

Preface

Why another book on preparative chromatography, what is our aim?

During the last decade, intensive research and development have considerably improved the acceptance and industrial applicability of chromatographic processes. An important key for this success has been the cooperation of different disciplines, in particular chemists and engineers, who have quite different views on preparative chromatography. A chemist, for instance, would say: „Look at the chromatogram, it tells you how to separate a mixture if you also know the chemical structure of the components." The engineer may say: „First of all I have to model the process to understand how it works. I can then explain and optimize the separation of a mixture of components that always have the names A, B and C". Faced with these statements an experienced chromatographer recommended that „We should drink mobile phase at night and discuss how to proceed". Having followed this advice we, a group of chemists and engineers, decided to write this book.

A first, early milestone in this process can be traced back to a 1995 symposium organized by DECHEMA e. V., which brought scientists from both industry and universities together to discuss future research in the field of preparative chromatography. This was the nucleus for further cooperation. Intense information exchange, joint research projects and, especially, sharing our different views on problems of preparative chromatography were the keys to a successful interaction. Very soon we also decided to offer a joint course on „Design and operation of preparative and production scale chromatographic processes".

Part of this book goes back to this course, which was continuously complemented and updated through discussions with colleagues from various disciplines in academia and industry and by the concurrent development of modern chromatographic process design.

Another stimulating factor was the breakthrough in the development in process-scale chromatography around this time. During 1993 the first scaled-down versions of SMB units had been presented for use in pharmaceutical product recovery, focusing on enantiomer separation. This triggered discussion on how to develop generic approaches for quick method development and implementation, and led to the

Preparative Chromatography. Edited by H. Schmidt-Traub
Copyright © 2005 WILEY-VCH Verlag GmbH & Co. KGaA, Weinheim
ISBN: 3-527-30643-9

insight that a solution could only be found by strong interactive and interdisciplinary cooperation between chemists and engineers. The tremendous success of SMB technology within less than one decade documents better than anything else the validity of this approach. Some of this may be attributed to the fact that the right people were in the right place at the right time to integrate equipment, methodology and personnel.

The authors of the various chapters have different backgrounds and different expertise. Reinhard Ditz and Michael Schulte have been with Merck KGaA for more than 20 years. Their responsibilities are in R&D of the Life Science Product Division. Joachim Kinkel was also with Merck KGaA before joining the University of Applied Science of Nürnberg, where he is now Professor of Chemistry and Environmental Analytics. Jules Dingenen is from Johnson & Johnson in Belgium and is responsible for preparative separation techniques. Klaus K. Unger was a Professor of Inorganic and Analytical Chemistry at the Johannes Gutenberg-Universität, Mainz before his retirement in 2001. Cédric du Fresne von Hohenesche was a member of the same institute before joining the ISIS Research Institute headed by Professor Lehn in Strasburg in 2003.

The research group at the Universität Dortmund, called AG CHROM, started in the early 1990s. Jochen Strube, the first PhD student, prepared the ground for very successful research projects at the Chair of Plant and Process Design. The next generation began to evaluate our theoretical ideas and findings by experiments. Achim Epping focused on batch chromatography while Andreas Jupke concentrated on the SMB and Jörg Fricke worked on integrated reaction and chromatography. Their research work contributed greatly to this book. They also formulated the concept of this book, but are now in industry. Therefore, the following team put these plans into effect: Klaus Wekenborg has applied SMB theory to the gradient separation of proteins and managed to verify his results by good agreement with measurements. However, his colleague Arthur Susanto demonstrated, by confocal microscopy, that our basic understanding of protein mass transfer in ion-exchange beads has still to be improved. Thomas Borren has run and simulated chromatographic SMB reactors for variable flowsheets, e.g. for the Hashimoto reactors, and is, therefore, known as the „lord of the valves". His colleague Mirko Michel has investigated the secrets of electrochemical reactors and combines them with chromatographic separation as well. Wolfgang Wewers research has focused on process design for the production of pharmaceutical agents by multi-step integration of chemical synthesis with chromatographic separation. In addition to the authors, other PhD students have contributed to improvements in the design and better understanding of chromatographic processes and, therefore, have their indirect part in this book. Silke Jünemann confronted our theorists with problems of real-life chromatography, forcing them to stick to the facts. Markus Meurer started the modeling of chromatographic reactors while Alexander Wiesel prepared the ground for further research on the chromatographic separations of bio-materials. Sven Felske was always a sought-after partner for discussing theoretical questions. I acknowledge their strong commitment and enthusiasm for the research work of AG CHROM.

The second group at the Universität Dortmund involved in research on chromatography is at the Chair of Process Control headed by Sebastian Engell. For the last eight years this group has developed and tested innovative control and optimization strategies for chromatographic processes. A highlight of this work is given in Chapter 9. The two groups have cooperated very closely in several successful research projects, and discussions of the complementary approaches to simulation, optimization and design have been very fruitful for the progress of research in both groups.

We also thank our students for their enthusiasm and support of our chromatographic work, e.g. by their master's theses or laboratory work as student assistants. Constantin Frerick, Sabine Frintrop and Günter Uebe prepared and improved the figures and, to our surprise, never complained about repeated changes.

Returning to the introductory question, naturally we want to summarize and publish recent results, which have been financed by industrial and public funding in a larger context. However, it is even more important to share what we perceive to be the key for success in making new approaches work. In this case it is the importance of addressing preparative and process chromatographic issues from the viewpoint of both chemists and process engineers in order to improve our understanding and to transfer the necessary knowledge between our disciplines. We want to address colleagues from both industries and universities who might find valuable additional information. Students and others who want to begin in this business and are looking for detailed information about the design and operation of preparative chromatography are another important target group. One message for them is: Chromatography is not as difficult and expensive as it is often said and perceived to be. However, it is not the solution for all separation problems in life sciences. It is rather an important and efficient operation within the sequence of other separation steps of down-stream processing. Further information about the content, and on how to read this book, is given in Chapter 1.

Again I thank all the authors for their contribution, and last but not least we thank our families and friends for their patience during completion of this book.

Henner Schmidt-Traub   Dortmund, December 2004

List of Contributors

Editor

Prof. Dr.-Ing. Henner Schmidt-Traub
University of Dortmund
Department of Biochemical and
Chemical Engineering
Chair of Plant and Process Design
44221 Dortmund
Germany

Authors

Dipl.-Ing. Thomas Borren
University of Dortmund
Department of Biochemical and
Chemical Engineering
Chair of Plant and Process Design
44221 Dortmund
Germany

Dr. Jules Dingenen
Johnson & Johnson
Pharmaceutical Research &
Development
Turnhoutseweg 30
2340 Beerse
Belgium

Dr. Reinhard Ditz
Merck KGaA
Life Science & Analytics
R & D/New Technologies
Frankfurter Str. 250
64293 Darmstadt
Germany

Dr. Cedric du Fresne von Hohenesche
Institut de Science et d'Ingénierie
Supramoléculaires
ISIS – BASF Laboratory
8, allée Gaspard Monge
67083 Strasbourg
France

Preparative Chromatography. Edited by H. Schmidt-Traub
Copyright © 2005 WILEY-VCH Verlag GmbH & Co. KGaA, Weinheim
ISBN: 3-527-30643-9

Prof. Dr.-Ing. Sebastian Engell
University of Dortmund
Department of Biochemical and
Chemical Engineering
Chair of Process Control
44221 Dortmund
Germany

Dr.-Ing. Achim Epping
BP Köln GmbH
Betriebstechnik
Postfach 75 02 12
50754 Köln
Germany

Dr.-Ing. Jörg Fricke
Bayer CropScience AG
Alfred-Nobel-Straße 50
40789 Monheim
Germany

Dr.-Ing. Andreas Jupke
Bayer Technology Services GmbH
Process Design
51386 Leverkusen
Germany

Prof. Dr. Joachim Kinkel
Georg Simon Ohm Fachhochschule
Nürnberg
Department of Applied Chemistry
90489 Nürnberg
Germany

Dipl.-Ing. Mirko Michel
University of Dortmund
Department of Biochemical and
Chemical Engineering
Chair of Plant and Process Design
44221 Dortmund
Germany

Prof. Dr.-Ing. Henner Schmidt-Traub
University of Dortmund
Department of Biochemical and
Chemical Engineering
Chair of Plant and Process Design
44221 Dortmund
Germany

Dr. Michael Schulte
Merck KGaA
Life Science & Analytics R & D
Frankfurter Str. 250
64293 Darmstadt
Germany

Dipl.-Ing. Arthur Susanto
University of Dortmund
Department of Biochemical and
Chemical Engineering
Chair of Plant and Process Design
44221 Dortmund
Germany

Dr.-Ing. Abdelaziz Tuomi
Bayer Technology Services GmbH
Process Management Technology
Advanced Manufacturing Solutions –
APC
51386 Leverkusen
Germany

Prof. Dr. Klaus K. Unger
Am Alten Berg 40
64342 Seeheim
Germany

Dipl.-Ing. Klaus Wekenborg
University of Dortmund
Department of Biochemical and
Chemical Engineering
Chair of Plant and Process Design
44221 Dortmund
Germany

Dipl.-Ing. Wolfgang Wewers
University of Dortmund
Department of Biochemical and
Chemical Engineering
Chair of Plant and Process Design
44221 Dortmund
Germany

Notation

Symbols

Symbol	Description	Units
a_i	Coefficient of the Langmuir isotherm	$cm^3\,g^{-1}$
a_s	Specific surface area	$cm^2\,g^{-1}$
A	Area	cm^2
A_c	Cross section of the column	cm^2
A_i	Coefficient in the Van Deemter equation	cm
A_s	Surface area of the adsorbent	cm^2
ASP	Cross section specific productivity	$g\,cm^{-2}\,s^{-1}$
b_i	Coefficient of the Langmuir isotherm	$cm^3\,g^{-1}$
B_i	Coefficient in the Van Deemter equation	s
Bo	Bodenstein-number	–
c_i	Concentration in the mobile phase	$g\,cm^{-3}$
$c_{p,i}$	Concentration of the solute inside the particle pores	$g\,cm^{-3}$
C	Annual costs	€
C_i	Coefficient in the Van Deemter equation	$cm^2\,s^{-1}$
$C_{DL,i}$	Dimensionless concentration in the liquid phase	–
$C_{p,DL,i}$	Dimensionless concentration of the solute inside the particle pores	–
C_{spec}	Specific costs	$€\,g^{-1}$
d_c	Diameter of the column	cm
d_p	Average diameter of the particle	cm
d_{pore}	Average diameter of the pores	cm
D_{an}	Angular dispersion coefficient	$cm^2\,s^{-1}$
$D_{app,i}$	Apparent dispersion coefficient	$cm^2\,s^{-1}$
$D_{app,pore}$	Apparent dispersion coefficient inside the pores	$cm^2\,s^{-1}$
D_{ax}	Axial dispersion coefficient	$cm^2\,s^{-1}$
D_m	Molecular diffusion coefficient	$cm^2\,s^{-1}$
$D_{pore,i}$	Diffusion coefficient inside the pores	$cm^2\,s^{-1}$

Preparative Chromatography. Edited by H. Schmidt-Traub
Copyright © 2005 WILEY-VCH Verlag GmbH & Co. KGaA, Weinheim
ISBN: 3-527-30643-9

Symbol	Description	Units
$D_{solid,i}$	Diffusion coefficient on the particle surface	$cm^2\,s^{-1}$
Da	Damkoehler number	–
EC	Eluent consumption	$cm^3\,g^{-1}$
F	Prices	$€\,l^{-1}$, $€\,g^{-1}$
f_i	Fugacity	–
h	Reduced plate height	–
hR_f	Retardation factor	–
Δh_{vap}	Heat of vaporisation	$kJ\,mol^{-1}$
H_i	Henry coefficient	–
H_p	Prediction horizon	–
H_r	Control horizon	–
HETP	Height of an equivalent theoretical plate	cm
$k_{ads,i}$	Adsorption rate constant	$cm^3\,g^{-1}\,s^{-1}$
$k_{des,i}$	Desorption rate constant	$cm^3\,g^{-1}\,s^{-1}$
$k_{eff,i}$	Effective mass transfer coefficient	$cm^2\,s^{-1}$
K_{eq}	Equilibrium constant	misc.
K_{EQ}	Dimensionless equilibrium coefficient	–
$k_{film,i}$	Boundary or film mass transfer coefficient	$cm\,s^{-1}$
k'_i	Retention factor	–
$\bar{k}'_i$	Modified retention factor	–
k_{reac}	Rate constant	misc.
LF	Loading factor	–
L_c	Length of the column	cm
$\dot{m}_i$	Mass flow	$g\,s^{-1}$
m_i	Mass	g
m_j	Dimensionless mass flow rate in section j	–
m_s	total mass	g
n_i	Molar cross section of component i	–
n_T	Pore connectivity	–
N	Column efficiency, number of plates	–
N_{col}	Number of columns	–
N_{comp}	Number of components	–
N_p	Number of particles per volume element	–
Δp	Pressure drop	Pa
Pe	Péclet number	–
Pr_i	Productivity	$g\,cm^3\,h^{-1}$
P_s	Selectivity point	–
Pu_i	Purity	%
q_i	Solid load	$g\,cm^{-3}$
q_i^*	Total load	$g\,cm^{-3}$
$\bar{q}_i^*$	Averaged particle load	$g\,cm^{-3}$
$q_{sat,i}$	Saturation capacity of the stationary phase	$g\,cm^{-3}$

Symbol	Description	Units
$Q_{DL,i}$	Dimensionless concentration in the stationary phase	–
r	Radial coordinate	cm
r_i	Reaction rate	misc.
r_p	Particle radius	cm
R_f	Retardation factor	–
R_i	Regulation term	–
R_s	Resolution	–
Re	Reynolds number	–
S_{BET}	Specific surface area	$m^2 g^{-1}$
Sc	Schmidt number	–
Sh	Sherwood number	–
St	Stanton number	–
t	Time	s
t_0	Dead time of the column (for total liquid hold-up)	s
$t_{0,int}$	Dead time of the column (for interstitial liquid hold-up)	s
t_{cycle}	Cycle time	s
t_g	Gradient time	s
t_{inj}	Injection time	s
t_{life}	Lifetime of adsorbent	h
t_{plant}	Dead time of the plant without column	s
$t_{R,i}$	Retention time	s
$t_{R,i,net}$	Net retention time	s
t_{shift}	Switching time of the SMB plant	s
t_{total}	Total dead time	s
T	Temperature	K
T	Degree of peak asymmetry	–
u_0	Velocity in the empty column	$cm\,s^{-1}$
u_{int}	Interstitial velocity in the packed column	$cm\,s^{-1}$
u_m	Effective velocity (total mobile phase)	$cm\,s^{-1}$
v_{sp}	Specific pore volume	$cm^3 g^{-1}$
V	Volume	cm^3
$\dot{V}$	Volume flow	$cm^3 s^{-1}$
V_{ads}	Volume of the stationary phase within a column	cm^3
V_c	Total volume of a packed column	cm^3
V_i	Molar volume	$cm^3 mol^{-1}$
V_{int}	Interstitial volume	cm^3
V_m	Overall fluid volume	cm^3
V_{pore}	Volume of the pore system	cm^3
V_{solid}	Volume of the solid material	cm^3
VSP	Volume specific productivity	$g\,cm^3 s^{-1}$
w_i	Velocity of propagation	$cm\,s^{-1}$
x	Coordinate	cm

Symbol	Description	Units
x_i	State of the plant	–
X_i	Mole fraction	–
X	Conversion	%
X_{cat}	Fraction of catalyst of the fixed bed	–
Y_i	Yield	%
Z	Dimensionless distance	–

Greek symbols

α	Selectivity	–
α_{exp}	Ligand density	$\mu\text{mol m}^{-2}$
β	Modified dimensionless mass flow rate	–
γ	Obstruction factor for diffusion or external tortuosity	–
Γ	Objective function	–
ε	Void fraction	–
ε^0	Solvent strength parameter	–
ε_p	Porosity of the solid phase	–
ε_t	Total column porosity	–
η	Dynamic viscosity	mPa s
Θ	Angle of rotation	°
λ	Irregularity in the packing	–
μ_t	First absolute moment	–
ν	Kinematic viscosity	$\text{cm}^2\text{ s}$
ν_i	Stoichiometric coefficient	–
π	Spreading pressure	Pa
ϱ	Density	g cm^{-3}
σ_t	Standard deviation	–
τ	Dimensionless time	–
ϕ	Bed voidage	–
ψ	Friction number	–
ψ_{reac}	Net adsorption rate	$\text{g cm}^{-3}\text{ s}^{-1}$
ω_j	Coefficient in the triangle theory	–
ω	Rotation velocity	$°\text{ s}^{-1}$

Subscripts

1, 2	Component 1/component 2
I, II, III, IV	Section of the SMB or TMB process
acc	Accumulation
ads	Adsorbent
c	Column
cat	Catalyst
conv	Convection

Symbol	Description
crude	Crude loss
des	Desorbent
diff	Diffusion
disp	Dispersion
DL	Dimensionless
eff	Effective
el	Eluent
exp	Experimental
ext	Extract
feed	Feed
het	Heterogeneous
hom	Homogeneous
i	Component i
in	Inlet
inj	Injection
j	Section j of the TMB or SMB process
l	Liquid
lin	Linear
max	Maximum
min	Minimum
mob	Mobile phase
mt	Mass transfer
opt	Optimal
out	Outlet
p	Particle
pore	Pore
pipe	Pipe within HPLC plant
plant	Plant
prod	Product
raf	Raffinate
reac	Reaction
rec	Recycle
sat	Saturation
sec	Section
shock	Shock front
SMB	Simulated moving bed process
solid	Solid adsorbent
spec	Specific
stat	Stationary phase
tank	Tank within HPLC plant
theo	Theoretical
TMB	True moving bed process

Definition of dimensionless groups

Bodenstein number $\quad \mathrm{Bo} = \dfrac{u_{\mathrm{int}} L_{\mathrm{c}}}{D_{\mathrm{ax}}}$ $\qquad$ Convection to dispersion (column)

Damkoehler number $\quad \mathrm{Da}_{\mathrm{solid}} = \dfrac{r_{\mathrm{feed}} \tau_{\mathrm{solid}}}{c_{\mathrm{feed}}}$ $\qquad$ Residence time solid to characteristic time of reaction

Péclet number $\quad \mathrm{Pe} = \dfrac{u_{\mathrm{int}} d_{\mathrm{p}}}{D_{\mathrm{ax}}}$ $\qquad$ Convection to dispersion (particle)

Reynolds number $\quad \mathrm{Re} = \dfrac{u_{\mathrm{int}} d_{\mathrm{p}} \rho_1}{\eta_1}$ $\qquad$ Inertial force to viscous force

Schmidt number $\quad \mathrm{Sc} = \dfrac{\eta_1}{\rho_1 D_{\mathrm{m}}}$ $\qquad$ Kinetic viscosity to diffusivity

Sherwood number $\quad \mathrm{Sh} = \dfrac{k_{\mathrm{film}} d_{\mathrm{p}}}{D_{\mathrm{m}}}$ $\qquad$ Mass diffusivity to molecular diffusivity

Stanton number (modified) $\quad \mathrm{St}_{\mathrm{eff},i} = k_{\mathrm{eff},i} \dfrac{6}{d_{\mathrm{c}}} \dfrac{L_{\mathrm{c}}}{u_{\mathrm{int}}}$ $\quad$ Mass transfer to convection

List of Abbreviations

ARX	Autoregressive exogenous
BET	Brunnauer–Emmet–Teller
BJH	Barrett–Joyner–Halenda
BR	Chromatographic batch reactor
CACR	Continuous annular chromatographic reactor
CD	Circular dichroism (detectors)
CEC	Capillary electro chromatography
CFD	Computational fluid dynamics
cGMP	Current good manufacturing practice
CIP	Cleaning in place
CLC	Column liquid chromatography
CLRC	Closed-loop recycling chromatography
CSEP®	Chromatographic separation
CSF	Circle suspension flow
CSP	Chiral stationary phase
CTA	Cellulose triacetate
CTB	Cellulose tribenzoate
DAC	Dynamic axially compressed
DAD	Diode array detector
DMF	Dimethyl formamide
DMSO	Dimethyl sulfoxide
DTA	Differential thermal analysis
DVB	Divinylbenzene
EC	Elution consumption
ECP	Elution by characteristic points
EDM	Equilibrium dispersive model
ELSC	Evaporating light scattering
EMG	Exponential modified Gauss (function)
FACP	Frontal analysis by characteristic points
FDM	Finite difference methods

Preparative Chromatography. Edited by H. Schmidt-Traub
Copyright © 2005 WILEY-VCH Verlag GmbH & Co. KGaA, Weinheim
ISBN: 3-527-30643-9

GC	Gas chromatography
GMP	Good manufacturing practice
GRM	General rate model
HETP	Height of an equivalent theoretical plate
HFCS	High fructose corn syrup
H-NMR	Hydrogen nuclear magnetic resonance (spectroscopy)
HPLC	High performance liquid chromatography
IAST	Ideal adsorbed solution theory
IEX	Ion exchange
IR	Infrared
ISEP®	Ion exchange separation
ISMB	Improved simulated moving bed
LC	Liquid chromatography
LOD	Limit of detection
LOQ	Limit of quantification
MPC	Model predictive control
MS	Mass spectroscopy
MW	Molecular weight
NMPC	Nonlinear model predictive control
NMR	Nuclear magnetic resonance (spectroscopy)
NN	Neural network
NP	Normal phase
NPLC	Normal phase liquid chromatography
OC	Orthogonal collocation
OCFE	Orthogonal collocation on finite elements
ODE	Ordinary differential equation
PDE	Partial differential equation
PEEK	Poly(ether ether ketone)
PES	Poly(ethoxy)siloxane
PLC	Programmable logic controller
PSD	Particle size distribution
R&D	Research and Development
RI	Refractive index
RMPC	Repetitive model predictive control
RP	Reversed phase
S/N	Signal-to-noise ratio
SEC	Size exclusion chromatography
SEM	Scanning electron microscopy
SFC	Supercritical fluid chromatography
SIP	Sanitization in place
SMB	Simulated moving bed process
SMBR	Simulated moving bed reactor
SOP	Standard operation procedure
SQP	Sequential quadratic programming
SSRC	Steady-state recycling chromatography

St-DVB	Styrene-divinylbenzene
TDM	Transport dispersive model
TEM	Trans electron microscopy
TEOS	Tetraethoxysilane
TFA	Trifluoroacetic acid
TG/DTA	Thermogravimetric/differential thermal analysis
THF	Tetrahydrofuran
TLC	Thin-layer chromatography
TMB	True moving bed process
TMBR	True moving bed reactor
USP	United States pharmacopoeia
UV	Ultraviolet
VSP	Volume-specific productivity

1
Introduction

R. Ditz

1.1
Liquid Chromatography – its History

For obvious reasons the 'official' birth date of chromatography is linked with the first use of the name 'Chromatography' (writing with colors) by M.S. Twsett in his article on chlorophyll substances published in 1903.

However, the existence and utilization of adsorptive methods for substance purification was practiced well before that time. Applications of liquid phase separations started as early as the turn of the 19th century, that is if one does not acknowledge wood as the first described adsorption medium (Moses 2, 15 (23–25))

Interesting or naturally enough the early history of liquid chromatography has been all preparative. Also, a significant amount of separation work was performed in an engineering environment long before the fundamental work of Michael Twsett, and this is a kind of parallelism, which has continued until today. For almost a century, separation methodology therefore has been developed and practiced in the two different fields without too much interaction and interfacing. In today's world, where as fast a transfer as possible from basic research to practical application has become mandatory, the integration of chemical research and process engineering is key to success. This is one of the major driving forces for producing this book, in order to describe the approaches to and the aspects of chromatography from the side of the chemist as well as from the side of the engineer.

Milestones in the development of Chromatography are linked to the work of many outstanding scientists, such as Ramsay, Langmuir, Berl and Schmidt, Kuhn, Martin and Synge, Cremer and others. It would expand the scope of this contribution to highlight the historical merits of these scientists and the reader is referred to special references (Wintermeier, s. Unger, 1990 and Ettre, 1996).

Also, as this book deals with preparative and process scale aspects and applications of chromatography, the vast efforts made in the analytical fields cannot be covered here.

After the first, and in many ways pioneering, work of Michael Twsett, further development of Column Liquid Chromatography (CLC) was hampered because his

Preparative Chromatography. Edited by H. Schmidt-Traub
Copyright © 2005 WILEY-VCH Verlag GmbH & Co. KGaA, Weinheim
ISBN: 3-527-30643-9

work had only limited public exposure as it was not published in English or, at least, German. Therefore, it took more than a decade, before Kuhn and Lederer in Heidelberg applied Twsett's approach to carotinoid separation in the late 1920s, which was followed by another dormant period until the late 1940s (Fig. 1.1).

Towards the end of the Second World War basic and directive studies on Gas Chromatography (GC) were performed. GC became a powerful separation method in the analysis of hydrocarbon mixtures obtained from petroleum fractions. It was followed by Thin Layer Chromatography (TLC) in the mid-1950s.

Between 1950 and 1960 Size Exclusion Chromatography (SEC) became a popular technique in two branches: the fractionation of synthetic polymers, described as gel permeation chromatography, and in the resolution of biopolymers, termed as gelfiltration. The former was performed on cross-linked porous synthetic polymers, the latter on cross-linked polysaccharides (Sephadex).

The real advancement of High-Performance Liquid Chromatography (HPLC) began during the late 1960s when small porous particles became available. It took several years before spherical porous particles were accepted over irregular chips. However, before porous particles took the floor porous layer beads with a solid core of 40 μm and a thin porous layer of 1–2 μm were the adsorbents of choice. The serious limitations of such packings were quickly recognized. The breakthrough of HPLC as a routine technique occurred in the mid-1970s with commercial availability of reversed phase silica packings that carried bonded n-octadecyl and n-octyl groups at the surface. One of the latest developments takes advantage of the thermodynamic properties of super-critical fluids and modified HPLC technology – so-called Supercritical Fluid Chromatography (SFC).

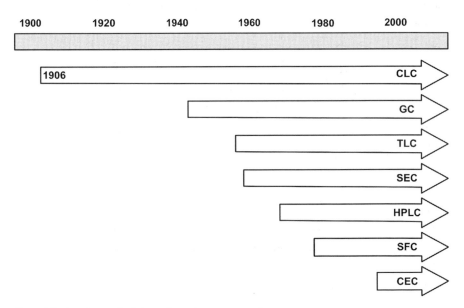

Figure 1.1 Development of chromatography over one century.

Finally, as a hybrid between pressure-driven chromatography and capillary electrophoresis, Capillary Electrochromatography (CEC) has emerged over the last decade.

In retrospect, the development of chromatography was not straightforward, following circles and dead ends before real breakthroughs.

The rapid development of HPLC columns for example was due to three major technical achievements: the ability to manufacture micro-particulate silicas, the invention of air elutriation technique as sizing technology and progress in the slurry technique for packing HPLC columns. However, it took more than ten years to build up sufficient know-how to produce stable, robust and reproducible HPLC columns that satisfied the continuously increasing demands of the chemical and pharmaceutical industry (Fig. 1.2).

The advancement of chromatographic separation techniques during the last century is closely linked to the synthesis of novel products and the isolation and purification of natural substances. The production needs for high-value compounds employed as pharmaceuticals, agrochemicals, food additives etc. put high demands on the purity of these products. The separation of racemates into pure enantiomers is one of the striking examples that highlight the importance and achievements of chromatography.

However, the 'Twsett' period of chromatography has also gone through a bifurcation during its maturization process. Originally conceived by Michael Twsett as a

Moses 2, 15 (23-25)	removal of bitter taste from waters of mara by addition of specific wood
turn of 19th century	adsorptive capacity of carbon for purification of beet juices
appr. 1850	*Runge's* capillary work with coloured chemicals on paper
1870's	ion exchange studies by *Eichhorn* and *Boecker*
1886	use of natural and synthetic ion exchangers in sugar production patented
1903	*Twsett* – chromatographic separation of plant pigments explained by adsorptive effects
1904	*Wislicenus* requests defined materials for adsorptive purposes
1930	*Lederer, Kuhn* – separation of carotin and zeaxanthin
1934	standardised aluminium oxides according to *Brockmann*
1941	*James* and *Martin* – gas liquid chromatography – trigger for development of chromatographic principles at analytical and preparative levels
since 1970's	liquid chromatography plays an ever increasing role
1981	1st process scale HPLC system (Kiloprep)
1986	1st preparative chromatography symposium in Paris
1993	1st scaled down SMB units for pharmaceutical applications
1996	1st example of large scale chiral purification process (*UCB*)
2000	1st 800 mm inner diameter SMB unit for contract purification (*Aerojet*)
2001	advanced SMB-applications (Multi-Component, VariCol, …)

Figure 1.2 Historic dates in preparative column liquid chromatography.

preparative method for the isolation and identification of plant pigments, the method still retained its preparative thrust during the reinvention in the 1930s by the Heidelberg research group of Kuhn, Lederle and Brockmann. Then, during the second rebirth in the 1960s as the extension of gas chromatography, it became reoriented for quite some time to serve as an analytical tool, driven not least by the instrument manufacturers. Preparative applications of liquid chromatography remained an exotic aspect for quite some time, served only by a minority of suppliers and estimated to be employed by a few hundred users throughout the world.

During the 'historical' chromatographic period, separations were developed mostly on a trial-and-error experimental basis due to lack of understanding of the underlying principles and the inability to address the complex interactions in fluid-phase systems by a computing approach.

Therefore, the 'chemical' scale-up approach was based on

1. Finding separation conditions for the mixture
2. 'Optimizing load' until elution profiles start to overlap (touching bands)
3. Go through a 'linear scale up' protocol by increasing the column geometry to deliver the required amount and purity of material while maintaining load and separation conditions.

While this protocol produced the 'predicted' product, its process economics were usually close to disastrous, allowing only extremely expensive materials to be processed. Therefore, preparative HPLC was long considered unsuitable for large-scale operation.

Both the 'understanding' as well as the 'computing' aspect have been significantly advanced during recent years by

- Applying more and more already well-known engineering concepts and approaches to chromatography and
- The availability of high-speed computing power at reasonable cost, allowing fast processing of complex data on-line to enable monitoring and control of chromatographic systems even under complex operating conditions.

1.2
Focus of the Book

The present book focuses on preparative chromatography, and is distinct from earlier approaches in that it will describe and develop access to chromatographic purification concepts through the eyes of both engineers and chemists. This is because, to develop a method that can be scaled up to a process environment, the earliest possible interaction and cooperation between chemist and engineer is required to achieve time and cost-effective solutions.

The differentiation between preparative and analytical chromatography has often been a point of heated discussion in the chromatographic community, especially over the last 30 or so years. Quite often, this differentiation has been based on size (column size, particle size of the packing or sample size). Unfortunately, this has often led to less than appropriate separation strategies and modes of operation (Fig. 1.3).

The difference between „analytical" and „preparative" work is not defined by the size of either sample or equipment. It is exclusively determined by the „GOAL" of the separation process. If „INFORMATION" is the goal of the separation, it is „analytical" chromatography. If the COLLECTION OF PRODUCTS is the intention, it is a „preparative" separation. This implies that „Sample Preparation" is always a „preparative" method. In an „analytical" mode, the sample can be processed, handled and modified in any way suitable to generate the required information, including degradation, labeling or otherwise changing the nature of the compounds under investigation, as long as a correct result can be documented. In a „preparative" mode, the sample has to be recovered in the exact condition that it was in before undergoing the separation, i.e. no degrading elution conditions, etc. This determines the whole separation strategy far more than any consideration of the size of the process or dimensions of columns ever would.

By this definition, one of the smallest scale operations could, and maybe should, be included in this context, which is the isolation, and subsequent characterization or identification of so-called low-abundant peptides in modern proteomics research.

At first glance, this is considered as nanoscale analytics, challenging the separation and detection capabilities of the most sensitive equipments and columns available

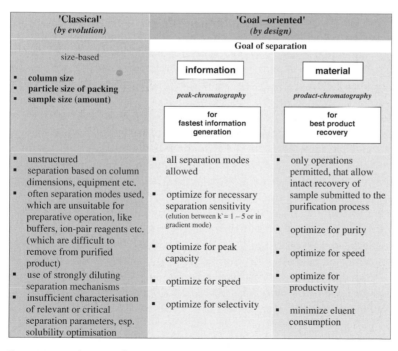

Figure 1.3 Development of a separation strategy.

today. At a second look, the boundary conditions for such isolations and characterizations under reproducible conditions put the same requirements on method development, sample handling and chromatographic packing materials as process-scale separations do.

1.3
How to Read this Book

It is not necessary to read all chapters of this book in sequence. For some readers the book may be a reference to answer specific questions. However, all readers should familiarize themselves with the content of Chapters 2–4 and 7 before they start solving chromatographic separations. Chapter 5 gives an overview of different chromatographic processes while Chapter 6 provides a detailed introduction to modeling and parameter estimation. These chapters, as well as Chapter 8 on chromatographic reactors and Chapter 9 on process control, should be read as the need arises. The book may not provide answers to all questions. In which case, the reader can obtain further information from the cited literature.

Brief contents of the individual chapters are: Chapter 2 presents the basic principles of chromatography and defines the most important parameters such as retention, retention factor, selectivity and resolution. It also explains the main model parameters, such as dispersion, porosity or void fraction, as well as different kinds of isotherm equations. Other topics are plate number, HETP values as well as their determination based on the first and second moments. Finally, the effects of linear and nonlinear chromatography, and the different kinds of column loading, are discussed. The experienced reader may pass quickly through this chapter to become familiar with definitions used. For beginners this chapter is a must in order to learn the general terminology and acquire a basic understanding. A further goal of Chapter 2 is the harmonization of general chromatographic terms between engineers and chemists, avoiding „slang terms", which are quite common in the literature.

Chapter 3 focuses on all aspects related to the chromatographic column. A major part explains and specifies the kinds of stationary phases as well as their chemical structure and properties. This part may be used as reference for special questions and will help those looking for an overview of attributes of different stationary phases. Packing procedures, as well as the design of the inlet and outlet heads, are very important for column efficiency. When deciding on the dimensions of a column the pressure drop also has to be taken into account.

Chapter 4 deals with the selection of a chromatographic system, i.e. the optimal combination of stationary phase, eluent or mobile phase for a given separation task. These key issues focus not only on solely scientific questions, but also take into account economy, speed, time pressure, hardware requirements, automation and legal aspects towards documentation, safety and others. Obviously, such rules of thumb may not cover all possible scenarios, but they may be useful in avoiding pitfalls.

The selection of chromatographic systems is critical for process productivity and thus process economy. On one hand, the selection of the chromatographic system offers the biggest potential for optimization but, on the other hand, it is a potential source of severe errors in developing separation processes.

Chapter 5 focuses on process concepts. The basis of every preparative chromatographic separation is the proper choice of the chromatographic system, as described in the previous chapters. Its implementation in a preparative process concept plays an important role in serving the different needs of substance production in terms of system flexibility and production scale. Depending on the operating mode, several features distinguish chromatographic process concepts:

- Batch-wise or continuous feed introduction
- Operation in single- or multi-column mode
- Elution under isocratic or gradient conditions
- Co-, cross- or counter-current flow of mobile and stationary phase
- Withdrawal of two or a multitude of fractions

Starting with the description of the main components of a chromatographic unit, Chapter 5 gives an overview of process concepts available for preparative chromatography.

In Chapter 6, modeling and determination of model parameters are key aspects. „Virtual experiments" by numerical simulations can considerably reduce the time and amount of sample needed for process analysis and optimization. To reach this aim, accurate models and precise model parameters for chromatographic columns are needed. Validated models can be used predictively for optimal plant design. Other possible fields of application for process simulation include process understanding for research purposes as well as training of personnel. This includes the discussion of different models for the column and plant peripherals. After a short explanation of numerical solution methods, a major part is devoted to the consistent determination of the parameters for a suitable model, especially those for the isotherm. Methods of different complexity and experimental effort are presented that allow a variation of the desired accuracy, on the one hand, and the time needed on the other hand. It is shown that an appropriate model can simulate experimental data within the accuracy of measurement, which permits its use for further process design as approached in Chapter 7. The necessity for optimization of chromatographic processes in the pharmaceutical and biotechnological industries results from high separation costs, which represent a major part of the manufacturing cost. However, due to the multitude of parameters and the complex dynamic behavior, pure empirical design and optimization of chromatographic processes are hardly possible. Mathematical modeling of the process is an essential requirement for this purpose.

Chapter 7, therefore, deals with model-based design and optimization of a chromatographic plant, where the already selected chromatographic system and concepts are applied. First, basic principles of the optimization of chromatographic processes will be explained. These include the introduction of the commonly used objective functions and the degrees of freedom. To reduce the complexity of the optimization and to ease the scale-up of a plant, this chapter will also emphasize the application of dimensionless parameters and degrees of freedom respectively. Examples for the scale-up of an optimized plant are also given. Subsequently, methods for the model-based design and optimization of batch and SMB (Simulated Moving Bed) processes are introduced. For this purpose dynamic simulations of an experimentally validated model are used. Especially in the design and optimization of the complex SMB processes, a quick design and optimization method for SMB will be introduced. Finally, the performance of batch and SMB processes are compared.

Chapter 8 introduces chromatographic reactors that combine the chromatographic separation, discussed in the earlier chapters of this book, and a chemical reaction. The elementary reaction type A $\rightleftarrows$ B + C is used to explain the operating principle of the different chromatographic reactors. An overview over the different reactor types as well as the investigated types of reactions is given, and their influence on process design and operation is discussed. Rigorous modeling is, so far, the only tool available to describe the behavior of chromatographic reactors. Therefore, the modeling approach presented in Chapter 6 is extended to chromatographic reactors. Finally, the design of chromatographic reactors is discussed, using the examples of the esterification of β-phenethyl alcohol with acetic acid as well as the isomerization of glucose.

The benefits of model-based control strategies for the operation of SMB processes are demonstrated in Chapter 9. This is a rather new concept as, in today's industrial practice, SMB processes are still „controlled" manually, based on the experience of the operators. A nonlinear model predictive (NMP) controller is described that can deal with the complex hybrid (continuous/discrete) dynamics of the SMB plant and takes hard process constraints (e.g. the maximal allowable pressure drop) and the purity requirements into account. The NMP controller employs a rigorous process model, the parameters of which are re-estimated online during plant operation, thus changes or drifting of the process parameters can be detected and compensated. The efficiency of this novel control concept is proven by an experimental study.

2
Fundamentals and General Terminology

M. Schulte and A. Epping

This chapter introduces fundamental aspects and basic equations for the characterization of chromatographic separations. Starting from the simple description of an analytical separation of different compounds the influences of fluid dynamics, mass transfer and thermodynamics are explained in detail. The important separation characteristics for preparative and process chromatography, e.g. the optimization of resolution and productivity as well as the differences compared with chromatography for analytical purposes, are described. Especially, the importance of understanding the behavior of substances in the nonlinear range of the adsorption isotherm is highlighted.

One further goal of this chapter is the harmonization of general chromatographic terms between engineers and chemists, avoiding „slang terms" that are quite common in the literature and daily use.

2.1
Principles of Adsorption Chromatography

Chromatography is part of the thermal separation processes used to separate homogeneous molecular dispersive mixtures. The separation itself can be divided into three steps (Sattler, 1995).

(1) In addition to the initially homogeneous mixture phase a second phase is generated by introduction of energy (e.g. during distillation) or an additive (e.g. during extraction or adsorption).

(2) Between the single phases, molecules and, in most cases, thermal energy are exchanged. The driving force for this molecular and heat transport is a thermodynamic imbalance between the two phases.

(3) After completion of the exchange procedures, the two phases, which now exhibit different concentrations to the starting situation, are separated. Combined with the phase separation is a partial separation of the homogeneous mixture.

In chromatography the homogeneous mixture phase is, in most cases, a fluid phase with dissolved substances. The additional second phase is a solid or a second

Preparative Chromatography. Edited by H. Schmidt-Traub
Copyright © 2005 WILEY-VCH Verlag GmbH & Co. KGaA, Weinheim
ISBN: 3-527-30643-9

immiscible fluid. The driving force between mixture phase and additional phase is the imbalanced adsorption between fluid and solid phase. The mixture is separated by the relative movement of the two phases. Usually, the solid phase is fixed as the so-called stationary phase. The fluid, which moves and is, therefore, responsible for the separation, is called the mobile phase. Chromatographic behavior is determined by the interaction of all single components in the mobile and stationary phases. The mixture of substances to be separated (solute), the solvent, which is used for their dissolution and transport (eluent), and the adsorbent (stationary phase) are summarized as the chromatographic system (Fig. 2.1). In laboratory practice a chromatographic system is selected by so-called method development.

According to the state of aggregation of the fluid phase chromatographic systems can be divided into several categories. If the fluid phase is gaseous the process is called gas chromatography (GC). If the fluid phase is a liquid the process is called liquid chromatography (LC). For a liquid kept at temperature and pressure conditions above its critical point the process is called supercritical-fluid chromatography (SFC). Liquid chromatography can be further divided according to the geometrical orientation of the phases. A widely used process for analytical purposes as well as rapid method development and, in some cases, even a preparative separation process is thin-layer chromatography (TLC). The adsorbent is fixed onto a support (glass, plastic or aluminium foil) in a thin layer. The solute is placed onto the adsorbent in small circles or lines. In a closed chamber one end of the thin-layer plate is dipped into the mobile phase, which then progresses along the plate due to capillary forces. Individual substances can be visualized by either fluorescence quenching or after chemical reaction with detection reagents. The advantages of TLC are the visualization of all substances, even those sticking heavily to the adsorbent as well as the easiness of parallel development.

In GC and LC the adsorbent is fixed into a cylinder that is usually made of glass, polymer or stainless steel (column). In this column the adsorbent is present as a porous or non-porous randomly arranged packing or as a monolithic block. Because of the high separation efficiency of packed columns made of small particles this type of chromatography is called high-performance liquid chromatography (HPLC).

2.1.1
Adsorption Process

Chromatography is an adsorptive separation process and is in general based on the deposition or accumulation of molecules on a surface. According to the nature of the molecules and the surface of the adsorbent the adsorption can be further divided into the following combination of phases:

- Gaseous molecules/solid surface
- Liquid molecules/solid surface
- Gaseous molecules/liquid surface
- Liquid molecules/liquid surface

Examples of industrial relevance for the first two combinations are the adsorption of pollutants from waste air or water onto activated carbon. Combinations three and four can be observed at the orientation of tensid molecules on water/air interfaces (foam formation, foam stabilization) or at the interface of two immiscible liquids. (e.g. oil and water, emulsion formation). This book deals mainly with the case of liquid molecules adsorbed onto solid surfaces. For this case the following definitions are made:

The solid onto which adsorption occurs is defined as the adsorbent. The adsorbed liquid molecule in its free state is defined as the adsorptive and in its adsorbed state as the adsorpt. To distinguish between the different molecules of a solute they are called here component A, B, etc. (Fig. 2.1).

On a molecular level the adsorption process is the formation of binding forces between the adsorbent's surface and the molecules of the fluid phase. The binding forces can be different in nature and, therefore, of different strength. Hence the energetic balance of the binding influences the adsorption equilibrium, which can also be very different in strength. Basically, two different types of binding forces can be distinguished (Atkins, 1990), (IUPAC, 1972):

1. Physisorption or physical adsorption, which is a weak binding based on van der Waals forces, e.g. dipole, dispersion or induction forces. These forces are weaker than the intramolecular binding forces of molecular species. Therefore, physisorbed molecules maintain their chemical identity.
2. The stronger binding type is chemisorption or chemical adsorption. It is caused by valence forces, equivalent to chemical, mainly covalent, bindings. The energy of the free adsorbent valences is strong enough to break the atomic forces between the adsorbed molecules and the adsorbent.

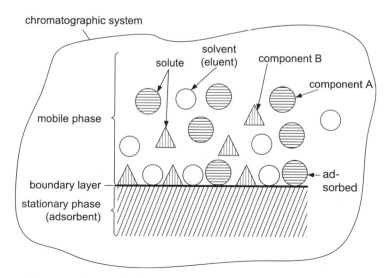

Fig. 2.1 Definitions of adsorption and the chromatographic system.

Table 2.1 Adsorption enthalpies.

	Physisorption	Chemisorption	
Gas phase	$1.5\Delta h_{vap}$	$2–3\Delta h_{vap}$	Ruthven, 1984
Liquid phase	$<50\,kJ\,mol^{-1}$	$\geq 60–450\,kJ\,mol^{-1}$	Kümmel and Worch, 1990

Table 2.1 gives estimated values for adsorption enthalpies. In gaseous systems they are proportional to the heat of vaporization h_{vap}.

As chromatographic processes require complete reversibility of the adsorption step, only adsorption processes based on physisorption can be used. The resulting energy is sufficient to increase the temperature of a gas due to its low volumetric heat capacity (Ruthven, 1984). Fluids, however, have a volumetric heat capacity 10^2 to 10^3 times higher; therefore the energy from the adsorption process has no influence on the temperature of the separation process and can be neglected. All of the following processes are, therefore, considered to be isothermal processes.

2.1.2
Chromatographic Process

In liquid phase chromatography the mobile phase is forced through a column, packed with a multitude of adsorbent particles (the stationary phase). Figure 2.2 shows the injection of a homogeneous mixture (represented by triangles and circles) into the system at the column inlet. The triangles represent the component B with a higher affinity to the stationary phase. The mean adsorption time on the stationary phase surface is, therefore, higher than that of component A with lower affinity (circles).

The difference in affinity, and thus adsorption time, results in a slower migration speed through the column of the more adsorbed component B. This delays its arrival at the column end compared with the less adsorbed component A. If the process

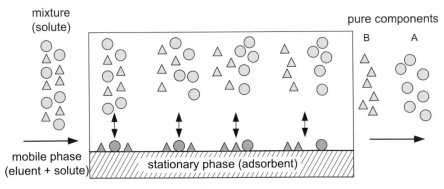

Fig. 2.2 Principle of adsorption chromatography.

conditions are chosen well the two substances can be completely separated and collected as pure components at the column outlet. Notably, the differences in retention are only based on adsorption. All molecules take the same time to pass through the liquid phase, as long as their molecular weights are of the same order of magnitude.

2.2
Basic Effects and Chromatographic Definitions

2.2.1
Chromatograms and their Parameters

Basic information for the development of a chromatographic separation process is obtained from the chromatogram. A typical chromatogram resulting from the injection of three different components in analytical amounts is shown in Fig. 2.3.

The interaction strength of each component with the stationary phase is proportional to its retention time $t_{R,i}$. Instead of the retention time some publications use the retention volume, which is obtained by multiplying $t_{R,i}$ by the flow rate $\dot{V}$. The unit „retention volume" must not be mistaken for the unit „column volume". The latter is based on the empty column volume V_c (Eq. 2.4).

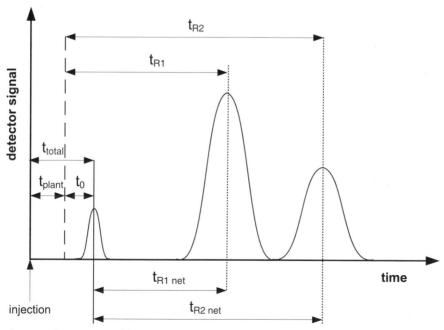

Fig. 2.3   Chromatogram of three components.

The retention time is determined from the peak maximum in the case of symmetrical peak shapes. For well-packed columns symmetrical peaks should be achieved as long as the amount injected into the column is in the linear concentration range of the adsorption isotherm. If increased amounts of substances in the nonlinear concentration range of the adsorption isotherm are injected the peak is often heavily distorted and asymmetric. In that case the retention time has to be calculated from the centroid following the momentum method (Eq. 2.27).

The total dead time t_{total} is the time a non-retained substance needs from the point of sample introduction to the point of detection. It depends on the hold up within the column itself as well as the hold up of the system or plant, which is, for instance, influenced by pipe length and diameter, pump head volume (if the sample is introduced via a pump) and the detector volume. Therefore, the dead time of the plant without the column t_{plant} has to be determined as well to get the correct dead time of the column t_0 by subtraction of t_{plant} from t_{total}. Tracer molecules are used to determine the dead time. These molecules should not be adsorbed by the solid phase but should have the same size as the components to be separated so as to penetrate the pore system in a comparable way.

The overall retention time $t_{R,i}$ of a retained component i is the sum of the deadtime t_0 and the net retention time $t_{R,i,net}$. The net retention time represents the time during which a substance is adsorbed onto the surface of the adsorbent. As t_0 depends heavily on the tracer substance used for its determination the calculation of the net retention time is very difficult. Therefore, this book always refers to the overall retention time, when the term retention time is used. This is in accordance with international convention.

Because the retention time of a substance depends on the column geometry and mobile phase flow rate the retention behavior has to be normalized. Consequently, the capacity factor k'_i, which is also called the retention factor, is defined according to Eq. 2.1:

$$k'_i = \frac{t_{R,i} - t_0}{t_0} \tag{2.1}$$

The capacity factor depends only on the distribution of the component of interest between the mobile and the stationary phase. It indicates the ratio of the time a component is adsorbed to the time it is in the fluid phase. It can also be expressed as the mole ratio of substance i in the stationary and mobile phase (Eq. 2.2):

$$k'_i = \frac{n_{i,stat}}{n_{i,mob}} \tag{2.2}$$

Both the capacity factor and the net retention time depend on the nature of the tracer substance, which is used to determine t_0. Therefore only k'_i values that are based on experiments with the same tracer substance should be directly compared and used to calculate selectivities.

Every chromatographic process aims to separate dissolved components, therefore the distance between the peak maximum of two or more components is of great importance. The selectivity α of a separation of two components is determined by dividing their capacity factors (Eq. 2.3):

$$\alpha = \frac{k'_j}{k'_i} = \frac{t_{R,j} - t_0}{t_{R,i} - t_0} \tag{2.3}$$

By convention, the more retained component is the denominator, thus the selectivity of two separated components is always >1. The selectivity is also called the separation factor.

2.2.2
Voidage and Porosity

Other important information that has to be derived from the standard chromatogram are the column porosities and the void fraction. These parameters have to be carefully determined as they will be the basis for detailed modeling and simulation of the purification process. As illustrated in Fig. 2.4, the total volume of a packed column (V_c) is divided into two sub-volumes: the interstitial volume of the fluid phase (V_{int}) as well as the volume of the stationary phase (V_{ads}) (Eq. 2.4)

$$V_c = \pi \frac{d_c^2}{4} L_c = V_{ads} + V_{int} \tag{2.4}$$

Again, V_{ads} consists of the volume of the solid material (V_{solid}) and the volume of the pore system, V_{pore} (Eq. 2.5):

$$V_{ads} = V_{solid} + V_{pore} \tag{2.5}$$

From these volumes different porosities can be calculated (Eqs. 2.6–2.8):

$$\text{Void fraction: } \varepsilon = \frac{V_{int}}{V_c} \tag{2.6}$$

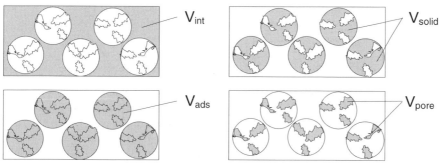

Fig. 2.4 Structure of packed beds.

Porosity of the solid phase: $\varepsilon_p = \dfrac{V_{pore}}{V_{ads}}$ (2.7)

Total porosity: $\varepsilon_t = \dfrac{V_{int} + V_{pore}}{V_c} = \varepsilon + (1 - \varepsilon)\varepsilon_p$ (2.8)

Practical determination of the porosities often suffers some difficulties. The most common method for determining the total porosity is the injection of a non-retained, pore-penetrating tracer substance (grey and small black circles in Fig. 2.5). In normal phase chromatography, toluene or 1,3,5-tri-*tert*-butylbenzene are often used, while in reversed phase chromatography uracil is the component of choice.

The interstitial velocity of the mobile phase is given by Eq. 2.9.

$$u_{int} = \frac{\dot{V}}{\varepsilon \, \pi \dfrac{d_c^2}{4}}$$ (2.9)

It is useful to measure the exact volume delivered by the pump during the determination of the column dead time. The total porosity is then calculated according to Eq. 2.10.

$$\varepsilon_t = \frac{t_0 \dot{V}}{V_c}$$ (2.10)

To obtain the void fraction or interstitial porosity of a column the same methodology is used with a high molecular weight substance that is unable to penetrate into the pores (large black circles in Fig. 2.5). Figure 2.5 also shows an ideal chromatogram obtained from the injection of two tracer substances of different molecular weight. For a non-penetrating molecule the dead time of the column reduces to Eq. 2.11,

$$t_{0,int} = \frac{L_c}{u_{int}}$$ (2.11)

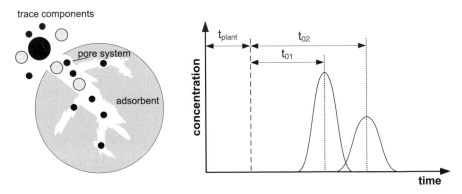

Fig. 2.5   Chromatography of tracer components.

Severe experimental problems often occur during the injection of a high molecular weight tracer. The ideal high molecular weight substance should be globular, exhibit no adsorption or penetration into any pore but should, however, be of high diffusivity. Obviously, from these prerequisites, there is no ideal high molecular weight volume marker. Therefore, alternative methods for determining the void fraction have been examined. If the adsorbent parameters can be determined with high accuracy from physical measurements the porosities can be calculated from the mass of the adsorbent m_{solid}, its density ρ_{solid} and the specific pore volume v_{sp} (determined by nitrogen adsorption or mercury porosimetry). The volume of the solid material is then calculated by Eq. 2.12:

$$V_{solid} = \frac{m_{solid}}{\rho_{solid}} \tag{2.12}$$

The pore volume is obtained from Eq. 2.13:

$$V_{pore} = v_{sp} m_{solid} \tag{2.13}$$

The sum of solid volume and pore volume gives the volume of the stationary phase, which, subtracted from the column volume, leads to the interstitial volume (Eq. 2.14):

$$V_{int} = V_c - (V_{solid} + V_{pore}) \tag{2.14}$$

The void fraction is calculated by Eq. 2.6. Subsequently, the other porosities can be obtained by Eqs. 2.7 and 2.8.

An alternative method for the easy and exact determination of total porosity can be used for small-scale columns. As long as a column can be weighed exactly, the mass difference of the same column filled with two solvents of different densities can be used to determine the porosity. The column is first completely flushed with one solvent and then weighed, afterwards the first solvent is completely displaced by a second solvent of different density. For normal phase systems methanol and dichloromethane can be used, for reversed phase systems water and methanol are quite commonly employed. The volume of the solvent, representing the sum of the interstitial volume and the pore volume, is determined by Eq. 2.15:

$$V_{int} + V_{pore} = \frac{m_{s,1} - m_{s,2}}{\rho_{s,1} - \rho_{s,2}} \tag{2.15}$$

The total porosity is again calculated by Eq. 2.8.

High efficiency adsorbents are very often spherical and monodisperse in order to reduce pressure drop. In this case the theoretical void fraction for spherical particles lies in the range $0.26 < \varepsilon < 0.48$ and a mean value of ε of approx. 0.37 can be roughly estimated (Brauer, 1971).

Particle porosity lies in range $0.50 < \varepsilon_p < 0.90$, meaning that 50 to 10 % of the solid is impermeable skeleton. In practice the total porosity is $0.65 < \varepsilon_t < 0.80$.

At large porosities, sufficient stability of the bed has to be ensured.

With monolithic columns (Chapter 3) the total porosity lies in the range $0.80 < \varepsilon_t < 0.90$.

2.2.3
Influence of Adsorption Isotherms on the Chromatogram

A chromatogram is influenced by several factors, such as the fluid dynamics inside the packed bed, mass transfer phenomena and, most importantly, the equilibrium of the adsorption at the temperature of the system.

The adsorption equilibrium is determined by the isotherm, which gives the correlation between the loading of the solute on the adsorbent q_i at different fluid phase concentrations c_i. Adsorption isotherms are discussed in more detail in Section 2.5.

The adsorption isotherm is the main parameter governing preparative-scale chromatographic separations. The influence of the isotherm type on the chromatogram is described in the following (Fig. 2.6).

The elution profile of an ideal chromatogram depends only on the thermodynamic behavior of the chromatographic system. In a real chromatogram additional mass transfer and fluid dynamic factors have to be taken into account.

In an ideal chromatogram the movement of an injected plug of a component i and thus its retention time is given by the basic equation of chromatography (Eq. 2.16) (Guiochon, 1994) (Seidel-Morgenstern, 1995).

$$t_{R,i}\left(c_i^+\right) = t_0 \left(1 + \frac{1 - \varepsilon_t}{\varepsilon_t} \frac{\partial q_i}{\partial c_i}\bigg|_{c_i^+}\right) \tag{2.16a}$$

$$t_{R,\text{lin},i}\left(c_i^+\right) = t_0 \left(1 + \frac{1 - \varepsilon_t}{\varepsilon_t} H_i\right) \tag{2.16b}$$

The differential term in Eq. 2.16a represents the slope of the isotherm at the concentration c_i^+. For the linear range of the adsorption isotherm the slope is equivalent to the Henry-constant of the isotherm H_i (Fig. 2.6-1a) and the retention time becomes independent of the fluid phase concentration (Eq. 2.16b). Each injected component then, theoretically, elutes at its retention time as an ideally rectangular pulse (Fig. 2.6-2a). The retention time is thus only influenced by the Henry constant.

For convex isotherms (Fig. 2.6-1b) the differential term depends on the solute concentration and each concentration c_i^+ corresponds to a different retention time. During elution of the solute all concentrations from zero to the maximum elution concentration can be observed. If the maximum elution concentration is outside the linear range of the adsorption isotherm the concentration profile of the chromatogram is determined by the convex shape of the isotherm. The slope of a convex isotherm, and thus the retention time, decreases with increasing concentration. According to Eq. 2.16a, regions of lower concentration with longer retention time are caught up with regions of higher concentration and thus shorter retention time. In total, the profile is concentrated and the front sharpened. The solute elutes in a compressed front. Fig. 2.6-2b shows the resulting chromatogram. The peak maximum is shifted towards the compressive front of the peak while the back of the peak is dispersed. Elution of a substance with a dispersed back of the peak is called tailing. This phenomenon is typical for convex adsorption isotherms of the Langmuir type. The opposite behavior is seen with concave adsorption isotherms (Fig. 2.6-1c). For concave

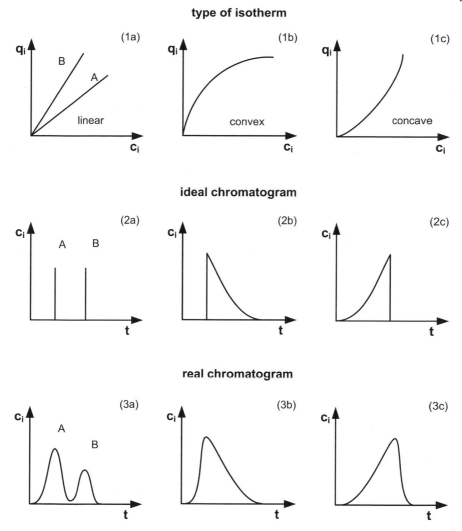

Fig. 2.6 Influence of the type of isotherm on the chromatogram. (Reproduced from Unger, 1994.)

adsorption isotherms lower concentrations move faster and, thus, the back of the late-eluting high concentrations are concentrated and sharpened. Fig. 2.6-2c shows the resulting chromatogram. The dispersed shape of the front part of the peak is called fronting.

Real chromatograms (Fig. 2.6-3) take into account the thermodynamic influences as well as the kinetics of mass transfer and fluid distribution. A rectangular concentration profile of the solute at the entrance of the column soon changes into a bell-shape Gaussian distribution, if the isotherm is linear. Figure 2.7a shows this distribution and some characteristic values, which will be referred to in subsequent chapters. With mass transfer resistance or nonlinear isotherms the peaks become asym-

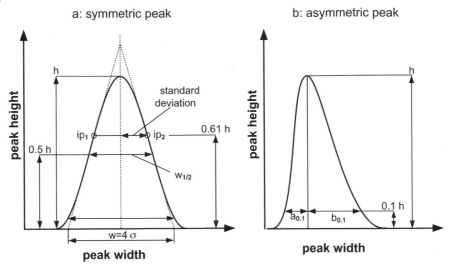

Fig. 2.7   Determination of peak asymmetry.

metric. The degree of peak asymmetry T is, in general, calculated from the difference between the two peak halves at $1/10^{th}$ of the peak height (Fig. 2.7b), (Eq. 2.17a).

$$T = \frac{b_{0.1}}{a_{0.1}} \qquad (2.17a)$$

A second calculation method for peak asymmetry can be found in the different pharmacopoeias, e.g. United States Pharmacopoeia (USP) or Pharmacopoeia Europea (PhEUR). These pharmacopoeias calculate the peak tailing at 5% of the peak height from the baseline and are therefore more stringent (Eq. 2.17b).

$$T_{USP} = \frac{a_{0.05} + b_{0.05}}{2a_{0.05}} \qquad (2.17b)$$

2.3
Fluid Dynamics

All preparative and production-scale chromatographic separations aim to collect target components as highly concentrated as possible. The ideal case would be a rectangular signal with the same time length and concentration as the pulse injected into the column. This behavior cannot be achieved in reality. In every chromatographic system non-idealities of fluid distribution occur, resulting in a broadening of the residence time distribution of the solute. All hydrodynamic effects that contribute to the total band broadening are summarized in the term axial dispersion. Figure 2.8 illustrates the effect of the axial dispersion. The rest of the band broadening results from mass transfer resistance.

Figure 2.8 shows that the rectangular pulse, which is introduced at the column inlet ($x = 0$), is symmetrically broadened as it travels along the column. In accor-

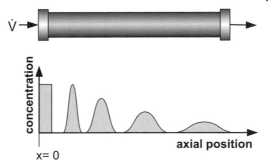

Fig. 2.8 Band broadening in a column due to axial dispersion.

dance with the band broadening the maximum concentration of the solute is decreased. This results in an unfavorable dilution of the target component fraction.

The main factors that influence axial dispersion are discussed below.

2.3.1
Extra Column Effects

Every centimeter of tubing, as well as any detector, between the point of solute injection and the point of fraction withdrawal contributes to the axial dispersion of the samples and thus decreases the concentration and separation efficiency. Obviously, the length of tubing should be minimized. More critical, however, are smooth connections of tubing and column, avoiding any dead space where fluid and, especially, sample can accumulate. The tubing diameter depends on the flow-rate of the system and has to be chosen in accordance to the system pressure. The tubing should contribute as little as possible to the system pressure drop, without adding additional large hold-up volumes.

2.3.2
Column Fluid Distribution

The most critical points for axial dispersion in column chromatography are the fluid distribution at the column head and the fluid collection at the column outlet. Especially with large-diameter columns, these effects have to be carefully considered. Fluid distribution in the column head is widely driven by the pressure drop of the packed bed, which forces the sample to be radially distributed within the inlet frit. It is, therefore, important to use high quality frits, which ensure an equal radial fluid distribution. In low-pressure chromatography with large dimensions, as well as with new types of adsorbents such as monolithic packings, which exhibit much lower pressure drops, the fluid distribution is of even greater importance. Several approaches to optimise the distribution have been made by column manufacturers to overcome this problem of low-pressure flow distribution. The introduction of specially designed fluid distributors has greatly improved the situation (Chapter 3.1.5).

2.3.3
Packing Non-idealities

The packed bed of a chromatographic column will never attain optimum hexagonal-dense packing due to the presence of imperfections. Those imperfections can be divided into effects due to the packing procedure and influences from the packing material. During the packing procedure several phenomena occur. The two most important are wall effects and particle bridges. Knox et al. (1976) determined a layer of about $30d_p$ thicknesses where wall effects influence the columns efficiency. Therefore, attention has to be paid to this effect with small diameter columns. The larger the column diameter the less severe is this effect. The second reason why imperfect packing may occur in columns is inappropriate pressure packing procedures. If the particles cannot arrange themselves optimally during the packing process they will form bridges, which can only be destroyed by immense axial pressure, with the danger of damaging the particles. One way to reduce particle bridging is to pack the column by vacuum and, afterwards, by axial compression of the settled bed. For more details see Chapter 3.3.

Insufficient bed compression, which leads to void volumes at the column inlet, is another source of axial dispersion. Preparative columns should therefore always exhibit a possibility for adjusting the compression of the packed bed. Fixed bed-length columns are not very stable at inner diameters >20 mm.

2.3.4
Sources for Non-ideal Fluid Distribution

The reasons for non-idealities in fluid distribution can be divided into macroscopic, mesoscopic and microscopic effects (Tsotsas, 1987). The different effects are shown in Fig. 2.9.

Microscopic fluid distribution non-ideality is caused by fluid dynamic adhesion between the fluid and the adsorbent particle inside the microscopic channels of the packed bed. Adhesion results in a higher fluid velocity in the middle of the channel than at the channel walls (Fig. 2.9a). Solute molecules in the middle of the channel thus have a shorter retention time than those at the channel walls.

Local inhomogeneities of the voidage are a second source of broadening the mean residence time distribution. For small particles, the formation of particle agglomerates, which cannot be penetrated by the fluid, is another reason for axial dispersion. Fig. 2.9b illustrates this mesoscopic fluid distribution non-ideality. This effect results in local differences in fluid velocity and differences in path-length of the solute molecules traveling straight through the particle agglomerates compared with those molecules moving around the particle aggregates. The second effect is also known as eddy-diffusion (Fig. 2.9d). Eddy-diffusion is a purely statistically effect. Due to its similar origin, this effect is summarized under mesoscopic fluid dynamic non-idealities.

Macroscopic fluid distribution non-idealities are caused by local non-uniformities of the voidage, which might occur especially in the wall region (Fig. 2.9c).

Fig. 2.9 Fluid distribution non-idealities according to Tsotsas (1987).

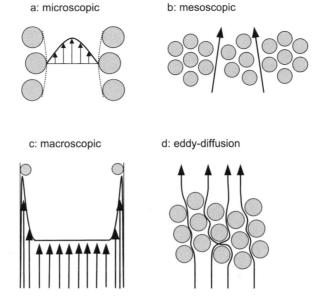

a: microscopic   b: mesoscopic

c: macroscopic   d: eddy-diffusion

All the above-mentioned effects cause a symmetrical increase of the peak width, which contributes besides mass transfer to the total band broadening.

2.4
Mass Transfer Phenomena

2.4.1
Principles of Mass Transfer

Most chromatographic production systems use particulate adsorbents with defined pore structures due to the higher loadability of mesoporous adsorbents compared with non-porous adsorbents. Adsorption of the target molecules on the inner surface of a particle has a tremendous influence on the efficiency of a preparative separation. Several factors with regard to mass transfer that contribute to the total band broadening effect in addition to axial dispersion can be distinguished (Fig. 2.10):

- Convective and diffusive transport towards the particle
- Film diffusion
- Pore (a) and surface (b) diffusion
- Kinetics of adsorption and desorption

Individual adsorbent particles within a packed bed are surrounded by a boundary layer, which is looked upon as a stagnant liquid film of the fluid phase. The thickness of the film depends on the fluid distribution in the bulk phase of the packed bed.

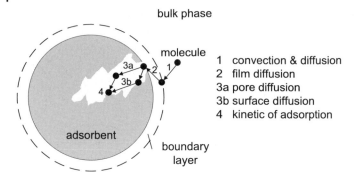

bulk phase

molecule

1 convection & diffusion
2 film diffusion
3a pore diffusion
3b surface diffusion
4 kinetic of adsorption

adsorbent

boundary
layer

Fig. 2.10 Mass transfer phenomena during the adsorption of a molecule.

Molecular transport towards the boundary layer of the particle by convection or diffusion is the first step (1) of the separation process.

The second step is diffusive transport of the solute molecule through the film layer which is called film diffusion (2). Transport of the solute molecules towards the adsorption centers inside the pore system of the adsorbent particles is the third step of the adsorption process. This step can follow two different transport mechanisms, which can occur separately or parallel to each other: pore diffusion (3a) and surface diffusion (3b).

If mass transport occurs by surface diffusion (3b) a solute molecule is adsorbed and transported deeper into the pore system by movement along the pore surface. During the whole transport process, molecules are within the attraction forces of the adsorbent surface. Notably, the attractive forces between the surface and absorbed molecules are so strong (Zhang et al, 2003) that, for many common adsorpt–adsorbent systems encountered in preparative chromatography, surface diffusion is physically implausible (especially in charged adsorpt – charged surface interactions).

Pore diffusion (3a) is driven by restricted Fickian diffusion of the solute molecules within the free pore volume. During the transport process the solute molecules are always outside the attraction forces of the adsorbent surface.

Once a solute molecule reaches a free adsorption site on the surface, actual adsorption takes place (4). If the adsorption processes (1–4) are slow compared with the convective fluid flow through the packed bed the eluted peaks show a non-symmetrical band broadening. Severe tailing can be observed, due to the slower movement of those solute molecules, which penetrate deeper into the pore system.

External mass transfer (1) and kinetic of adsorption (4) are normally very fast in the fluid phase (Guiochon, 1994) (Ruthven, 1984). The speed limiting processes are the film diffusion (2) and transport inside the pore system (3a, 3b).

For particle diameters >5 μm, which is the case for most preparative chromatographic separations, transport resistance in the pore is the dominant contribution to the total mass transfer resistance (Ludemann-Hombourger et al., 2000).

The transport factors described above are influenced by the morphology of the adsorbent particles as well as by their surface chemistry. Pore sizes influence chromatographic properties in two main ways. The first is through the overall pore size. Molecules can only penetrate into the pores above a certain threshold. For bigger molecules the theoretical surface area is only partly accessible. The second factor is determined by the equivalence of all adsorption sites in terms of the diffusion pathways and unhindered accessibility. Because transport within a mesoporous system is always diffusive, band broadening due to mass transfer resistance increases dramatically when pores with large differences in sizes are present within an adsorbent backbone. This is because molecules that adsorb in the larger pores have a significantly shorter residence time in the fluid phase than molecules adsorbed in the smaller pores.

One of the most important factors influenced by the physical adsorbent parameters is the available surface area. Within an ideal adsorbent the surface area is high so as to offer a multitude of adsorption sites. All adsorption sites should be equal and accessible within short distances, thus keeping diffusion pathways short. This prerequisite leads to the consequence that all pores should be of equal size, that no deep dead-end pores are present and that the accessibility of the pores from the area of convective flow within a packed bed should be equal for all pores.

The above-mentioned factors indicate that an ideal chromatographic adsorbent would exhibit a high surface area but a moderate pore volume and pore size (which is large enough to ensure good accessibility for the adsorptive molecules). As these three physical parameters are interlinked they cannot be independently chosen. Their interdependence is summarized in the Wheeler equation, which assumes a constant pore diameter within the particle. In Eq. 2.18, d_{pore} is the pore diameter (nm), V_{pore} the pore volume (ml g^{-1}) and S_{BET} the specific surface area (m^2 g^{-1}).

$$d_{pore} = 4 \times 10^3 \left(\frac{V_{pore}}{S_{BET}} \right) \qquad (2.18)$$

The design of a preparative adsorbent is always a compromise between a high surface area, with good loadability and limited mechanical strength, and a high mechanical strength and reduced surface area.

2.4.2
Efficiency of Chromatographic Separations

One of the parameters that characterize a preparative chromatographic system is the plate number N, which is also called the column efficiency.

The plate number as well as the corresponding height of an equivalent theoretical plate (HETP) is well known in chemical engineering as a common measure for ideal mass transfer within packed beds (Eq. 2.19).

$$\text{HETP} = \frac{L_c}{N} \qquad (2.19)$$

However, the definitions related to specific processes have to be kept in mind. In chromatography the plate height is a measure that lumps together the contribution of the fluid dynamic non-idealities (axial dispersion) and the mass transfer resistance to the total band broadening. It is defined as the rate of increase of the Gaussian peak profile per unit length of columns (Knox and Pardur, 1969) (Eq. 2.20).

$$\text{HETP} = \frac{\partial \sigma_z^2}{\partial z} \tag{2.20}$$

For uniform packed columns and incompressible eluents this becomes Eq. 2.21,

$$\text{HETP} = \frac{\sigma_z^2}{L_C} \tag{2.21}$$

In practice the peak profile is not measured in axial coordinates but detected with time. Therefore, the standard deviation of the time dependent peak σ_t has to be introduced (Eq. 2.22).

$$\frac{\sigma_t}{t_R} = \frac{\sigma_z}{L_C} \tag{2.22}$$

The plate height is then calculated by Eq. 2.23.

$$\text{HETP} = \left(\frac{\sigma_t}{t_R} \right)^2 L_C \tag{2.23}$$

The higher the efficiency of a column the closer is the peak shape to the ideal rectangular elution profile of an ideal chromatogram and the narrower is the peak width. Owing to the non-idealities of mass transfer and fluid dynamics every peak has only a certain, limited efficiency. A narrow peak results in a good peak resolution, a small elution volume and thus a high outlet concentration. All these facts give favorable conditions for preparative chromatography.

For the quantitative determination of the column efficiency a small amount of a retained test component is injected into the column. From the resulting chromatogram the column efficiency can be calculated.

For high efficiency adsorbents with symmetrical peaks the efficiency can quickly be calculated if the width of the Gaussian distribution w (Fig. 2.7) and Eq. 2.19 are introduced in Eq. 2.23:

$$N_i = \left(\frac{t_R}{\sigma_t} \right)^2 = 16 \left(\frac{t_{R,i}}{w_i} \right)^2 \tag{2.24}$$

The index i indicates that the plate number is different for each component. Another equation for the same plate number uses the peak width at half-height $w_{1/2,i}$ (Eq. 2.25).

$$N_i = 5.54 \left(\frac{t_{R,i}}{w_{1/2,i}} \right)^2 \tag{2.25}$$

To compare the efficiencies of different length columns the efficiency per metre (N_L) is often used (Eq. 2.26):

$$N_{L,i} = \frac{N_i}{L} = \frac{1}{\text{HETP}} \tag{2.26}$$

For asymmetric peaks the mean retention time of the chromatogram is calculated by the first absolute moment μ_t (Eq. 2.27). The variance σ_t^2 is determined by the second central moment according to Eq. 2.28.

$$\mu_t = \frac{\displaystyle\int_0^\infty t\,c(t)\,\mathrm{d}t}{\displaystyle\int_0^\infty c(t)\,\mathrm{d}t} \tag{2.27}$$

$$\sigma_t^2 = \frac{\displaystyle\int_0^\infty (t - \mu_t)^2 c(t)\,\mathrm{d}t}{\displaystyle\int_0^\infty c(t)\,\mathrm{d}t} \tag{2.28}$$

The plate number for a certain asymmetric peak is then calculated by Eq. 2.29.

$$b = \frac{H}{q_{\text{sat}}} \tag{2.29}$$

For asymmetrical peaks another empirically derived equation has been introduced by Bidlingmayer and Warren (1984). According to Eq. 2.30 and the parameters illustrated in Fig. 2.7b the plate number of non-symmetrical peaks can be calculated as

$$N_i = 41.7 \left[\frac{\left(\dfrac{t_{R,i}}{w_{t0.1}} \right)^2}{1.25 + \dfrac{b_{0.1i}}{a_{0.1i}}} \right] \tag{2.30}$$

The influence of the different mass transfer parameters on the overall efficiency of a column is shown in Fig. 2.11, where the efficiency represented by the plate height is plotted versus the mobile phase velocity.

This relationship is best expressed by the HETP curve, which is often called the van Deemter equation (van Deemter et al., 1956) (Eq. 2.31).

$$H_i = A_i + B_i\, u_{\text{int}} + \frac{C_i}{u_{\text{int}}} \tag{2.31}$$

This equation shows the dependence of the plate height of a column on the mobile phase velocity. The three terms of Eq. 2.31 describe different effects that have to be taken into account when selecting a stationary phase in preparative chromatography. The *A* term, which is almost constant over the whole velocity range, is mainly gov-

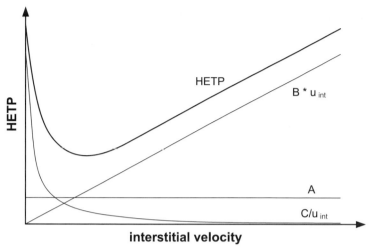

Fig. 2.11 Relationship between HETP and interstitial phase velocity.

erned by eddy-diffusion. It results from packing imperfections as well as adsorbents with very broad particle size distributions. The absolute height of the A term is proportional to the mean particle diameter. The plate height can therefore be decreased by using smaller particles.

The linear increase of the plate height at high velocities is summarized in the B term (B_i u_{int}). It is caused by the increasing influence of mass transfer resistance at higher velocities. Mass transfer resistance itself, especially inside the pores, is nearly independent of fluid velocity. However, if the fluid velocity increases, the relation between the convective axial transport in the bulk phase of the column and mass transfer into and in the stationary phase dominates. The slope of the B term depends on the nature of the packing material. The more optimised the adsorbent in terms of pore accessibility and diffusion pathlength the lower the slope of the B term and the higher the column efficiency at high mobile phase velocities.

The hyperbolic C term, C_i/u_{int}, expresses the influence of axial diffusion of the solute molecules in the fluid phase. It can only be observed in preparative systems with large diameter adsorbents operated at very low flow rates. In most cases of preparative chromatography it can be neglected, as the velocity of the mobile phase is rather high. As the longitudinal diffusion depends on the diffusion coefficient of the solute, it can be influenced by changing the mobile phase composition. To achieve high diffusion coefficients, low viscosity solvents should be preferred, which will, in addition, result in lower column pressure drops and thus will be favorable.

2.4.3
Resolution

In contrast to analytical chromatography, where it is possible to deconvolute overlapping peaks and to obtain even quantitative information from non-resolved peaks, preparative chromatography requires complete peak resolution if the components of interest are to be isolated with 100% purity and yield. The chromatographic resolution R_S is a measure how well two adjacent peak profiles of similar area are separated. R_S is defined as (Eq. 2.32):

$$R_S = \frac{2(t_{R,j} - t_{R,i})}{w_i + w_j} \tag{2.32}$$

where $t_{R,j}$ and $t_{R,i}$ are the retention times of components j and i, of which component i elutes first $(t_{R,j} > t_{R,i})$; w_j and w_i are the peak widths for components j and i at their base line $(w = 4\sigma)$.

A resolution of 1.5 corresponds to a baseline separation at touching band situation. At a resolution of 1.0 there is still an overlap of 3% of the peaks.

The assumption of equivalent peaks, where the peak width is equal for both components, leads to the following transformation of Eq. 2.33.

$$R_S = \left(\frac{\alpha - 1}{\alpha}\right)\left(\frac{k'_j}{k'_j + 1}\right)\frac{\sqrt{N_j}}{4} \tag{2.33a}$$

Note the index for component j, which has the longer retention time. The equivalent expression for resolution based on the less retained component i is

$$R_S = (\alpha - 1)\left(\frac{k'_i}{k'_i + 1}\right)\frac{\sqrt{N_i}}{4} \tag{2.33b}$$

The resolution of a separation system depends on three important parameters: the selectivity, capacity factors and column efficiency (Eq. 2.33a, b).

In theory, the equation above offers three possibilities for increasing the resolution. In fact increasing the capacity factor k' is only of limited use because it is linked to a higher retention time and thus increases the cycle time. This leads not only to lower throughput but also to more severe band broadening, which reduces the column efficiency.

The two real options for resolution optimisation are illustrated in Fig. 2.12a–c.

Fig. 2.12a shows a separation with a resolution factor of 0.9. The column efficiency is given by a plate number of 1000 N m^{-1}, while the selectivity is equal to 1.5. The first approach optimizes the column fluid dynamics, for example, by improving the column packing quality, reducing the particle diameter of the adsorbent or optimizing the mobile phase flow rate. The resulting chromatogram is shown in Figure 2.12b. The resolution has been increased to 1.5 due to the higher plate number of 2000 N m^{-1} while the selectivity is kept constant at 1.5. Conversely, the system thermodynamics can be changed by optimizing the temperature or altering the chromatographic system by changing the mobile phase and/or the adsorbent. Figure 2.12c shows the effect of the thermodynamic approach. The plate number is still constant at 1000 N m^{-1} but the selectivity has increased to 2.0, resulting in an increased resolution of 1.5.

The above two optimisation approaches have different effects on the cost of preparative chromatography. Especially, if efficiency is increased by decreasing the particle

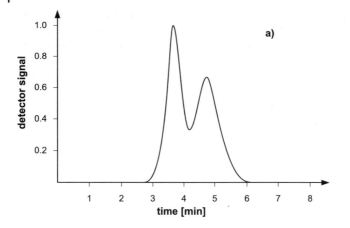

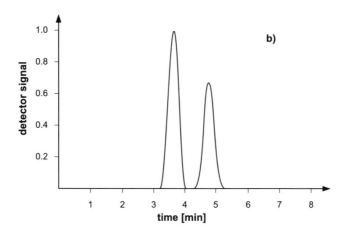

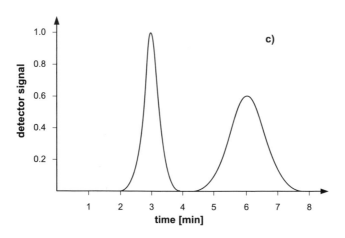

Fig. 2.12 Optimisation of peak resolution.

diameter of the stationary phase then the cost of the phase rapidly increases, along with the cost of the equipment, which needs to be more pressure-stable for operation with smaller particle diameters. It is, therefore, worth looking for a system with optimised selectivity.

With Eq. 2.33 it is also possible to calculate the plate number necessary for a given resolution (Eq. 2.34):

$$N_j = 16R_S^2 \left(\frac{\alpha}{\alpha - 1} \right)^2 \left(\frac{1 + k'_j}{k'_j} \right)^2 \tag{2.34}$$

Figure 2.13 shows the necessary plate numbers for a resolution of 1.5 and 1.0 versus the selectivity. For separation systems with selectivities <1.2 the necessary plate number clearly increases drastically. As higher plate numbers are linked to high-pressure systems with small particles, these separations obviously have much higher cost than separations operated at higher selectivities. This, once again, illustrates that rigorous screening for the most selective separation system is necessary to decrease operational costs.

In summary, for optimised preparative chromatographic systems some consideration should be taken into account right from the beginning:

1. System efficiency: good packing quality and low band-broadening due to axial dispersion and mass transfer resistance are necessary
2. Resolution vs. load: good resolution in the nonlinear range of the adsorption isotherm has to be achieved
3. Pressure drop: should be as low as possible to allow operation at maximum linear velocity

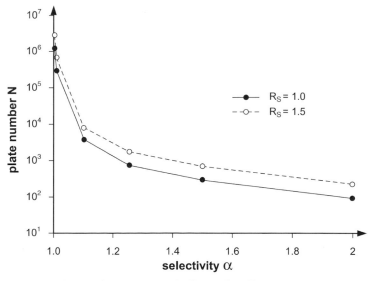

Fig. 2.13 Influence of selectivity on peak efficiency for different resolution.

2.5
Equilibrium Thermodynamics

2.5.1
Definition of Isotherms

The most dramatic difference between analytical and preparative chromatography is the extension of the working range of the adsorption isotherm into its nonlinear region. The behavior of single components as well as their mixtures over the complete range of the adsorption isotherm has, therefore, to be determined with great accuracy.

As with any other phase equilibrium the adsorption equilibrium is, according to Gibbs, defined by the equality of the chemical potential of all interacting phases. A more detailed description of the thermodynamic fundamentals can be found in the literature (Ruthven, 1984), (Guiochon, 1994).

The adsorption of molecules on a solid surface can reach significant loads. The surface concentration may be defined as the quotient of the adsorbed amount of substance and the surface of the adsorbent in $mol\,m^{-2}$. However the effective internal surface area of an adsorbent is especially difficult to determine because it depends on the nature and size of the solutes. It is therefore advisable to use the mass or the volume of the adsorbent instead of the internal surface. The loading is then expressed as $mol\,g^{-1}$ or $g\,l^{-1}$ adsorbent. Adsorbent volume can be expressed as total adsorbent volume V_{ads} or as solid-phase volume ($V_{ads} - V_{pore}$). According to Eq. 2.35 both values can be transformed into each other. In Eq. 2.35 $c_{p,i}$ represents the concentration of component i within the pore system. The total load of the adsorbent is given as q_i*and the pure solid load as q_i.

$$q_i^* = \varepsilon_p c_{p,i} + (1 - \varepsilon_p) q_i \qquad (2.35)$$

Graphical representation of the solid load q_i or the total adsorbent load q_i* versus the concentration of the solute in the fluid phase c_i at constant temperature provides the adsorption isotherm. In the following the pure solid load, q_i, will be preferred.

In the literature (Kümmel and Worch, 1990) different types of isotherms are described (Fig. 2.14).

In contrast to the well-developed thermodynamic methods for determining gas/liquid equilibriums the theoretical determination of adsorption isotherms is not yet feasible. Only approaches to determining multi-component isotherms from experimentally determined single-component isotherms are known. Such approaches are explained in more detail in Section 2.5.2.3. Careful experimental determination of the adsorption isotherm is therefore absolutely necessary. The different approaches for isotherm determination are discussed in Chapter 6.5.7.

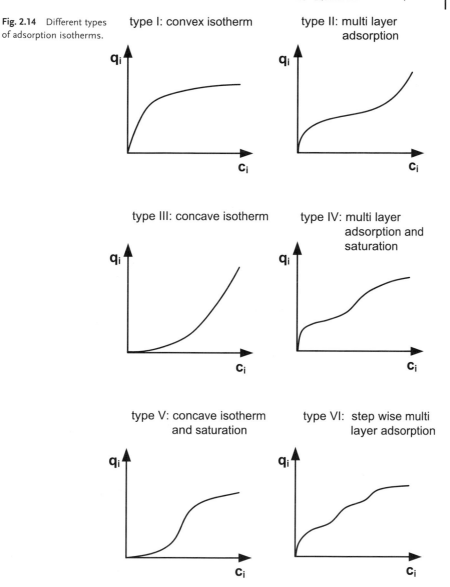

Fig. 2.14 Different types of adsorption isotherms.

2.5.2
Models of Isotherms

For modeling and simulation of preparative chromatographic systems the experimentally determined data have to be represented by mathematical equations. From the literature a multitude of different isotherm equations is known. Many of these equations are derived from equations developed for gas phase adsorption. Detailed literature can be found in the textbooks of Guiochon et al. (1994), Everett (1984) and Ruthven (1984) or articles of Seidel-Morgenstern and Nicoud (1996) or Bellot and Condoret (1993).

Isotherm models have to be divided into single-component and multi-component models. Because most of multi-component models are directly derived from single-component models the latter will be presented first.

2.5.2.1 Single-component Isotherms

Figure 2.15 shows a single-component adsorption isotherm of the Langmuir-type, the most common type in preparative chromatography. While increasing the concentration of the solute in the mobile phase the amount adsorbed onto the stationary phase no longer increases linearly. Only the very first region, with very low mobile phase concentrations, shows a linear relationship. This region is used for quantitative analysis in analytical chromatography because only this working region ensures that no retention time shifts take place if different amounts are injected.

In the linear range of the adsorption isotherm the relationship between the mobile and the stationary phase concentration c and q is expressed by Eq. 2.36

$$q = Hc \qquad (2.36)$$

Nevertheless this region is also of importance for preparative chromatography as the Henry coefficient is obtained from it. For its determination the retention time of a substance, the column dead time and the total porosity are necessary (Eq. 2.16b):

$$H = \left(\frac{t_R}{t_0} - 1\right) \frac{\varepsilon_t}{1 - \varepsilon_t} \qquad (2.37)$$

The higher the Henry coefficient for a substance the stronger is its adsorption and thus the longer its retention time. This definition shows that for two components to be separated their Henry coefficients have to differ. According to Eqs. 2.3 and 2.37 the quotient in the Henry coefficients expresses the selectivity of a separation system.

$$\alpha = \frac{H_j}{H_i} \qquad (2.38)$$

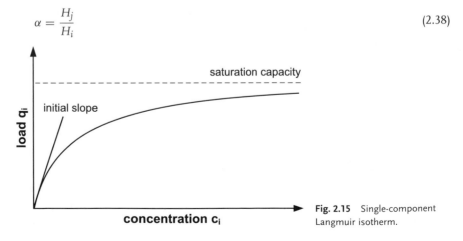

Fig. 2.15 Single-component Langmuir isotherm.

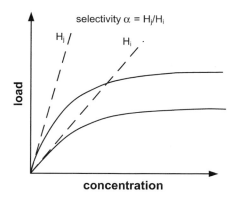

Fig. 2.16 Initial slopes of the isotherms for two different components.

Figure 2.16 shows the relationship between the isotherms of two different components and their Henry coefficients.

The complete Langmuir isotherm is represented by Eq. 2.39. The following assumptions are the theoretical background of Langmuir type isotherms:

- All adsorption sites are considered energetically equal (homogeneous surface)
- Each adsorption site can only adsorb one solute molecule
- Only a single layer of adsorbed solute molecules is formed
- There are no lateral interferences between the adsorbed molecules

$$q = q_{sat}\frac{bc}{1+bc} = \frac{Hc}{1+bc} \tag{2.39}$$

The Langmuir factor b in Eq. 2.39 represents the term

$$b = \frac{H}{q_{sat}} \tag{2.39a}$$

Experimental data are often fitted more precisely to the adsorption isotherm when an additional term is introduced that covers the non-specific adsorption of the solute to the adsorbent. This modifies the equation to the single-component, modified Langmuir isotherm (Eq. 2.40):

$$q = \lambda c + q_{sat}\frac{bc}{1+bc} = \lambda c + \frac{Hc}{1+bc} \tag{2.40}$$

An even more flexible equation is given by the Bi-Langmuir isotherm (Graham, 1953) (Eq. 2.41):

$$q = q_{sat,1}\frac{b_1c}{1+b_1c} + q_{sat,2}\frac{b_2c}{1+b_2c} = \frac{H_1c}{1+b_1c} + \frac{H_2c}{1+b_2c} \tag{2.41}$$

Equation 2.41 represents an additive extension of the modified Langmuir isotherm. The second Langmuir term covers the adsorption of the solute molecules on a second, completely independent group of adsorption sites of the adsorbent. Those completely independent adsorption sites can occur on reversed phase adsorbents with remaining silanol groups or on chiral stationary phases with a chiral selector coated or bonded to a silica surface (Guiochon, 1994).

One drawback of Langmuir-type adsorption isotherms is the conjunction between the initial slope of the isotherm and its curvature. This can be overcome by the Toth isotherm model (Toth, 1971) (Eq. 2.42):

$$q = q_{sat} \frac{c}{\left(\frac{1}{b} + c^e\right)^{1/e}} \tag{2.42}$$

The model shown in Eq. 2.42 has three independent fitting parameters, q_{sat}, e and b, which allow independent control of slope and curvature. For $e = 1$ it approaches the Langmuir model.

2.5.2.2 Multi-component Isotherms

If mixtures of solutes are injected into a chromatographic system, not only interferences between the amount of each component and the adsorbent but also between the molecules of different solutes occur. The resulting displacement effects cannot be appropriately described with independent single-component isotherms. Therefore, an extension of single-component isotherms that also takes into account the interference is necessary. For the modified Langmuir isotherm the extension term is shown in Eq. 2.43, representing the multi-component Langmuir isotherm.

$$q_i = q_{sat} \frac{B_{i,i} c_i}{1 + \sum_{j=1}^{n} b_{i,j} c_j} = \frac{H_i c_i}{1 + \sum_{j=1}^{n} b_{i,j} c_j} \tag{2.43}$$

The coupled isotherm equation takes into account the displacement of one component by the other with the term $(b_{i,j} c_j, j \neq i)$. Equation 2.43 is the general form of the multi-component Langmuir isotherm and is also called asymmetric as for each component i specific parameters $b_{i,j}$ have to be determined. With symmetric coefficients only a set of b_j parameters is taken into account and Eq. 2.43 reduces to Eq. 2.44,

$$q_i = q_{sat} \frac{b_i c_i}{1 + \sum_{j=1}^{n} b_j c_j} = \frac{H_i c_i}{1 + \sum_{j=1}^{n} b_j c_j} \tag{2.44}$$

A prerequisite for Eq. 2.43 is the equality of the maximum loadability q_{sat} of all solutes. In case of different loadabilities ($q_{sat,i} \neq q_{sat,j}$) for the different solutes Eq. 2.43 no longer accords with the Gibbs–Duhem equation and is thus thermodynamically inconsistent (Broughton, 1948), (Kemball et al. 1948). This is, for example, the case if solutes with substantially different molecular masses are separated on sorbents where the pore accessibility is hindered for large molecules.

The selectivity α_i of a separation is directly linked to the isotherms of the solutes. For linear chromatography the selectivity can be expressed as the relationship between the quotients of load and concentration of the solute components i and j (Eq. 2.45) (Seidel-Morgenstern, 1995), (Guiochon, 1994).

$$\alpha = \frac{q_j/c_j}{q_i/c_i} \tag{2.45}$$

With symmetrical Langmuir coefficients ($b_{i,i} = b_{j,i}$), Eq. 2.45 results in a selectivity that is constant and independent of the relative concentrations of the solutes. The selectivity is equivalent to the initial slope H of the adsorption isotherms of the components i and j (Eq. 2.46).

$$\alpha = \frac{q_{sat,j}b_j}{q_{sat,i}b_i} = \frac{H_j}{H_i} \tag{2.46}$$

Equations 2.45 and 2.46 show that simple multi-component Langmuir isotherms obviously can not explain the decrease in selectivity observed under increased loading factors. The extended form of the modified Langmuir isotherm, however, can represent these phenomena. The modified multi-component Langmuir isotherm is shown in Eq. 2.47 (Charton and Nicoud, 1995).

$$q_i = q_{sat,i}\frac{b_{i,i}c_i}{1+\sum\limits_{j=1}^{n}b_{i,j}c_j} + \lambda_i c_i = \lambda_i c_i + \frac{H_i c_i}{1+\sum\limits_{j-1}^{n}b_{i,j}c_j} \tag{2.47}$$

The Bi-Langmuir isotherm can be extended in the same way to give the multi-component Bi-Langmuir isotherm. (Eq. 2.48) (Guiochon, 1994).

$$q_i = q_{sat,1,i} \cdot \frac{b_{1i,i}c_i}{1+\sum\limits_{j=1}^{n}b_{1i,j}c_j} + q_{sat,2,i} \cdot \frac{b_{2i,i}c_i}{1+\sum\limits_{j=1}^{n}b_{2i,j}c_j} \tag{2.48}$$

$$= \frac{H_{1i}c_i}{1+\sum\limits_{j=1}^{n}b_{1i,j}c_j} + \frac{H_{2i}c_i}{1+\sum\limits_{j=1}^{n}b_{2i,j}c_j}$$

Modified multi-component Langmuir and multi-component Bi-Langmuir isotherms offer a maximum flexibility for adjustment to measured data if all coefficients are chosen individually. But in the same way as for multi-component Langmuir isotherms (Eq. 2.43) it is possible to use, for Eqs. 2.47 and 2.48, constant Langmuir terms ($b_{i,j} = b_{i,i}$, $b_{1i,j} = b_{1i,i}$, $b_{2i,j} = b_{2i,i}$) as well as constant adjustment terms ($\lambda_i = \lambda$) or equal saturation capacities ($q_{sat,1,i} = q_{sat,2,i} = q_{sat,i}$).

2.5.2.3 Ideal Adsorbed Solution (IAS) Theory
Presently, there are no methods that allow the theoretical calculation of isotherms. However, there are methods available that allow the calculation of mixture isotherms

from single component data. This can significantly reduce the necessary number of experiments and therefore the effort to estimate competitive isotherms. Several approaches are documented in the literature. The most successful approach is based on the theory of Myers and Prausnitz (1965), which was developed for gas phase adsorption: the Ideal Adsorbed Solution (IAS) theory. This theory was subsequently extended by Radke and Prausnitz (1972) to dilute liquid solutions. It assumes an ideal solution as well as an eluent, which is inert and does not influence the adsorption of the solutes.

According to Gibbs theory, adsorption is a three-phase system consisting of the stationary phase (adsorbent), the mobile phase and the adsorbed phase (adsorbt) (Fig. 2.1). The adsorbt is viewed as a 2-dimensional boundary layer between the other phases. It is also assumed that the activated surface of the adsorbent is identical for all components and does not change its properties. This reduces the adsorption to a two-phase system with an interaction between the adsorbt and the mobile phase.

In case of ideal adsorption of a mixture the fugacity f_i^{ad} of the adsorbt component i is proportional to the fugacity $f_i^{ad,0}$ of a one-component system and the mole fraction x_i of the adsorbed solutes.

$$f_i^{ad}(T, \pi, x_i) = f_i^{ad,0}(T, \pi)x_i \tag{2.49}$$

In Eq. 2.49 T is the temperature and π stands for the spreading pressure, which characterizes the difference in surface tension between a clean surface and a surface covered with adsorbate. The mole fraction of the ideal system is

$$x_i = \frac{q_i}{\displaystyle\sum_{i=1}^{N_{comp}} q_i} = \frac{q_i}{q_{tot}} \tag{2.50}$$

According to Radke and Prauznitz the total load of the mixture, q_{tot}, is only a function of the load $\tilde{q}_i^0$ for single-component adsorption of component i (Eq. 2.51a) at the hypothetical pure component concentration $\tilde{c}_i^0$ (Eq. 2.51b). In Eq. 2.51 $\tilde{c}_i^0$ is a hypothetical concentration of component i of a single-component system that causes a spreading pressure π in the adsorbed phase, which is equal to the spreading pressure of the mixture (Eq. 2.52).

$$\frac{1}{q_{tot}} = \sum_{i=1}^{N_{comp}} \frac{x_i}{\tilde{q}_i^0} \tag{2.51a}$$

$$\tilde{q}_i^0 = q_i(\tilde{c}_i^0) \tag{2.51b}$$

$$\pi_{mix} = \pi_i(\tilde{c}_i^0) \tag{2.52}$$

In case of equilibrium between fluid phase and adsorbt their chemical potentials are equal. For an ideal diluted mobile phase the concentration of component i is

$$c_i = c_{tot} x_i \tag{2.53}$$

$$c_i = \tilde{c}_i^0(T, \pi)x_i \tag{2.54}$$

These equations can be rearranged to give x_i

$$x_i = \frac{c_i}{\tilde{c}_i^0(T, \pi)} \tag{2.55a}$$

Note that the sum of the mole fractions of all solutes is equal to one:

$$\sum_i^{N_{comp}} x_i = 1 \tag{2.55b}$$

Together with Eq. 2.55 and the definition of the spreading pressure (Eq. 2.56), adsorption data of the mixture can be calculated from single-component adsorption data. The spreading pressure of a multi-component mixture corresponds to the spreading pressure calculated for each single component at the hypothetical concentration $\tilde{c}_i^0$ (Eq. 2.52) and is defined as:

$$\pi_{mix} = \pi_i(\tilde{c}_i^0) = \int_0^{\tilde{c}_i^0} \frac{q_i^0}{c_i^0} \, dc_i^0 \tag{2.56}$$

To calculate the adsorption equilibrium of a binary mixture for the given concentrations c_1 and c_2 the hypothetical concentrations $\tilde{c}_1^0$ and $\tilde{c}_2^0$, which fulfill Eqs. 2.55 and 2.56, have to be determined.

The procedure is shown in Fig. 2.17. After the single-component data are fitted to isotherm equations, the spreading pressure integrals (Eq. 2.56) can be calculated for pure component isotherms. During the calculation of the multi-component isotherm, it has to be distinguished whether the pressure spreading integral can be explicitly solved for $\tilde{c}_i^0$. If not, the integral must be solved iteratively. As an example for the former case, the Langmuir isotherm Eq. 2.39 is considered. In this case Eq. 2.56 can be solved and rearranged to give $\tilde{c}_i^0$,

$$\pi_{mix} = \pi_i(\tilde{c}_i^0) = \int_0^{\tilde{c}_i^0} \frac{H_i c_i^0}{(1 + bc_i^0)c_i^0} \, dc_i^0 = \frac{H_i}{b_i} \cdot \ln(1 + b_i \tilde{c}_i^0) \tag{2.57a}$$

$$\tilde{c}_i^0 = \frac{1}{b_i}\left[\exp\left(\frac{\pi_i(\tilde{c}_i^0)b_i}{K_i}\right) - 1\right] \tag{2.57b}$$

As shown in Fig. 2.17, the point wise determination of the multi-component isotherm at given mixture compositions is performed iteratively to find the hypothetical concentrations ($\tilde{c}_i^0$s) of all components that fulfill all necessary equations. Afterwards, the respective hypothetical loadings ($\tilde{q}_i^0$s) are calculated with the single-component isotherm model. The real loadings q_{tot}, q_1 and q_2 are then calculated with Eqs. 2.50 and 2.51.

For practical applications the resulting points of the multi-component isotherm may be fitted to a suitable model equation.

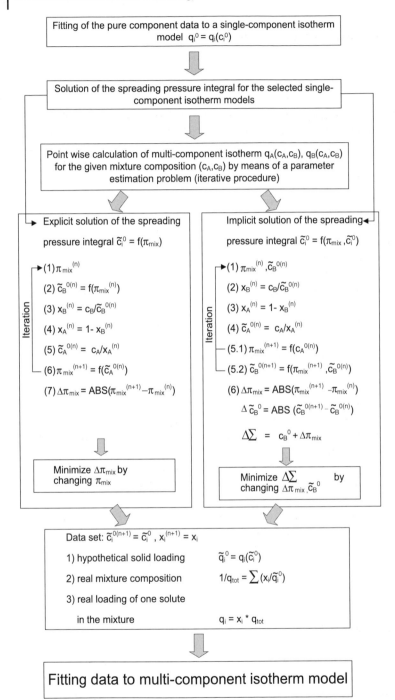

Fig. 2.17 Procedure to determine the multi-component isotherms from pure component data with the IAS theory for binary mixtures.

2.6
Thermodynamic Effects on Mass Separation

2.6.1
Mass Load

In analytical chromatography the column is typically run with very dilute sample mixtures. Therefore the chromatographic parameters generally remain within the range of the linear isotherm and are independent of the mass of sample loaded. The concentration profiles are symmetric and Gaussian (Fig. 2.18a). The sample mass m is given by Eq. 2.58,

$$m_{inj} = c_{inj} V_{inj} \tag{2.58}$$

where c_{inj} is the sample concentration and V_{inj} the sample volume. Typically, 10 to 100 µl of a sample mixture are injected into an analytical column 100 mm long and 4 mm in diameter. The concentration of the solute is approximately 1 mg ml^{-1}. When one increases the sample concentration at constant sample volume, the peak profile changes as indicated in Fig. 2.18b. A differently shaped profile is obtained when increasing the sample volume at constant sample concentration (Fig. 2.18c). The first case is called concentration overload or mass overload and the second is volume overload. In a third case both sample concentration and sample volume are changed. This situation is called column overload and is characterized by a decrease in the retention coefficients of solutes, a decrease in plate number or increase in plate height and in a loss of chromatographic resolution.

The term „overload" has, notably, been introduced by analytical chemists. Here a column should not be „overloaded" in order to achieve a constant retention time for a reproducible analytical detection of each component peak. Preparative chromatography has a different aim, which is called „productivity". To achieve this goal the columns are operated under so-called „overloaded" conditions. From the engineering view point overloaded systems are nonlinear because of nonlinear isotherms as well as dispersion and mass transfer effects.

A column is called overloaded by the convention of analytical chemists when the sample mass per unit mass of packing causes a 10% decrease in the retention factor k' or a 50% decrease of the plate number N. For visualization purposes one usually plots these quantities against the sample mass per gram of packing.

2.6.2
Linear and Nonlinear Isotherms

In the simplest case the adsorption isotherm is of the Langmuir type with a steep slope at the initial part and reaching a saturation value at higher amounts of adsorbed solute. The slope of the isotherm corresponds to the distribution coefficient K, discriminating two parts of the isotherm: the linear and nonlinear parts. In the linear range K is equal to the Henry constant. For the nonlinear part K becomes

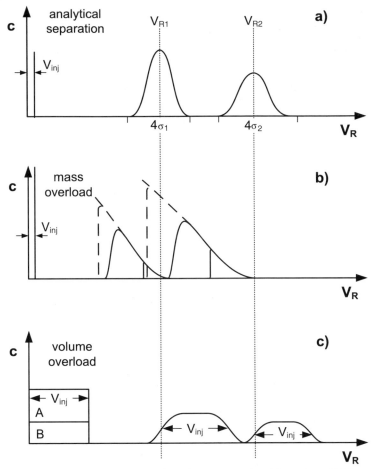

Fig. 2.18 Overloaded chromatographic columns. (Reproduced from Unger, 1994.)

smaller and reaches a limiting value at high solute concentrations. The subsequent part of the isotherm is called the overload regime whereas the linear part corresponds to the non-overload regime.

Analytical chromatography aims to achieve an adequate, not necessarily a maximum, resolution of solute bands to identify and to quantify the analytes based on their retention coefficient, peak height and peak area. Information on the analytes is the target.

In preparative chromatography the goal is to purify and to isolate compounds at a high yield, high purity or high productivity. Productivity is the major goal, being the mass of isolate (target compound) per unit mass of packing and per unit time. To increase productivity samples are applied on the column with much higher concentrations than in analytical chromatography. One should keep in mind that at higher column loading the selectivity coefficients change as well as the column plate num-

Fig. 2.19 Asymmetric multi-component Langmuir isotherm for different mass ratios (H_1 = 1.84, H_2 = 3.00, b_{11} = 0.01, b_{12} = 0.05, b_{21}= 0.025, b_{22}= 0.02).

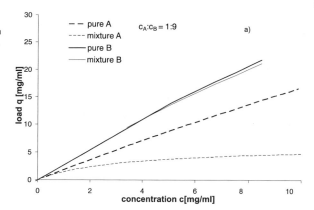

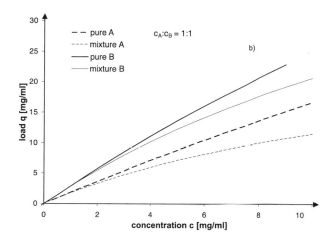

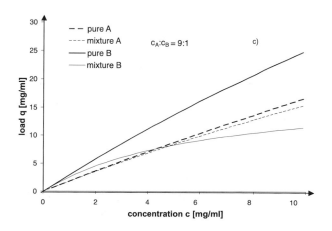

ber. The gain in selectivity is the major objective, which is controlled by the choice of eluent and packing.

There are additional effects occurring during the elution of a binary mixture at higher concentration depending on thermodynamics described by the isotherms and their coefficients as well as the relative mass ratio of the two components in the mixture. Two of the most prominent effects are the decrease of retention time at high load combined with the sharpening of the elution front and increased tailing of the disperse front (displacement effect) as well as plateau formation of the second peak on the disperse front (tag-along effect).

Figure 2.19a–c presents the multi-component isotherms of a two-component mixture for different concentrations (Eq. 2.43). In Fig. 2.19a the mass ratio of the first to the second eluting component is 1:9. In this case the isotherm for component B changes drastically while the equilibrium for component A is nearly unchanged. This relation changes for different mass ratios of 1:1 (Fig. 2.19b) and 9:1 (Fig. 2.19c).

Competition between molecules A and B for the interaction sites of the adsorbent results in different elution profiles (Fig. 2.20a–c).

The amount of one component adsorbed at equilibrium with a given mobile phase composition onto an adsorbent is constantly decreased, while the concentration of a second component is constantly increased. If a component is adsorbed at one point of the column and the second, more strongly adsorbed, component arrives at that point, the molecules of the first component are desorbed from the adsorbent surface and displaced into the mobile phase. The tuning of such systems by adding displacers with well-defined adsorption characteristics is used in displacement chromatography. An easier and more economic way is the self-displacement effect, which can occur when a small amount of the first eluting component is displaced by a large amount of the second eluting component (Fig. 2.20a). Adjusting the chromatographic system (components to be separated as well as mobile and stationary phase) to self-displacement conditions can increase productivity tremendously.

The second, so-called tag-along effect can occur when a high concentration of the first eluting component is injected together with a small amount of the second eluting component. Under these conditions the high concentration of the first component desorbs the molecules of the second component, thus forcing them into the mobile phase and resulting in shorter retention times of the second component. This effect causes a long plateau of the second small-amount component under the elution profile of the first eluting large-amount component (Fig. 2.20c).

Figure 2.20b represents an intermediate state, where both isotherms are affected by the other component. Notably, the isotherms shown in Fig. 2.19a–c are valid for constant mass ratios of the solutes A and B and represent the situation at the column head. During the chromatographic separation the mass ratio changes locally and the local equilibrium data are calculated by Eq. 2.43.

These effects stem from thermodynamic principles. If in a separation with a given purity requirement the peak resolution is not high enough due to these effects there is no way to overcome this problem by increasing column efficiency, which means optimizing the fluid-dynamic term of the resolution equation (Eq. 2.33). If a tag-along effect occurs the use of a smaller particle diameter has no effect, as long as the

Fig. 2.20 Elution profiles for different mass ratios of the feed concentration.

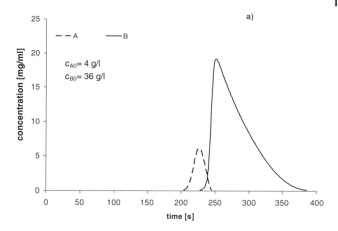

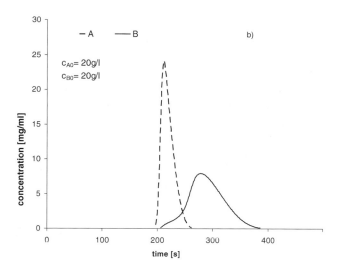

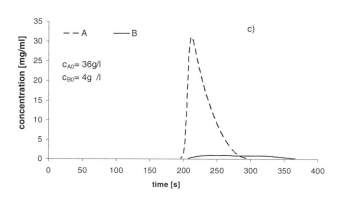

surface chemistry of the stationary phase and the mobile phase composition are kept constant. For such systems only a systematic screening approach for an optimised chromatographic system can lead to a production process with high enough productivity.

2.6.3
Elution Modes

Two elution modes can be applied in operating column chromatography: isocratic elution or gradient elution. In isocratic elution the composition of the eluent is kept constant throughout the operation. In gradient elution the eluent composition is changed during the separation, such that one starts with a weak solvent A and constantly adds volumes of a stronger solvent B, e.g. 5 % B min^{-1}. In this case a linear gradient is accomplished. Alternatively, a step-wise gradient can be applied by increasing the volumes of solvent B. Solvent strength is a parameter describing the strength of a solvent for a given adsorbent using standard solutes.

At isocratic elution the peaks broaden and become flatter with increasing retention, and might disappear in the base line noise. As a rule of thumb the retention factor k' of the last eluting solute in isocratic elution should not be larger than 10. This corresponds to an elution volume of less than 10 column volumes. Very strongly retained solutes may not be eluted and, hence, the column has to be washed with a strong solvent from time to time.

At gradient elution the solvent strength is changed continuously or stepwise, as described above and shown in Fig. 5.12 in Chapter 5.3.5, by increasing the amount of the stronger solvent. As a result of the gradient the peaks become sharper and higher and one can cover several orders of magnitudes of the retention coefficient. Gradient elution is commonly applied at the screening of a complex mixture to receive information on the number and polarity of the components. A linear gradient expands the chromatogram in its first part and compresses it for the late-eluting solutes. An example is given in Fig. 2.21. After running the gradient one has to go back to the initial conditions, which means that the column has to be washed and reconditioned. Parameters to adjust a gradient with respect to optimum resolution are: the starting and final eluent composition and the gradient time and steepness. Problems may occur with respect to the reproducibility of the eluent composition at low amounts (volumes) of solvent A and B, respectively, in a binary mixture. Gradient elution is the preferred technique for analytical HPLC and is also commonly used at preparative scale. One disadvantage is that one ends up with solvent mixtures, which makes re-use, by distillation, difficult and expensive.

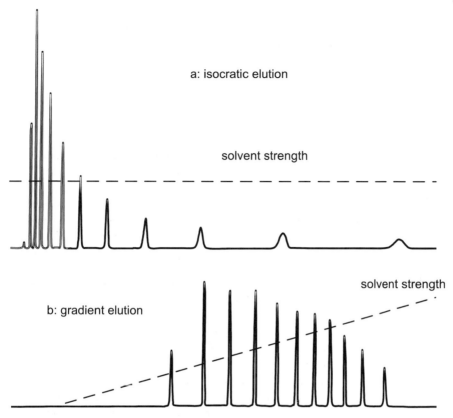

a: isocratic elution

solvent strength

solvent strength

b: gradient elution

Fig. 2.21   Chromatogram for isocratic as well as gradient elution.

2.7
Practical Aspects of Parameter Determination

The parameters defined in this chapter are divided into model parameters and evaluation parameters. Model parameters are: porosity, voidage and axial dispersion coefficient, type and parameters of the isotherm as well as mass transfer and diffusion coefficient. All of them are decisive for the mass transfer and fluid flow within the column. They are needed for process simulation and optimisation. Therefore their values have to be valid over the whole operation range of the chromatographic process. Experimental as well as theoretical methods for determining these parameters are explained and discussed in Chapter 6.

Evaluation parameters are: retention time, selectivity, efficiency or plate number, HETP values and resolution. They characterize the performance of a certain chromatographic separation and are indicators for the selection (or better definition) of a chromatographic system. Strictly speaking they are defined for linear chromatography only.

2.7.1
Linearized Chromatography

In most cases isotherms are nonlinear. Theoretically the range for linear chromatography is restricted to the tangent of the isotherm at its origin, where the slope is equal to the Henry coefficient. For practical applications the lower part of the isotherm can be linearized with satisfactory accuracy. If a chromatographic system has to be selected the evaluation parameters are derived from measured chromatograms, which are more or less accurate and include measurement failures and noise. Before this background the following procedures and recommendations should be considered.

No overload: The chromatogram may be asymmetric, but make sure that the column is not overloaded. This can be tested by two or three pulses with different concentrations. Section 2.6.1 describes the characteristics of an overloaded column.

Retention time and selectivity: With symmetric peaks the retention time is given by the peak maximum. For slightly asymmetric peaks the peak maximum may be taken as well from the maximum, otherwise it is determined by calculation of the first absolute moment (Eq. 2.27). When the dead time of the column as well as the retention times of the components is known the k' and the selectivity α can be calculated.

HETP and plate number: The determination of these parameters, which characterize the efficiency, is more complex. The basic equations are explained in Section 2.4.2. The problem is to determine the peak width or variance from a measured chromatogram. Especially, if the second central moment has to be calculated the result depends very much on the extension of the base line, which has to be chosen for the integration interval. Figure 2.22 gives an example for a slightly asymmetric peak.

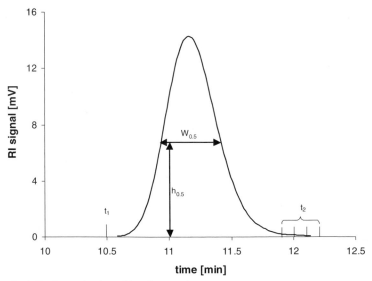

Fig. 2.22 Determination of the plate number.

Table 2.2 Residence time and plate number for different integration intervals.

	1	2	3	4	5
t_1 (min)	10.5	10.5	10.5	10.5	10.5
t_2 (min)	11.8	11.9	12	12.1	12.2
μ_t	11.20	11.20	11.20	11.21	11.21
σ_{t_2}	0.0400	0.0416	0.0428	0.0438	0.0444
$N_i = f\,(h_{0.5})$	3276	3277	3278	3279	3279
$N_i = f\,(\mu_t, \sigma_t)$	3138	3020	2933	2864	2828

Table 2.2 presents the first absolute moment (Eq. 2.27) and the second central moment (Eq. 2.28) for five different integration intervals.

The changes in mean residence time are minimal while the second moment alters very much. Consequently, in this example, the number of plates varies by up to 10% if it is calculated with the second moment. The differences are much smaller if the number of plates is determined on the basis of the peak width at 0.5 h.

One recommendation is to measure the peak width at half peak height. The plate number is then calculated by Eq. 2.25. Alternatively, the peak width of an asymmetric peak is measured at 0.1 of the peak height (Fig. 2.7b). In this case Eq. 2.30 will be used to calculate the plate number. When k_i', α and N_i are known the resolution is calculated by Eq. 2.33a,b.

Please note that different procedures may be used to determine the evaluation parameters. Therefore it has to be ensured that the same methods have been applied if different chromatographic systems are compared. If the values are calculated by automated HPLC-systems the algorithms implemented in the software should be known.

2.7.2
Nonlinear Chromatography

In nonlinear chromatography the thermodynamic effects of the isotherm can be seen in the chromatogram as indicated by Eq. 2.16a and shown in Fig. 2.6. Therefore the mean retention time cannot be calculated by the first absolute moment (Eq. 2.27). A not exact but more reliable value is, for example, the position of the compressed front in case of Langmuir isotherms (Fig. 2.6.3b). This might be used as a qualitative value to characterize and compare different chromatographic systems.

Sometimes selectivity is calculated by Eq. 2.45. But in most cases of nonlinear chromatography this is only a qualitative indicator. As the selectivity depends on the local gradient of the isotherm (Eq. 2.16a) its value varies with the local composition of the mobile phase.

The plate number characterizes the influence of fluid dynamics and mass transfer on the chromatogram for a given column but not the thermodynamic effects of nonlinear isotherms. Therefore this characterization is not applicable in the range of nonlinear chromatography. Consequently, the resolution is not defined as well.

However, as will be shown in Chapter 7.1.2, the evaluation parameters, especially the plate number for the linear regime, are well established for process optimisation.

3
Columns, Packings and Stationary Phases

K. K. Unger, Cédric du Fresne von Hohenesche, M. Schulte

3.1
Column Design

3.1.1
Column Hardware and Dimensions

The column is a cylindrical tube with a given inner diameter d_c and length, L_c. Stainless steel, glass and plastic materials (cross-linked organic polymers) are employed as tube materials. Stainless-steel tubes have a mirror finish inside with a surface roughness smaller than 1 μm. The column material should be mechanically stable towards high flow rates and high pressure and chemically resistant to aggressive eluents. Typical column dimensions for preparative and large-scale chromatography (process chromatography) are summarized in Tab. 3.1.

Figure 3.1 shows a scheme of an HPLC plant. The column contains a packing that is a dense array of porous or non-porous particles. Alternatively, the column is composed of a continuous bed, i.e. a porous piece called a monolith.

Figure 3.2 shows a photo of a real preparative chromatography system that corresponds to the scheme of Fig. 3.1.

Table 3.1 Typical column dimensions and operation conditions in preparative and process chromatography.

Type of column	Column dimensions [$L_c \times d_c$ (mm)]	Flow-rate range (ml min^{-1})	Max. pressure drop (bar)
Preparative	100 × 20	20–40	300
	300 × 50	100–200	200
Large scale	300 × 100	500–1000	150
	200 × 300	8 000–12 000	100
	100 × 600	32 000–48 000	70
	100 × 1000	80 000–13 0000	50

Preparative Chromatography. Edited by H. Schmidt-Traub
Copyright © 2005 WILEY-VCH Verlag GmbH & Co. KGaA, Weinheim
ISBN: 3-527-30643-9

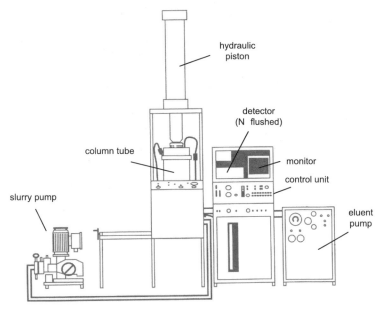

Fig. 3.1 Scheme of a preparative chromatography system.

There is a major difference in maintaining the stability of the column bed between analytical and process columns packed with particles. Analytical columns with d_c = 4 mm (4000 μm) packed with 5 μm particles contain 800 particles (4000/5) over the column diameter. The wall of the column stabilizes the column bed; however, the packing density at the column wall is less than at the core. The column wall region extends to approximately 30 particle diameters.

Fig. 3.2 Photo of a preparative chroma-tography system.

A preparative column of $d_c = 50\,mm$ ($50\,000\,\mu m$) packed with $15\,\mu m$ particles contains approximately 3000 particles over the column diameter. For column stability the column wall plays only a minor role and the packing depends on the mechanical stability of the bed. While the packing pressure to pack analytical columns is often raised to 550 bar (stainless steel microbore columns can only be efficiently packed at 2000 bar according to Roumeliotis et al., 1984), the maximum packing pressure for preparative columns seldom exceeds 100 bar.

When considering a packed bed the interparticle or interstitial porosity amounts to approximately 40%. The average diameter of these interstitial voids calculates for a fairly regular packing to 0.4 times of the particle diameter, e.g. a regularly packed bed with $15\,\mu m$ particle creates interstitial pores of $6\,\mu m$. A bimodal particle size distribution of silicas employed for preparative columns can often be found, e.g. a major distribution around $15\,\mu m$ and a minor peak at $4\,\mu m$. These small particles are thought to occupy the interstices between the $15\,\mu m$ diameter particles and thus to stabilize the bed.

One of the main tasks in preparative chromatography is the scale-up and scale-down of separations from one column dimension to another one. It is of great importance to obtain correct values for all process parameters when changing the column geometry. The basic equation allowing for comparison between different columns is Eq. 3.1.

$$\frac{X_i}{\frac{\pi}{4}d_{c,i}^2} = \frac{X_j}{\frac{\pi}{4}d_{c,j}^2}\frac{L_{c,i}}{L_{c,j}} \tag{3.1}$$

In most cases the scale-up and -down factors are calculated by using Eq. 3.2:

$$F = \frac{d_{c,j}^2}{d_{c,i}^2} \cdot \frac{L_j}{L_i} \tag{3.2}$$

Scaling factors between typical column dimensions for preparative and large-scale chromatography are summarized in Appendix A1.

To scale the properties of preparative chromatographic separations it has to be kept in mind that the different factors are scaled according to column diameter, length or volume. Table 3.2 summarizes the relevant scaling factor for single system parameters.

Table 3.2 Scaling factors according to column diameter, length and volume.

Scaling according to diameter	Scaling according to length	Scaling according to volume
$F = \dfrac{d_{c,j}^2}{d_{c,i}^2}$	$F = \dfrac{L_{c,i}}{L_{c,j}}$	$F = \dfrac{d_{c,j}^2}{d_{c,i}^2} \cdot \dfrac{L_{c,j}}{L_{c,i}}$
• Flow rate • Eluent consumption	• Retention time • Cycle time • Plate number	• Amount of adsorbent • Feed amount injected

Table 3.3 Commercial suppliers for preparative columns.

Company	Contact details
Armen Instruments	www.armen-instrument.com
Bayer Technology Services	www.bayertechnology.com
Dan Process A/S	www.dan-process.dk
Eastern Rivers Inc.	easternriv@aol.com
Euroflow Ltd. (only glass columns)	www.euroflow.net
Infraserv Hoechst Technik	www.infraserv-hoechst-technik.com
Merck KGaA	www.merck.de
(United States: EMD Chemicals Inc.)	www.emdchemicals.com
Millipore Corporation	www.millipore.com
MODcol Corporation (now Grace Vydac)	www.modcol.com
Novasep Inc.	www.novasep.com
Omnichrom.	www.omnichrom.com
Varian Inc.	www.varianinc.com
Waters Corporation	www.waters.com

Table 3.3 provides a survey of commercially available column systems made of stainless steel for high-pressure applications. Only column systems that allow the packing to be performed by the user are listed.

3.1.2
Columns with Particles (Particulate Column Beds)

Special processes have been developed to manufacture spherical packing materials with a given particle size distribution. The size distribution is narrowed by a size classification process, e.g. by air elutriation. Typically, the average particle size of a packing for an analytical column lies between 3 and 5 µm, the d_p of a packing for preparative columns ranges from 10 to 50 µm.

Analytical columns are packed by the slurry technique, where a dilute suspension of the packing is pumped at a high flow rate and a high pressure through the column. Particles are retained by a porous frit at the end of the column. Preparative columns with d_p between 10 and 30 µm are, preferably, packed by the dynamically axial compression technique (Unger, 1994). The column contains a movable piston that keeps the packing under an external pressure during operation. The operation pressure should be always less than the piston pressure.

The particles of the bed of an analytical column are held by the column wall and by the porous frits at the column top and column end. Specific frit systems have been developed by column manufacturers to enable a homogeneous distribution of the flow across the column. Wide bore preparative columns contain distributors at both ends for optimum sample distribution. Both the quality of the frits and the distributors significantly affect the performance of a chromatographic column. While for analytical columns the bed is supported by friction between the column wall and the packed particles, the particles of wide bore preparative columns are subjected to a much higher mechanical stress.

Packed bed particles are porous (for definitions see Chapter 2.2.2). Particle porosity ε_p ranges from 50 to 90%. This means that 50 to 10% are solid impermeable skeleton. Rigid particles made of silica show a permanent porosity while particles made of cross-linked organic polymers might change their porosity depending on the liquid of immersion. The total column porosity ε_t ranges from 65% to 80%. Interstitial porosity ε amounts to approximately 26 to 48%. For comparison the densest packing of ideal spheres in a hexagonal closed array calculates to 26%. In other words, the particles in a packed column retain some degree of mobility, which might change the bed under pressure fluctuations and other influences.

Notably, the flow passes through the interstitial cross section, because the pores of the particles are 200 times smaller (approximately 10 nm) than the width of the interstitial cross section (ca. 2000 nm at a particle size of 5 μm). Transport within the pores is governed by diffusion.

The total porosity of a small column is most accurately measured by weighing the column filled with the mobile phase followed by re-weighing after the eluent has been removed by displacement of helium or nitrogen with subsequent drying.

3.1.3
Columns with a Continuous Bed (Monolithic Columns)

Monolithic columns are composed of a single piece of highly porous adsorbent. The total porosity reaches up to 90%, i.e. 10% remains as a solid skeleton. Silica monoliths possess a bimodal pore size distribution: Macropores with average pore diameters between 1 and 10 μm, so-called through-pores, and mesopores of 10 to 30 nm average pore diameter. The latter generate the necessary specific surface area for the solute–surface interactions. Mesopores are termed as diffusional pores because their access occurs through diffusion. More than 70% of the total porosity belongs to the macropore porosity and 25% to the porosity of mesopores. Thus, the macroporosity is remarkably higher than the interstitial porosity of a particulate column. The most striking feature of the monolithic columns is the high connectivity of the macropores and of the mesopores. This highly connected macropore system generates a much higher permeability than particulate columns and, hence, a lesser column pressure drop. The high connectivity of mesopores creates a fast mass transfer, which is reflected by the maintenance of the column efficiency at high flow rates of the eluent. The outstanding pore connectivity arises because two separate coherent bulk systems are formed during synthesis, i.e. an organic polymeric phase and a siliceous phase – the former phase is removed during synthesis to generate the macropores. Figure 3.3 shows a scanning electron micrograph of a cross-section of a Chromolith™ of Merck KGaA, Darmstadt, Germany. The high inter-connectivity of the macropores can be easily seen. Mesopores are not visible due to the low magnification.

Fig. 3.3 SEM picture of a monolithic silica packing (Chromolith™, Merck KGaA, Darmstadt, Germany).

3.1.4
Column Pressure Drop

In chemical engineering the Ergun equation (Eq. 3.3) is well known for the calculation of pressure drops for fixed beds with granular particles.

$$\psi = \frac{150}{\text{Re}} + 1.75 \tag{3.3}$$

It covers the broad span from fine particles to coarse materials (Brauer, 1971). For chromatographic columns very small particles, mostly spherical, are used. Therefore, the Re numbers are very small and inertial forces can be neglected. Equation 3.3 then reduces to its first term, which represents the pressure drop because of viscous forces only.

The friction number is defined (Eq. 3.4) as

$$\psi = \frac{\varepsilon^3}{(1-\varepsilon)^2} \frac{\Delta p}{\rho u_0^2} \frac{d_p}{L_c} \tag{3.4}$$

where u_0 is the velocity in the empty column (superficial velocity) (Eq. 3.5),

$$u_0 = \frac{\dot{V}}{A_c} = \frac{\dot{V}}{\pi d_c^2 / 4} \tag{3.5}$$

Introducing Eq. 3.4 in the reduced Eq. 3.3 leads to Eq. 3.6

$$\Delta p = 150 \frac{(1-\varepsilon)^2}{\varepsilon^3} \frac{\eta u_0 L_c}{d_p^2} \tag{3.6}$$

This equation is identical to Darcy's law (Eq. 3.7) which is well known for chromatography (Guiochon et al., 1994b),

$$\Delta p = \frac{1}{k_0} \frac{\eta u_0 L_c}{d_p^2} \tag{3.7}$$

with

$$k_0 = \frac{\varepsilon^3}{150(1 - \varepsilon)^2} \tag{3.8}$$

The coefficient k_0 lies between 0.5×10^{-3} and 2×10^{-3}. This agrees with Eq. 3.8 where a void fraction of 0.4 results in a corresponding value of 1.2×10^{-3}. For practical applications k_0 has to be measured for a given packing.

Another expression derived from Darcy's equation that is often used in the literature is the column permeability B (Eq. 3.9).

$$B = \frac{\eta u_0 L_c}{\Delta p} \tag{3.9}$$

Comparing Eqs. 3.7 and 3.9, the permeability is related to k_0 by

$$B = k_0 d_p^2 \tag{3.10}$$

Pressure drop measurements are used to check the stability of a column bed as a function of eluent flow rate. Deviations from linearity at ascending and descending flow rates serve as a strong indication of irreversible changes in the column bed. It is also important to notice the linear dependency between pressure drop and viscosity. The pressure drop for high-performance columns may increase up to 200 bar. Therefore the viscosity is also a strong argument for the choice of the eluent. To pack column beds properly a liquid with low viscosity has to be chosen. If ethanol for instance is the eluent, methanol may be used for the packing procedure as its viscosity is 1.9 times lower. Equation 3.6 indicates the influence of the void fraction on the pressure drop. For particles with varying diameter the void fraction of the column can decrease tremendously. Therefore, the span of the particle size distribution for chromatographic adsorbents should be chosen within the range 1.7–2.5 for analytical columns (Section 3.2.6.1). For preparative columns packed with 10–20 μm particles the distribution can be much broader or can even be bimodal.

3.1.5
Frit Design

The heart of every preparative chromatographic system is the column packed with the adsorbent. If all other parts of the equipment are well designed with regard to minimum hold-up volume, the column is responsible for the axial dispersion of the separation. The column has therefore to be designed in an optimal way. Tremendous work has been done to obtain good preparative columns. Typical column design is of cylindrical form due to the ease in manufacturing. Quite obviously, the bigger the column diameter the more severe the problems of equal distribution and collection are, e.g. with a flow coming from a 2 mm tube diameter into a column with a diameter of more than 1000 mm and collecting the effluent again in a 2 mm tube (Fig. 3.4). To optimise the fluid flow in a chromatographic column several attempts have been made with regard to the fluid distributor and collector systems and the shape of the column tube itself.

The easiest means of fluid distribution at the column inlet is by using a high-pressure drop of the packed bed, which forces the fluid inside the inlet frit into the radial dimension (Fig. 3.5).

The layer design of the frits and the quality of the radial flow characteristics of the frit are of great importance to obtain a good distribution. Distribution by using the column pressure drop has nevertheless a severe drawback: every HPLC plant should be designed to show the lowest possible system pressure drop (i.e. to allow the maximum linear velocity with a given maximum pressure drop). Therefore, efficient pressure-less distribution systems had to be designed. One approach is the integration of distribution plates into the column inlet. Nowadays, their design is optimised by computational fluid dynamics (CFD), as shown by Boysen et al. (2002). The most advanced approach uses distribution channels with engineered fractal geometries as an alternative to random free turbulence (Kearney 1999 and 2000).

The second possibility for influencing the fluid distribution is the design of the column inlet and outlet geometry. Geometry optimisation aims to equalize all fluid streams regardless of their radial position. A conical shape was first proposed by Stahl (1967) for the zone focusing in thin-layer chromatography (Fig. 3.6).

It was successfully adapted to column chromatography with the Lobar® glass column system by, e.g., Henke and Ruelke (1987). The conical inlet and outlet made these columns superior to cylindrical glass tubes with regard to column efficiency and fraction concentration (Fig. 3.7).

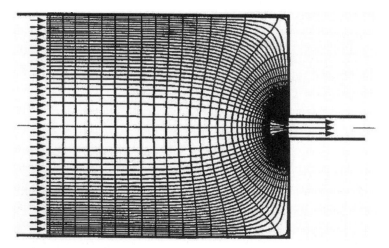

Fig. 3.4 Isobars and streamlines for a cylindrical column outlet.

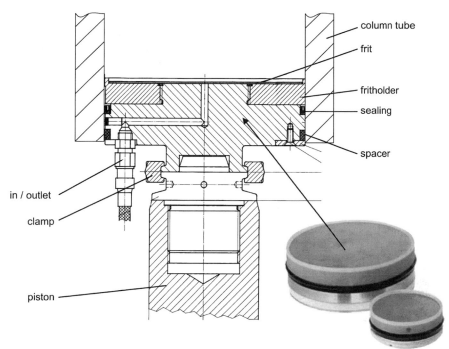

column tube
frit
fritholder
sealing
spacer
in / outlet
clamp
piston

Fig. 3.5 Cross section of a column inlet with frit system.

This concept was not taken into consideration when the era of high-pressure stainless steel columns was introduced. At that time fluid distribution by high-pressure drop, as explained above, was dominant. Only more recently, with the advent of low-pressure high-performance chromatography, has the focus been more and more on

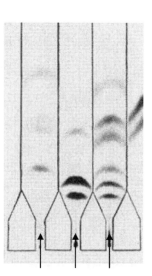

Fig. 3.6 Conical concentration zone of a thin-layer plate according to Stahl (1967).

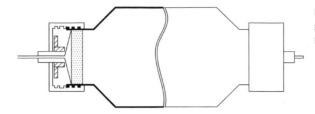

Fig. 3.7 Scheme of a Lobar® glass column for efficient low-pressure chromatography.

column design. Pelz et al. (1998) have compared different inlet and outlet geometries (cylindrical, conical, parabolic and hyperbolic funnel) for large-scale columns (Fig. 3.8).

The result of such column inlet and outlet design is shown in Fig. 3.9. The concentration profile is plotted against dimensionless eluted volume for the cylindrical and exponential funnel. As can be seen, the exponential funnel having the same volume exhibits a much steeper peak.

Lisso et al. (2000) have designed elliptic frits that exhibit the same effect with higher manufacturing challenges.

Whether the open space in the column inlet or outlet has to be filled (by packing or frit material) is not yet decided, but it seems to be much easier to operate the column with an open funnel-type column inlet and outlet. However, today's machining possibilities allow the manufacturing of the stainless steel parts with high precision at nearly all diameters.

An alternative approach to the conical form of column inlet and outlet is the conical column. Here the whole diameter of the column is narrowed in the axial direction (Jiping et al., 2003).

A totally different approach for optimum fluid distribution is the column inlet spray concept proposed by Müller et al. (1997), which shows favorable effects for very large columns. The mobile phase flow is stopped during the feed introduction, the feed is sprayed on top of the packed bed through several nozzles and, when

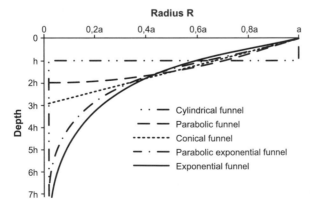

Fig. 3.8 Comparison of different inlet geometries for large-scale columns. (Reproduced from Pelz et al. 1998.)

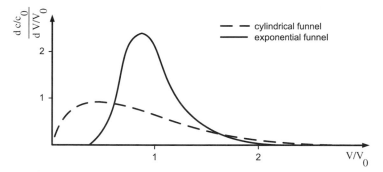

Fig. 3.9 Improvement of axial dispersion of a tracer solute as a consequence of different outlet geometries. (Reproduced from Pelz et al., 1998.)

the appropriate amount is introduced into the column, the mobile phase flow is restarted, pushing the feed as a real plug flow distribution through the column.

The final approach worth mentioning for optimizing the geometry of the column with respect to throughput is the radial column design (Fig. 3.10).

In this column the flow is distributed over the cylindrical segment of the column radius and then passed radially through the packed bed towards the inner cylinder, where it is collected and pushed out of the column in the axial direction. This concept can be used if the efficiency requirement in terms of plates per column is low so that the radial passage through a short packed bed is sufficient for separation. Thus, the concept is more advisable for adsorption/desorption processes than for high efficiency chromatographic separations.

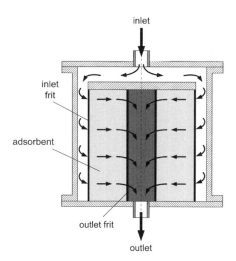

Fig. 3.10 Scheme of radial column design.

3.2
Column Packings

3.2.1
Survey of Packings and Stationary Phases

Adsorbents can be grouped according to their chemical composition. In general, one distinguishes between inorganic types such as active carbon, zeolites, porous glass and porous oxides, e.g. silica, alumina, titania, thoria and magnesia, and the families of cross-linked organic polymers. The former are classical adsorbents and possess a crystalline or amorphous bulk structure. They possess hydrophobic as well as hydrophilic surface properties. The main criterion that distinguishes them from cross-linked polymers is their high bulk density and the porosity, which is permanent except under certain conditions, e.g. very high pressures. The texture of inorganic adsorbents resembles more closely a corpuscular structure rather than a cross-linked network.

The mechanical strength of polymers is achieved through extensive cross-linking whereby a three-dimensional network of hydrocarbon chains is formed. Depending

Table 3.4 Survey on column packings and stationary phases in chromatographic techniques.

Designation	Polarity	Types	Mobile phases
Normal phase packings	Polar	Varying in pore size and specific surface area	Non-polar , moderately polar mobile phases
Reversed phase packings	Non-polar	Reversed phase silicas with n-alkylsilyl groups, reversed phase silicas with hydrophobic polymer coatings, hydrophobic cross-linked organic polymers, porous carbon	Aqueous/organic mobile phases
Medium polar packings	Between polar and non-polar	Silicas with bonded propylcyano groups, silicas with bonded diol groups, silicas with bonded amino groups	Buffered aqueous /organic mobile phases (reversed phase conditions) moderately polar organic mobile phases
Enantioselective packings	Polar or non-polar	Packings with enantioselective cages or enantioselective surfaces, microcrystalline cellulose triacetate, cellulose ester or cellulose, carbamate/silica composites, optically active poly(acrylamide)/silica composites, chemically modified silicas (Pirkle phases), cyclodextrine modified silicas	Operated either with normal phase or reversed phase mobile phases

on the extent of cross-linking, soft gels, e.g. agarose, and highly dense polymer gels are obtained. The bulk density is much lower than that of inorganic adsorbents. The hydrophilicity and hydrophobicity of cross-linked polymer gels is tuned by the chemical composition of the backbone polymer as well as by the surface chemistry. Organic chemistry provides enormous scope in designing the functionality of the surface of polymeric adsorbents. Even at a high degree of cross-linking, polymers, still show a swelling porosity, i.e. the porosity of a polymeric adsorbent depends the type of solvent. For example, the volume of a soft gel can be ten times higher when immersed in a solvent than in the dry state. For both types of adsorbents the porosity and pore structure can be manipulated by additives such as templates, volume modifiers (so-called porogens) and other additives.

A note to the terms hydrophobicity and hydrophilicity: these are frequently employed in characterizing adsorbents. Hydrophilicity/hydrophobicity is a qualitative measure of an adsorbent, characterizing its behavior towards water. The term lipophilicity/lipophobicity is applied to characterize the polarity of a compound. Table 3.4 summarizes the different column packings and stationary phases used in preparative chromatography.

3.2.2
Generic, Designed and Tailored Adsorbents

Another classification of adsorbents that considers their use in preparative chromatography is the division into generic, designed and tailored stationary phases. Generic stationary phases include non-sophisticated adsorbents that are made available in large quantities and serve for rather simple purification processes. In contrast, tailored stationary phases represent adsorbents with high specificity and are employed in small-scale isolations. Designed stationary phases are located in-between the former groups. „Design" usually refers to modification of the adsorbents surface chemistry to match a given isolation process. Reversed phase silica is a predominant example of designed stationary phases.

The basic features for preparative chromatography are their applicability, selectivity and specificity, which affect the column performance, productivity and cost. Furthermore, it is important to know how flexible the materials have to be concerning integration into a given process chain and which bulk quantities are available when a scale-up of the purification and isolation is considered.

The next section briefly surveys the most relevant types of generic adsorbents and then focuses on two groups of designed adsorbents, i.e. porous silicas and synthetic cross-linked polymers. Detailed discussion of tailored adsorbents is beyond the scope of this chapter as they are used in the purification of enzymes and other biopolymers, which is not the focus of the book.

3.2.2.1 Generic Adsorbents
Generic adsorbents usually represent bulk materials of low cost produced in several thousand metric tons per year. They are employed in industrial purification pro-

Table 3.5 Survey of the generic types of inorganic packings for chromatography and their characteristic properties.

Designation	Bulk composition and bulk structure	Surface characteristics
Activated carbons	Microcrystalline carbon	Hydrophobic, terminating polar groups such as hydroxyl, carboxyl etc.
Zeolites	Crystalline alumosilicates, three-dimensional network of silica and alumina with adjusted silica to alumina ratio	Hydrophobic or hydrophilic depending on the silica/alumina ratio, cation exchanger, Brønsted and Lewis acidity
Porous glass	Amorphous silica	Hydrophilic, cation exchanger, Brønsted acid sites
Porous silica	Amorphous, partially crystalline	Hydrophilic, cation exchanger, Brønsted acid sites, point of zero charge at pH 2–3, pK_a of Brønsted acid sites 7
Porous titania	Anatas	Brønsted and Lewis acid and basic sites, cationic and anionic exchange groups, point of zero charge at pH 5, pK_a of Brønsted acid sites 0.2–0.5
Porous zirconia	Crystalline, monoclinic	Brønsted and Lewis acid and basic sites, cationic and anionic exchange groups, point of zero charge at pH 10–13, pK_a of Brønsted acid sites 7
Porous alumina	γ-alumina	Brønsted and Lewis acid and basic sites, cationic and anionic exchange groups, point of zero charge at pH 7, pK_a of Brønsted acid sites 8.5

cesses, e.g. in cleaning drinking water or in drying air or removing pollutants. Typical examples are active carbons and bentonites. Various types of materials have been surveyed by Kurganov et al. (1996) and Nawrocki et al. (1993) and are summarized in Tab. 3.5.

Activated Carbons
Activated carbons are the most widely used adsorbents in gas and liquid adsorption processes. They are manufactured from carbonaceous precursors by a chain of chemical and thermal activation processes. The temperature for carbonization and activation reaches up to 1100 °C in thermal processes. Activated carbons develop a large surface area, between 500 and 2000 $m^2 g^{-1}$, and micropores with an average pore diameter <2 nm. Mesoporosity and macroporosity is generated by secondary procedures such as agglomeration. The products are shaped as granules, powders and pellets depending on their application.

The bulk structure is predominantly amorphous and the surface is hydrophobic. During activation with oxygen polar functional groups such as hydroxyl, both carbonyl and carboxyl are formed that act as acidic and basic surface sites. As a result, hydrophilicity is introduced to the surface. The world production of activated carbons in 2002 was estimated to be ca. 750 000 metric tons. One discriminates between gas- and liquid-phase carbons. Typical liquid phase applications are: potable water treatment, groundwater remediation, and industrial and municipal waste water treatment and sweetener decolorization. Gas adsorption applications are solvent recovery, gasoline emission control and protection against atmospheric contaminants.

Synthetic Zeolites

Zeolites represent a family of crystalline alumosilicates with a three-dimensional structure. They possess regularly shaped cavities with eight-, ten- and twelve-membered silicon–oxygen rings. The pore openings range from 0.6 nm (10-ring) to 0.8 nm (12-ring). Zeolites are cation exchangers. The exchange capacity is controlled by the aluminium content. Silica-rich zeolites, e.g. MFI-type, possess a hydrophobic surface. Zeolites are made from water glass solutions under alkaline conditions. An amorphous gel is first formed, which is subjected to hydrothermal treatment. The amorphous gel then converts into crystallites 0.1 to 5 μm in size. Silica-rich zeolites are manufactured in the presence of a structure-directing template, e.g. tetraalkylammonium salts. Table 3.6 gives a survey of the type of zeolites and their applications (Ruthven, 1997).

The most important applications are the UOP Sorbex processes (Ruthven, 1997). These are automated continuous processes that separate hydrocarbon mixtures mainly in the liquid phase."

Porous Oxides: Silica, Activated Alumina, Titania, Thoria, Zirconia

Porous inorganic oxides are made through a sol–gel process. The sol is converted into a hydrogel that is subjected to dehydration to form a porous xerogel. Special techniques have been developed to combine the sol–gel transition with a shaping to spherical particles (Fig. 3.11).

Table 3.6 Survey of the type of zeolites and their applications.

Structure	Cation	Window Obstructed	Window free	Effective channel diameter (nm)	Applications
4A	Na$^+$	8-ring		0.38	Desiccant; CO_2 removal; air separation (N_2)
5A	Ca^{2+}		8-ring	0.44	Linear paraffin separation; air separation (O_2)
3A	K$^+$	8-ring		0.29	Drying of reactive gases
13X	Na$^+$		12-ring	0.84	Air separation (O_2);
10X	Ca^{2+}	12-ring		0.80	removal of mercaptans
Silicalite			10-ring	0.60	Removal of organics in aqueous systems

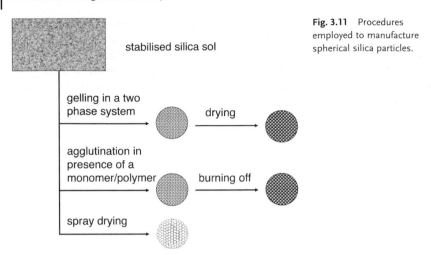

stabilised silica sol

gelling in a two phase system

drying

agglutination in presence of a monomer/polymer

burning off

spray drying

The starting material for processing porous alumina is crystalline hydrargyllite (gibbsite), Al(OH)$_3$, which is subjected to a heat treatment, under controlled conditions, between 500 and 800 °C. During this process water is released, leaving crystalline porous alumina (γ-alumina) (Unger, 1990).

Porous Glasses
Porous glass is manufactured from glass by special heat treatment and leaching processes where the porosity and pore size is adjusted and controlled (Janowski and Heyer, 1982).

Styrene-divinylbenzene Copolymers and Derivatives
Copolymers of styrene-divinylbenzene are the classical polymeric adsorbents, manufactured for producing synthetic ion exchangers. These types of materials are discussed in more in detail in Section 3.2.4.

3.2.2.2 Tailored Adsorbents
Tailored adsorbents are those synthesized for a purification of a specific component. They are made by sophisticated chemistry whereby biospecific or biomimetic ligands are bonded via a linker to a support. Examples of tailored adsorbents are affinity adsorbents and immuno adsorbents. Such materials serve for the isolation of components of high value in milligram amounts.

3.2.2.3 Designed Adsorbents
As named, this type show high flexibility in achieving and adjusting distinct properties during manufacture so as to be easily adapted to the various application scenarios in preparative chromatography. This means that desired properties of the adsorbents, which are mandatory to successfully execute purifications, are adjusted by an appropriate design during synthesis or by an after-treatment of the adsorbent. Such

designed adsorbents enable the user to quickly elaborate and to accomplish a purification procedure and protocol without a high investment of time and cost.

The idea of designing adsorbents for chromatography arose at the beginning of the 1970s when HPLC was developed as a powerful separation technique. There was a need for size classified microparticulate adsorbents. Initial processes were based on milling and sizing of larger grains, e.g. of porous silicas of technical grade. The particles obtained were of irregular shape and contained fines of an average particle diameter <1 µm that were adhered at the surface of the larger ones. Irregularly shaped particles did not allow one to pack highly efficient HPLC columns with acceptable reproducibility and stability as compared with those packed with beads. Thus, spherical particles became the materials of choice. The beads can be formed in different ways, as shown in Fig. 3.11 for porous silicas. The term design not only refers to the particle morphology but also to other parameters such as the average particle diameter and particle size distribution, the porosity, the specific surface area and the pore size distribution. All these properties are directly related to the performance of the stationary phase packings and columns in chromatography in a predictable manner.

3.2.3
Reversed Phase Silicas

Reversed phase silicas are discussed in depth here as this type of packing materials is often applied in preparative chromatography. According to the classification used above, reversed phase silica is an example of a designed adsorbent.

The term „reversed phase" stands for a hydrophobic packing. Reversed phase packings are operated with polar mobile phases, typically aqueous mobile phases containing organic solvents as a second solvent, such as methanol, acetonitrile or dioxane. For hydrophobic solutes the retention time increases with the hydrophobic character. The hydrophobic character of a solute is proportional to its carbon content, its number of methylene groups in the case of a homologue series, its number of methyl groups in the case of alkanes, or its number of aryl groups in the case of aromatic compounds.

The term reversed phase packing is a synonym for a packing with a hydrophobic surface: the most common reversed phase packings are silicas with surface-bonded long-chain n-alkyl groups, also termed reversed phase silicas. The same term is used to describe silicas with hydrophobic polymer coatings. Reversed phase packings are also hydrophobic cross-linked organic polymers (cross-linked styrene-divinylbenzene copolymers) and porous graphitized carbons. These reversed phase packings differ in the degree of hydrophobicity in the relative sequence:

porous graphitized carbon > polymers made from cross-linked styrene/divinalbenzene > n-octadecyl (C18) bonded silicas > n-octyl bonded (C8) silicas > phenyl bonded silicas > n-butyl (C4) bonded silicas > n-propylcyano bonded silicas > diol bonded silicas.

We will demonstrate the synthesis of n-alkyl bonded silicas by chemical surface modification and their properties.

3.2.3.1 Silanisation of the Silica Surface
Objectives
Chemical surface modification of the silica serves to:

- Chemically bind desired functional groups (ligands) at the surface, mimicking the structure of solutes and thus achieving retention and selectivity (group-specific approach).
- Deactivate the original heterogeneous surface of the silica surface to avoid matrix effects. As silica has a weak acidic surface basic solutes are strongly adsorbed, which should be minimized by surface modification.
- Enhance the chemical stability of silica, particularly at the high pH range above pH 8.

Silanisation
The silica surface bears 8–9 $\mu mol\, m^{-2}$ of weakly acidic hydroxyl groups (~ 5 silanol groups per nm^2) when silica is in its fully hydroxylated state. The hydroxyl groups react with halogen groups, OR groups and other OH groups, leaving acids, alcohols and water respectively. The most suitable approach is the use of organosilanes with reactive groups X. Thus, the surface reaction can be written as

$$\equiv SiOH + X\text{-}SiR_3 \longrightarrow \equiv Si\text{-}O\text{-}SiR_3 + HX \qquad (3.11)$$

As a result a siloxane bond is formed and the functional group R is introduced by the organosilane. Silanisation can be performed in many ways and thus the products differ in surface chemistry, which is reflected in the chromatographic behavior.

Starting Silanes
Silanes differ in the type of the reactive group X and in the type of the organic group R. At constant R and X one can discriminate three types of silanes: Monofunctional, bifunctional and trifunctional. Monofunctional silanes undergo a monodentate reaction, i.e. they react with one hydroxyl group, bifunctional silanes may react with one or two groups X. Trifunctional silanes react with a maximum of two groups X per molecule in case of anhydrous reaction conditions. When water is present in the reaction mixture trifunctional silanes hydrolyze and form oligomers by intramolecular condensation. The starting compounds as well as the intermediate then perform a condensation with surface hydroxyl groups. The organic group R is an n-alkyl group with a chain length of 8 or 18. There are also silanes employed with terminated polar groups such as diol, cyano and amino groups with a short n-alkyl spacer such as n-propyl. The latter are employed for consecutive reactions at the terminal polar group.

Depending on the surface modification reversed phase silicas can be grouped into (a) monomeric reversed phase silicas chemically modified with monofunctional silanes and (b) polymeric reversed phase silicas with a polymeric layer made by surface reaction with trifunctional silanes.

Parent Porous Silica

The parent silica is usually subjected to activation prior to silanisation, e.g. treatment with diluted acids under reflux. In this way the heterogeneous surface becomes smoother with homogeneously distributed surface hydroxyl groups to ensure batch to batch reproducibility. Simultaneously, the treatment extracts traces of metals that would otherwise affect the chromatographic separation due to a secondary interaction mechanism, e.g. ionic interaction of the solute with the adsorbent. Special care has to be devoted to the average pore diameter of the parent silica in relation to the size of the silanes and to the size of solute molecules to be resolved. For example, long-chain silanes reduce drastically the pore opening of the modified adsorbent, which leads to hindered diffusion of the components to be isolated. Peak broadening, reduced capacity and low resolution are the resulting effects. Commonly, 10 nm pore diameter materials are recommended as starting silicas because a reduction of the specific surface area, a diminution of the specific pore volume and a decrease of the average pore diameter occur by the silanisation (see also Tab. 3.8).

Reaction and Reaction Conditions

Silanisation is a heterogeneous reaction. Silanes can be in the gas or liquid phase or in solution. The reaction is carried out at elevated temperatures, depending on the volatility of the silane and solvent, in a vessel under gentle stirring or in a fluidized bed reactor. To enhance the kinetics, catalysts are added. With chlorosilanes, organic bases are added as acid scavengers; acids are employed in case of alkoxysilanes as reagents. By-products must be carefully removed by extraction with solvents.

Endcapping

The term endcapping originates from polymerization chemistry, when reactive groups after polymerization are removed by a specific reaction. After primary silanisation the maximum ligand density amounts to 3.5–4.5 $\mu mol\,m^{-2}$ for monofunctional silanes. As the initial hydroxyl group concentration is ca. 8 $\mu mol\,m^{-2}$, only half of the hydroxyl groups have reacted. The large size of the silanes makes is almost impossible to convert all hydroxyl groups due to steric reasons. The remaining are still present at the surface and provide the surface with a partially hydrophilic character. As a result, the chromatographic separation will show significant peak tailing due to the weak ion-exchange properties of the hydroxyl groups present. Reversed phase silicas, even bonded with C_{18} groups, are operated with aqueous eluent up to approximately 70 % v/v water/organic solvent, i.e. the C_{18} bonded phases are not completely hydrophobic. To diminish the hydroxyl groups and the so-called silanophilic activity the silanised materials are subjected to a second silanisation with reactive short-chain silanes. Hexamethyldisilazane (HMDS) and others are the preferred reagent. Figure 3.12a represents the surface of a C_8-modified silica after endcapping. Endcapped reversed phase silica packings exhibit a different selectivity towards polar solutes than non-endcapped materials. Unfortunately, these „base deactivated" phases possess low polarity and therefore similar selectivity towards polar compounds. To overcome the lack of selectivity, a new type of base deactivated stationary phase with polar groups, such as amides or carbamates, „embedded" in the bonded

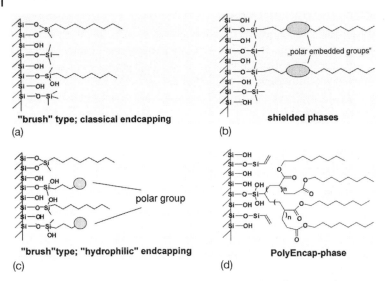

Fig. 3.12 Types of RP columns. (Reprinted from Engelhardt et al. (2001) with permission.)

phase (Fig. 3.12b) have been developed. These polar embedded phases provide polar selectivity without the poor chromatographic performance associated with stationary phases that have high silanol activity. The use of polar or hydrophilic endcapping (Fig. 3.12c) along with bonding of longer alkyl chains such as C_{18} is a successful approach for stationary phases that can retain polar analytes reproducibly under highly aqueous conditions. These polar or hydrophilic endcapping chemicals allow the silica surface to be wetted with water and allow the full interaction with the longer alkyl chains, i.e. even 100 % water can be applied as solvent.

Depending on the parent silica and the way the reversed phase silica was modified with silanes, reversed phase columns exhibit a distinct selectivity towards hydrophobic and polar solutes (Engelhardt et al., 2001).

Example: C₁₈ Bonded Spherical Silica Packing

In the following, the synthesis of the most often employed stationary phase is discussed: spherical silica with an n-octadecyl modification. The synthesis route has been chosen because all synthesis steps are well characterized and documented in standard operation procedure (SOP) protocols. The objective of this work was to develop a manufacturing process for a reversed phase C_{18}-bonded silica column for HPLC according to standardized and validated procedures and to perform certification of the column, the tests and the mobile phases (du Fresne von Hohenesche et al., 2004). Figure 3.13 shows a scheme of the whole manufacturing process, and Table 3.7 summarizes the main steps.

Starting with acid-catalyzed hydrolysis and condensation (step 1) of the silica precursor (tetraethoxysilane, TEOS), the viscous product (poly(ethoxy)siloxane, PES) is converted into silica hydrogel beads in a stirring process under basic conditions (step 2). Both the viscosity of the sol–gel derived PES and the stirring speed directly influence

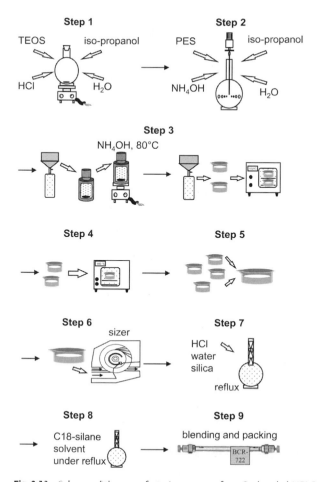

Fig. 3.13 Scheme of the manufacturing process for a C_{18}-bonded HPLC phase.

the particle size and size distribution and are, therefore, critical parameters. As the average particle size and size distribution can never reach an ideal value during synthesis, the products have to be after-treated. After washing and sedimentation (step 3), very few fines (small particles) can be found. The ageing process is performed at elevated temperature and results in the densification and stiffening of the siloxane framework and the complete reaction of the accessible functional groups. However, alkoxy groups from the monomers still remain even after the conversion of the hydrogel into the xerogel by drying. Calcination at temperatures above 673 K has to be performed (step 4). Size classification yields a material with a narrow size distribution (step 6). The broadness and mean value of the latter can be adjusted by varying the parameters of the sizing equipment. At this stage, the loss in silanol groups due to calcination is significant and metal contaminants are often observed. Treatment with hydrochloric acid (step 7) tackles both problems, as siloxane bonds are

Table 3.7 Synthesis and post-synthetic steps of an analytical 5 μm spherical particulate silica with a C_{18}-modified silica surface.

Step number	Synthesis steps	Notes
1	Synthesis of polyethoxysiloxane (PES)	Acid-catalyzed condensation and hydrolysis of monomers, control of viscosity
2	Converting PES into silica hydrogel beads	Stirring process, base-catalyzed
3	Washing, ageing, drying and blending	Removal of impurities
4	Calcination	Densification of material, removal of remaining functional groups from monomer
5	Blending of individual batches	Homogenization
6	Size classification	Narrowing the particle size distribution
7	HCl treatment	Removal of inorganic impurities, increase of surface silanol group concentration
8	Silanisation	Surface modification
9	Blending and packing	Application of material in HPLC

converted into silanol groups and metals are dissolved. The fully hydroxylated spherical silica batch that possesses the desired pore structure as well as a defined particle size distribution can be subjected to silanisation (step 8).

3.2.3.2 Reversed Phase Packings with Polymer Coatings (Types of Polymer Coatings)

There are several variants in immobilizing polymers on supports. Figure 3.14a shows one of the most common coating variants in chromatography, which is generated by physisorption of reactive prepolymers with well-defined chemical composition on the support and subsequent immobilization by thermal treatment or γ-radiation. The procedure was developed by Schomburg (1988) and Figge et al. (1986) on porous silica and alumina. Hanson et al. (1990) have optimised this technique and applied it to non-porous, microparticulate silica, thereby combining polymer selectivity with support efficiency. This method is advantageous, especially in bio-chromatography, since it allows the design of stationary phases with well-defined properties such as hydrophobicity and denaturation potential against polypeptides. Furthermore, chromatographers can design tailor-made polymers, without restriction of the surface properties of the support material, so that any support can be coated. In addition, the following advantages should be mentioned: Retention behavior of the phase can easily be varied by coating the surface with variable amounts of prepolymer before immobilization. Also, more than one polymeric layer can be deposited; the multilayer so-obtained shows special mixed polarities. However, in this so-called „dry" or bulk polymerization, some oligomers may behave like non-wetting liquids on the support surface so that the support material is not completely encapsulated during the cross-linking procedure. This results in inhomogeneous coatings and may lead in some cases to a poor liquid chromatographic performance.

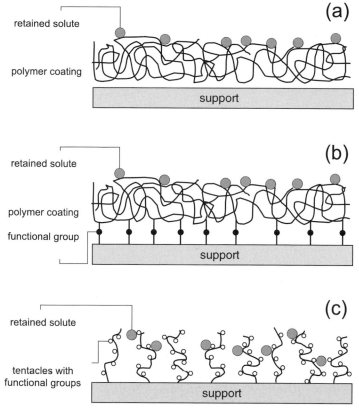

Fig. 3.14 Polymer coating without (a) and with (b) chemical bond to the support surface, and flexible polymeric or oligomeric ligands as stationary phases (c).

An alternative is a chemically-bond coating (Fig. 3.14b) where the polymer layer is connected to the surface by functional or reactive groups, such as vinyl. Polymerization may be carried out in situ in a monomer or oligomer solution, in which the support material is dispersed and the resulting, chemically-bond precursor is subsequently cross-linked. Alternatively, an externally synthesized prepolymer with functional groups is reacted with the support surface, thereby generating an „anchored coating". This type of polymer coating is considered to be more homogeneous, but is limited to the surface chemistry. Therefore, inorganic supports, except silica, are barely described in the literature when this method is reported.

A further polymeric modification is the so-called tentacle type (Fig. 3.14c), where the polymer or oligomer does not encapsulate the bead. This stationary phase is useful for mild separation or purification of proteins by ion-exchange chromatography, as described by Müller (1986, 1990). Chang et al. (1985) and Chang and An (1988) reported a hydrophobic interaction chromatography (HIC) variant where long, flexible side-chains can surround the proteins and thereby interact only with their hydrophilic and/or charged outer surface so that the ternary or quaternary structure of proteins is maintained.

Coating the silica surface with a polymer has become a fashionable technique in surface modification. Polymers and their precursors can be grafted to the silica surface or they can be physically adsorbed. Polymers offer a wide variety of backbones in terms of chemical composition – from hydrophilic to hydrophobic ones. They also provide a unique opportunity to introduce terminating functional groups. Polymer coating has the particular advantage that it is applicable to any kind of inorganic support and not restricted to silica. Polymer coating techniques have been adapted to thoria, zirconia and titania.

3.2.3.3 Physico-chemical Properties of Reversed Phase Silicas

The carbon content, which can be derived from elemental analysis, is used to calculate the ligand density, α_{exp}, when the specific surface area of the starting silica is known.

Figure 3.15 shows a comparative thermogravimetric (TG/DTA) measurement between native (rehydroxylated) silica and the corresponding modified sample. The mass loss of the silica materials upon annealing is recorded. With native silica, the mass loss can be attributed to (a) removal from the surface of physisorbed water that is hydrogen bonded to silanols and (b) evaporation of water molecules formed by the condensation of surface silanol groups, thus leaving siloxane groups. The latter value is used to calculate the silanol group concentration of the material. For chemically modified samples, the residual surface hydroxyl groups can not be assessed by thermo gravimetry because the mass loss arises not only from the condensation of silanol groups. For example, the exothermic DTA peak in Fig. 3.15 indicates combustion of the hydrocarbon surface modification, and endothermic peaks (condensing silanols) are not detected due to the broad range of combustion temperature of the reversed phase bonding.

A better solution is to couple the method with mass spectrometry. Here, the signal for water molecules is clearly separated from the combustion products, leading to a better quantification. Measurement of the silanol group density by deuterium exchange with CF_3COOD followed by ^1H NMR spectroscopy is a chemical method combined with a spectroscopic measurement and gives reliable values.

IR and NMR spectroscopy are powerful techniques for following the modification of silica surfaces.

Nitrogen adsorption/desorption experiments have been performed with silanised silica samples using n-alkyldimethylchlorosilanes with varying chain lengths (Fig. 3.16).

Table 3.8 gives the corresponding data on the pore structural parameters. With increasing silane chain length the specific surface area, as well as the specific pore volume, is evidently reduced by almost a factor of two for the C_{18} modification compared with the native silica sample. In the same fashion, the initial pore diameter of 13.3 nm is diminished to 10.2 nm after silanisation. The decrease in surface coverage (α_{exp}) stresses the increasing steric problems upon silanisation with long-chain silanes.

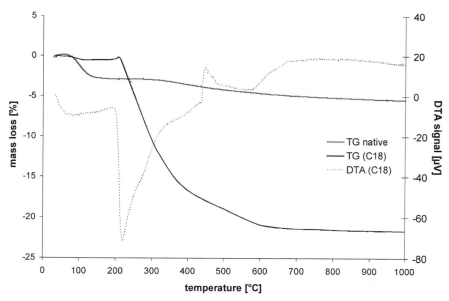

Fig. 3.15 Comparative TG/DTA experiments of native and C_{18}-modified silica.

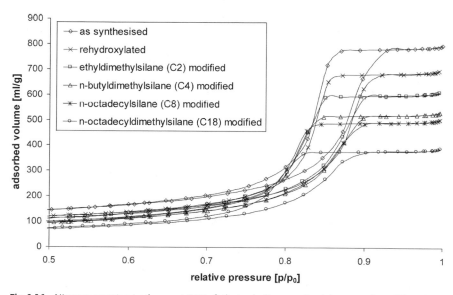

Fig. 3.16 Nitrogen sorption isotherms at 77 K of silanised silica samples (p/p_0 range from 0.5 to 1).

Table 3.8 Pore structural parameters and ligand densities of silanised silica samples.

Modification	a_s (BET) (m^2 g^{-1})	v_{sp} (G) (cm^3 g^{-1})	d_{pore} (BJH) (nm)	α_{exp} (μmol m^{-2})
Rehydroxylation	303	1.06	13.3	0
C2-modified silica	250	0.88	11.6	2.9
C4-modified silica	276	0.83	11.5	2.6
C8-modified silica	224	0.76	11.4	2.6
C18-modified silica	178	0.59	10.2	2.5

3.2.3.4 Chromatographic Characterization of Reversed Phase Silicas

Surface-modified silica-based stationary phase packings in chromatography are mostly characterized under isocratic conditions. The employed tests help to assess chromatographic parameters and make it possible to compare different stationary phases. Robustness, reproducibility and easy handling are the requirements for such tests. It is also important to separate extra-column effects in order to be able to evaluate the column itself rather than the whole HPLC plant system (Chapter 6.3). The following tests give information on hydrophobic properties (retention of non-polar solutes), silanol activity (retention of base solutes), performance, purity and shape selectivity towards selected solutes of modified materials in reversed phase HPLC. It is impossible to find one single suitable test that covers the whole range of chromatographic properties. In addition, the following tests are performed under analytical chromatography conditions. Column tests characterizing normal phase stationary phases are presented in Section 3.4.

Chromatographic Performance

The number of theoretical plates N is a measure of the peak broadening of a solute during the separation process (for definitions see Chapter 2). The efficiency of a column can be given for any solute of a test mixture but is strongly dependent on the retention coefficient of the solute.

Hydrophobic Properties

A dependency on the type of ligand, its density, the eluent used and temperature is found when evaluating hydrophobicities of stationary phases. This property can be assessed by the retention factor of a hydrophobic solute or by the ratio of the retention factors of two non-polar solutes. The latter is called selectivity; for example, when the components differ only in one methyl group, the term methylene selectivity coefficient is applied. Hence, hydrophobic properties describe the polarity of a column and its selectivity towards solutes with only small differences in polarity. This becomes rather important when endcapped stationary phases are compared (Section 3.2.3.1) as some new types of adsorbents allow separation with 100% water as eluent.

Shape Selectivity

Molecular recognition of the solute by the stationary phase with respect to its geometrical dimension is called shape selectivity. For this test, one can employ aro-

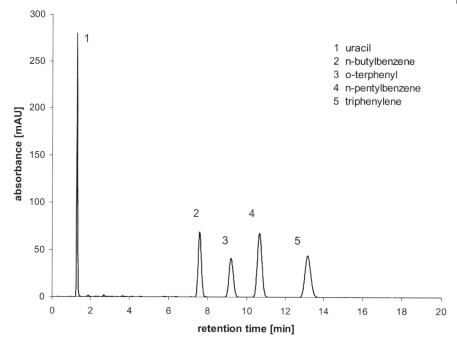

Fig. 3.17 Chromatogram of a test mixture to assess hydrophobic properties, efficiency and shape selectivity of a RP material (eluent: methanol–water, 75:25 v/v).

matic components with identical hydrophobicity that differ only in their three-dimensional shape. The chromatographic selectivity of o-terphenyl/triphenylene or tetrabenzonaphthalene/benzo[a]pyrene are commonly used and show dependencies on several features of the phase, e.g. pore structure, ligand type and density. Figure 3.17 shows a chromatogram of a test mixture of uracil (t_0 marker), n-butylbenzene and n-pentylbenzene (to assess hydrophobic properties and efficiency), and o-terphenyl and triphenylene (to assess shape selectivity). The test mixture was chosen to provide a short analysis time and to facilitate calculation of parameters from baseline separated peaks.

Silanol Activity

As already mentioned, a certain amount of silanol groups remain unreacted on the surface after silanisation. To suppress the resulting secondary interactions in HPLC, buffers can be applied. The selectivity of two basic compounds is a measure of silanol activity. Another way to gain information on this property is to assess the peak symmetry of a basic solute, which is defined as the tailing factor by the USP (United States Pharmacopeia) convention. Figure 3.18 shows a chromatogram of a test of uracil (t_0 marker), benzylamine and phenol using a non-endcapped stationary phase. As can be seen, benzylamine shows peak tailing, indicating strong interaction with residual hydroxyl groups of the silica surface. Some novel adsorbents with hydrophilic endcapping have been developed that reduce peak tailing of base components while retaining high selectivity towards polar and non-polar solutes.

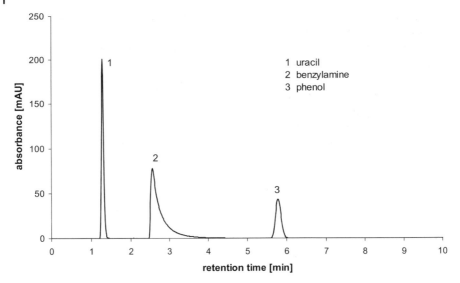

Fig. 3.18 Chromatogram of a test mixture to assess the silanol activity of a RP material (eluent: methanol–buffer, 30:70 v/v, pH 7.0).

Purity

Metals present at the surface of the phase increase the number of secondary interactions with basic substances. One of the proposed tests using 2.2'-bipyridine (complex forming type) and 4.4'-bipyridine (inactive type) shows a good correlation between the metal content and the peak symmetry of the complexing base, starting at impurity levels around 100 ppm. However, the results change with increasing life time of the column as metal ions are accumulated during use.

3.2.4
Cross-linked Organic Polymers

Organic-based supports for use in chromatography have appeared mostly through applications in biochromatography and in size exclusion applications for organic polymers. Such applications also represent the two different basic 'sources' of organic polymers for separation purposes: natural polymers, like agaroses and dextranes with varying degrees of cross-linking, and synthetic organic polymers such as hydrophobic styrene divinyl benzene-copolymers as well as more hydrophilic materials like poly(vinylacetates), synthesized in the late 1960s by Heitz (1970) predominantly for use in size exclusion chromatography. In biochromatography research focused on the increase in biocompatibility and alkaline stability of dextrane and agarose gels, allowing the uninhibited use of sodium hydroxide for cleaning and sterilization protocols. However, the relative softness of such gels has always been a limitation in large-scale applications.

Synthetic organic porous polymers in chromatography suffered for a long time from structural problems. One was the diffusion hindrance in the porous structures – mostly traced back to a considerable amount of micropores generated during the synthesis. The other issue interfering with its more widespread use was the compressibility of the beads, limiting their use significantly, especially in high-pressure applications above around 5 MPa. Several approaches have helped significantly to overcome many of these drawbacks, allowing better utilization of their useful properties in loading, selectivity and cleaning. One approach was the development of more selective synthesis procedures, generating better defined pore structures with significantly reduced micropores together with the synthesis of macroporous material with much higher rigidity, making the use of porous polymers possible even in large-scale preparative chromatography environments. Another was the control of particle size during synthesis, leading to a much better particle size distribution and even to monodispersity, resulting in reduced back-pressure during operation and longer maintenance of the initial system pressure by lack of generation of fines during operation. These improvements, together with an increasing demand for more selective adsorption systems, have already made organic polymers a valuable asset to modern chromatographic supports, and will do so even more in the future. Among other things, their enhanced properties will also allow their broader use in hybrid materials for upcoming separation tasks, especially in the booming life sciences.

3.2.4.1 **General Aspects**

Synthetic cross-linked organic polymers were introduced as packings in column liquid chromatography one decade later than oxides. The first organic-based packings were synthetic ion exchangers made by condensation polymerization of phenol and formaldehyde (Adams and Holmers, 1935). In the 1960s, procedures were developed by Moore (1964) to synthesize cross-linked polystyrenes with graduated pore sizes for size exclusion chromatography. The synthesis of cross-linked dextrane (Porath and Flodin, 1959, Janson, 1937) and agarose (Hjerten, 1964) were milestones in the manufacture of polysaccharide-type packing. At the same time, polyacrylamide packings were synthesized from acrylamide and *N,N*-bismethylene acrylamide by Lea and Sehon (1962).

All the above products, except the first, served as packings in size exclusion chromatography. The major breakthrough in the synthesis of cross-linked organic polymers with tailor-made properties for column liquid chromatography occurred in the decade 1960–1970 (Seidl et al., 1967). Since then, cross-linked organic polymers have maintained a leading position as adsorbents in ion exchange and size exclusion chromatography while their use in column liquid adsorption chromatography to resolve non-polar and polar low molecular weight compounds has been rather limited and bare silica has dominated the market. Currently, the situation seems to be changing slightly, but distinctly. It is the authors' belief that organic-based packings will gain greater importance.

Textbooks often treat the structure of organic- and oxide-based materials according to different aspects (Epton, 1978). On viewing a particle of an organic polymer and an oxide, its structure is best described by a coherent system either of three-

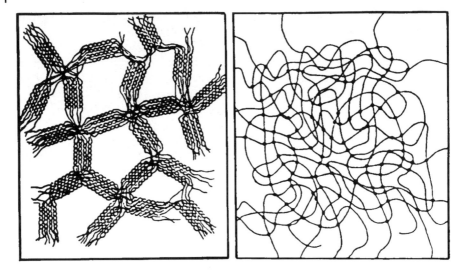

Fig. 3.19 Scheme of the agarose gel network (left) compared with a network formed by random chains of Sephadex or Bio-Gel P (right) at similar polymer concentrations. (Reprinted from Hjerten (1983) with permission.)

dimensionally cross-linked chains or of a three-dimensional array of packed colloidal particles as limiting cases. Thus, previous classifications into xerogel, xerogel-aerogel hybrid, and aerogel appear to be inadequate. To provide a sufficient rigidity of particles, the chains or colloidal particles should be linked by chemical bonds rather than by physical attraction forces.

Some examples serve to illustrate the structure of organic-based packings. Figure 3.19 (Hjerten, 1983) shows a scheme of the structure of a cross-linked polyacrylamide gel (right) and a cross-linked agarose gel of an equivalent polymer concentration (left). Both polymers possess a random coil structure. In the agarose, the double-helix-shaped chains are collected in bundles that generate a quite open structure (Arnott et al., 1974). The structure is stabilized by hydrogen bonds between the chains. When agarose is subjected to cross-linking, links are formed among the chains in these bundles (Porath et al., 1975).

Macroporous, macroreticular or isoporous polymer packings exhibit another type of structure (Fig. 3.20). As the name implies, these polymers contain so-called macropores (>100 nm) and micropores (<2 nm), the latter being inaccessible to large solutes. In other words, macroporous polymer particles constitute an agglomerate made of secondary particles that themselves represent agglomerates of microspheres. This structure resembles that of porous silica particles, which are composed of agglomerates of spherical colloidal silica particles. With respect to the mechanical rigidity of the polymeric packings, cross-linking becomes an essential means in the synthesis. Other requirements that must be met are insolubility, resistance to oxidation and reduction, and a defined, controllable, and reproducible pore structure.

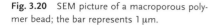

Fig. 3.20 SEM picture of a macroporous polymer bead; the bar represents 1 μm.

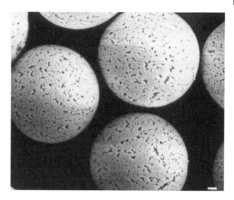

Polymerization is performed either by condensation or addition polymerization, depending on the type of starting monomer. For cross-linking, co-monomers such as divinylbenzene (styrene), ethylene glycol dimethacrylate, epichlorohydrin, 2,3-dibromopropanol, and divinylsulfone (saccharides) are added (Gethie and Schell, 1967, Porath et al., 1971, Laas,1975). The cross-linking reagent can amount to as much as 70 wt.%. Macroporous copolymers are synthesized in the presence of an inert solvent that functions as a volume modifier. Both the cross-linker and the inert solvent have a substantial impact on both the polymerization kinetics and the resulting properties of the copolymer. The decisive parameters relevant for the synthesis of macroporous copolymers have been reviewed by Mikcš et al. (1976). As in the synthesis of silica packings, specific processes must be chosen in polymerization to manufacture polymeric packings with beads of controlled size (Bangs, 1987). Emulsion polymerization starts with a solution of a detergent to which the monomers are added. As a result, micelles swollen with the monomer are formed. After a water-soluble initiator is added (for styrene as a monomer), polymerization leads to particles of exactly the same size as the swollen micelles. Emulsion polymerization processes generate particles of up to 0.5 μm in one step.

Suspension polymerization is usually designed to prepare larger beads, > 5 μm mean particle diameter. The monomer or co-monomer solution is vigorously agitated in water in the presence of a colloidal suspending agent. The colloidal agent coats the hydrophobic monomer droplets (in the case of, e.g., styrene or divinylbenzene). Coalescence of the droplets is prevented by their surface. Adding a lipophilic catalyst or initiator starts the polymerization in the droplets and this continues until the beads are solidified in bulk. The size of the beads is thus controlled by the size of the droplets via the stirring speed.

A third variant in polymerization technology is the swollen emulsion polymerization pioneered by Ugelstad et al. (1980). The procedure is performed in two steps. First the polymerization is started by adding a swelling agent, which causes the submicrometre polymer particles to swell by large volumes of the monomer. The increase in volume can reach a factor of 1000. Secondly, the monomer-swollen beads of defined size are polymerized in a consecutive step.

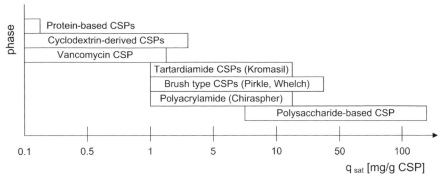

Fig. 3.22 Typical saturation capacity of the most used commercially available CSPs. (Reproduced from Francotte, 2001.)

Table 3.10 Product features of different classes of chiral stationary phases. Loadability data from Francotte (2001).

CSP	Application range	Solvent use	Loadability ($mg\,g^{-1}$ CSP)	Stability	Availability (>10 kg)
CTA, CTB	Broad	Alkanes, alcohols, water	10–100	Good	Yes
Brush-type columns	Small, but focused	Unlimited	1–40	Very good	Major types: yes
Cellulose-, Amylose-derivatives coated on silica	Very broad	Alkanes, alcohols, acetonitrile	5–100	Good, but solvent restrictions	Yes
Cyclodextrin-CSP	Broad		0.1–3	Limited	No
Antibiotic-CSP	Broad	Alcohol + water	0.1-1	Limited	Yes
Polymer-silica composites (Chiraspher, Kromasil)	Limited	Mainly alkanes + alcohols	1–20	Very good	Yes

sure stability requirements for the SMB systems could be reduced. In that context the new monolithic stationary phases may offer some possibilities.

3.2.6
Properties of Packings and their Relevance to Chromatographic Performance

The relevant packing properties are divided into bulk and column properties. The former pertain to the bulk powder before it is packed into the column, the latter characterize the chromatographic properties of the packed column.

3.2.6.1 Chemical and Physical Bulk Properties

The bulk composition and the bulk structure depend on the type and chemical composition of the adsorbent and are largely determined by the manufacturing process. Parameters that characterize the bulk structure are the phase composition, phase purity, degree of crystallinity, long- and the short-range order and defect sites etc. Special care has to be taken with regard to the purity of the adsorbents. Metals incorporated in the bulk and present at the surface of oxides often give rise to additional and undesired retention of solutes. Remaining traces of monomers and polymerization catalysts in cross-linked polymers are leached during chromatographic operation and may affect column performance. Thus, high purity adsorbents are aimed for during the manufacturing processes for the isolation of value-added compounds, e.g. pharmaceuticals.

Among the physical properties of stationary phases the skeleton density and the bulk density of an adsorbent are of major interest, in particular when the column packing is considered. Skeleton density is assessed as the apparent density due to helium from helium penetration measurements. For silica the skeleton density varies between 2.2 and 2.6 g cm^{-3} depending on the bulk structure. The bulk density of a powder is simply determined by filling the powder in a cylinder under tapping until a dense bed is obtained. The bulk density is inversely proportional to the particle porosity: the higher the specific pore volume of the particles the lower is the bulk density. In column chromatography the packing density is a commonly used parameter, expressed in gram packing per unit column volume. The packing density varies between 0.2 and 0.8 g cm^{-3}. The low value holds for highly porous particles, the high value for particles with a low porosity.

Morphology

The morphology of the adsorbent controls the hydrodynamic and kinetic properties in the separation process, e.g. the pressure drop, mechanical stability and column performance in terms of number of theoretical plates. In classical applications granules of technical products were applied with irregularly shaped particles. The desired particle size was achieved by milling and grinding and subsequent sieving. Irregular chips were successively replaced by spherical particles. The latter generate a more stable column bed with adequate permeability. For crude separations, low-cost irregular particles can be used, e.g. in flash chromatography.

Although columns packed with beads can be supplied by manufacturers with highly reproducible properties, the mechanical stability of columns needs to be improved, particularly in terms of the maintenance of a constant flow rate in automated chromatographic systems such as simulated moving bed processes or high-throughput separations. This problem is solved by the design of monolithic columns composed of polymeric materials or silica. Monoliths are constituted of a bimodal, highly interconnected pore system: flow-through pores in the size range of a few micrometres and mesopores in the 10 nm pore size range. Figure 3.3 shows the highly interconnected macropore system, providing a much higher permeability and lower pressure drop than particulate columns. Monoliths are mechanically stable and show neither bed-settling nor channeling if the correct cladding is performed.

One of the greatest challenges in monolith production is to increase the diameter. Presently, silica monoliths with a diameter of up to 50 mm can be produced. Commercial monolithic silica columns e.g. Chromolith columns of Merck KGaA possess a column permeability equivalent to a column packed with 15 μm particles, but show a column performance of a column packed with 5 μm beads. The major benefit of monolithic columns, however, is their robustness in use, their tailored pore structure and the tunable surface chemistry.

Particulate Adsorbents: Particle Size and Size Distribution

Especially when beading processes are applied, the final material has to be subjected to a sizing process to obtain a narrow size distribution. The sized factions are characterized by the average particle diameter and the size distribution. Figure 3.23 shows a differential size distribution based on the volume averaged particle diameter of the as-made product (Fig. 3.23a) and the sized product (Fig. 3.23b). Significantly, the amount of small and very large particles is reduced by the sizing process. As a result, the particle size distribution of the sized material has a near Gaussian appearance."

Usually, the width of the distribution is expressed by d_{p90}/d_{p10}, which should be lower than 2.5 for the chromatographic application; d_{p90} (d_{p10}) equals the value at 90% (10%) of the cumulative size distribution. Figure 3.24a and b visualize the effect of size classification and efficient removal of fine particles.

The particle size is given by an average value: d_p. The average particle diameter can be expressed as a number average, $d_{p(n)}$, a surface area average, $d_{p(s)}$, a weight average, $d_{p(w)}$ and a volume average, $d_{p(v)}$. Based on statistics the following sequence is achieved $d_{p(n)} < d_{p(s)} < d_{p(w)} < d_{p(v)}$.

Usually, in chromatography the volume average particle diameter is employed. For comparison the particle size distribution based on the number and the volume average is shown for the same silica measured by the same technique (see Tab. 3.11).

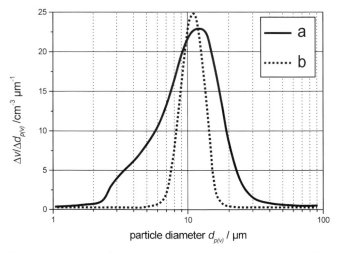

Fig. 3.23 Particle size distribution of silica spheres before (a) and after (b) size classification.

Fig. 3.24 SEM pictures of silica spheres before (a) and after (b) size classification.

Table 3.11 Comparison of the particle size distribution data of a LiChrospher Si 100 silica. (Kindly supplied by Dr. K.-F. Krebs, Merck KGaA, Darmstadt, Germany.)

	Volume statistics	Number statistics
Calculation range	0.96% to 33.5%	0.96% to 33.5%
Volume	100%	100%
Mean	8.267 µm	6.240 µm
Median (d_{p50})	8.236 µm	7.138 µm
Mean/median ratio	1.04	0.874
Mode	8.089 µm	7.341 µm
Standard deviation	0.091	0.221
Variance	0.0082	0.049
Skewness	462.5 left skewed	-2.010 left skewed
Size analysis		
d_{p5}	6.067	1.545
d_{p10}	6.475	2.865
d_{p50}	8.236	7.138
d_{p90}	10.740	9.344
d_{p95}	11.540	10.090
d_{p90}/d_{p10}	1.66	3.52

Apart from the respective value the distribution expressed as the ratio of $d_{p(90)}/d_{p(10)}$ is different. Based on the volume average, $d_{p(90)}/d_{p(10)}$ calculates to 1.66, for the number average it is 3.52.

The particle size of a packing affects two major chromatographic properties: the column pressure drop and the column performance in terms of plate number. For simplicity:

- Δp is inversely proportional to the square of the average particle diameter of the packing and
- the plate number is inversely proportional to the particle diameter

The optimum average particle size of preparative stationary phases with respect to pressure drop, plate number and mass loadability is between 10 and 15 µm.

Fig. 3.25 TEM image of a silica xerogel (primary particles 10 nm).

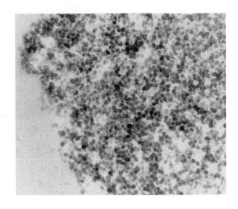

Pore Texture

The pore texture of an adsorbent is a measure of how the pore system is built. The pore texture of a monolith is a coherent macropore system with mesopores as primary pores that are highly connected or accessible through the macropores. Inorganic adsorbents often show a corpuscular structure; cross-linked polymers show a network structure of inter-linked hydrocarbon chains with distinct domain sizes. Porous silicas made by agglutination or solidification of silica sols in a two-phase system are aggregates of chemically bound colloidal particles (Fig. 3.25).

The size of the colloidal non-porous particles determines the specific surface area of an adsorbent (Eq. 3.12) and the porosity of particles is controlled by the average contact number of non-porous primary particles.

$$a_s = \frac{6}{d_p \rho_{app(He)}} \tag{3.12}$$

The average pore diameter is given by Eq. 3.13 and shows that pore structural parameters such as the specific surface area, pore diameter and specific pore volume are interrelated and, therefore, can be varied independently only to a certain extent.

$$d_{pore} = \gamma \left(\frac{v_p}{a_s} \right) \tag{3.13}$$

As an example, spray drying of nanosized porous particles yields spherical agglomerates with a bimodal pore size distribution: the pores according to the primary particles (intraparticle pores) and the secondary pores (interparticle pores) made by the void fraction of the agglomerated nanobeads. Figure 3.26 shows spherically agglomerated nanoparticles that build up a porous bead. For a given size d_p of the nanoparticles the pore diameter of the interstitial pores corresponds to about 40% of d_p. According to porosimetry experiments this rule of thumb was verified, showing a mean pore size at 300 nm in the case of 750 nm nanoparticles.

Pore Structural Parameters

The pore structure of an adsorbent is characterized by the dimensionality of the pore size (unidimensional channels, three-dimensional pore system), pore size distribu-

Fig. 3.26 SEM picture of a 20 µm spherical agglomerate consisting of 750 nm particles.

tion, pore shape, pore connectivity and porosity. The specific surface area is related to the average pore size. Micropores generate high specific surface areas in excess of $500 \, m^2 \, g^{-1}$, mesopores have values between 100 and $500 \, m^2 \, g^{-1}$ and macropores of d_{pore} larger than 50 nm possess values of less than $50 \, m^2 \, g^{-1}$. Micropores of d_{pore} smaller than 2 nm possess low specific pore volumes of smaller than $0.2 \, cm^3 \, g^{-1}$. Mesopores have specific pore volumes of the order of 0.5 to $1.5 \, cm^3 \, g^{-1}$ whereas macropores generate much higher porosities. Porosity can change due to swelling with cross-linked organic polymers. Porous oxides exhibit a permanent porosity at common conditions. The pore size distribution of adsorbents with a permanent porosity is determined either by nitrogen sorption at 77 K (micropore size and mesopore size range) or by mercury intrusion (macropore size and mesopore size range). The pore size distribution can be expressed as a number or volume distribution. However, the pore size distribution does not give any indication of how the pores are interconnected across a porous particle. The most useful parameter in this context is the (dimensionless) pore connectivity n_T, which is derived from the pore network model applied to the experimental nitrogen sorption isotherm (Meyers and Liapis, 1998 and 1999). Pore connectivity describes the number of pore channels meeting at a node at a fixed lattice size. Low η_T values indicate a low interconnectivity; n_T can be as high as 18 for highly interconnected pore systems. Such a high interconnectivity leads to fast mass transfer kinetics and favorable mass transfer coefficients (Unger et al., 2002).

Surface Chemistry

In *normal-phase* chromatography the native, non-modified adsorbent is employed with organic solvent mixtures as eluents. Normal phase chromatography was the classical chromatography mode performed with native silica or alumina, i.e. the adsorbent's surface is hydrophilic and the interaction with the solutes takes place via the hydroxyl groups on the surface. As an example the surface of silica consists of

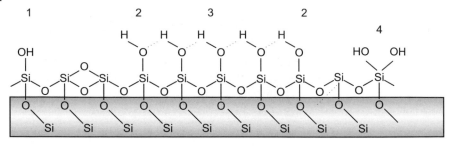

Fig. 3.27 Types of silanol groups: isolated (1), terminal (2), vicinal (3) and geminal (4) species.

silanol groups of different types. One can distinguish between free (isolated, non-hydrogen bonded), terminal, vicinal (hydrogen-bonded), geminal and internal hydroxyl groups (Fig. 3.27).

Knowledge of the materials surface chemistry is crucial. The required adsorbents are silicas with a chemically modified surface carrying bonded n-octadecyl or n-octyl groups. The long n-alkyl ligands invert the surface polarity from hydrophilic (native adsorbent) to hydrophobic. Silica-based reversed phase columns are operated with aqueous/organic eluents and inversion of the elution order takes place (hydrophilic components elute first). More details are given in Tab. 3.12.

Table 3.12 Interrelationship between adsorbent characteristics and chromatographic properties.

Method	Parameter derived	Chromatographic properties affected
Light microscopy, scanning electron microscopy	Particle morphology, particle size distribution	Stability of packed bed, column performance
Light scattering coulter counter	Particle size distribution, volume average particle diameter, number average particle diameter	Stability of packed bed, hydrodynamic column properties, column performance
Gas sorption (nitrogen at 77 K), mercury intrusion (mercury porosimetry)	Specific surface area (BET), pore size distribution, average pore diameter, specific pore volume, particle porosity	Retention of solutes, mass loadability, column regeneration, column performance, mass loadability, pore and surface accessibility for solutes of given molecular weight, mechanical stability, column pressure drop, pore connectivity
Atomic absorption spectroscopy, neutron activation analysis, ICP-optical emission spectroscopy	Inorganic bulk and surface impurities of packings	Chemical stability of packing, retention of solutes, peak tailing

3.3
Column Packing Technology

Several column packing processes have been developed and applied, depending on the column design and on column dimensions. The procedures have been reviewed in depth by Majors (2003b). The most common technique is the dynamic axial compression (DAC) process (Fig. 3.28). The core of the equipment is a cylindrical column with a movable piston at the lower end. The column is approximately three times longer than the actual column length. Separately, a slurry is prepared using the packing and a slurry liquid or a mixture of liquids. The suspension is either poured or pumped into the column. Then, the column is closed at the top by a frit or distribution system. Pressure is applied by the piston pressing the particles into a dense bed from downside. The packing pressure is increased from 30 and 70 bar until the column is packed. After packing the column is conditioned with the eluent while the piston pressure is maintained. This leads to a further densification of the column bed under column operation (in case channels and holes were formed). After usage the column bed can be pressed out as a cylinder and the equipment can be reused.

Appendix A2.2 gives a detailed description of different column packing techniques and their standard operating procedures.

Decisive parameters in the dynamic axial compression technique with respect to optimum column performance and column stability are:
- Quality (permeability) and fit of frit at the column top
- Choice of the solvent and solvent composition for the slurry
- Slurry concentration
- Packing pressure
- Piston pressure

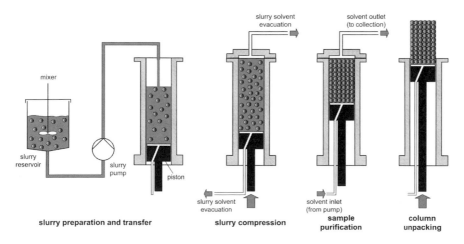

Fig. 3.28 Scheme of the dynamic axial compression technique. (Courtesy of H. Colin, formerly Promochrom, Champigneulles, France.)

In most cases the packing manufacture supplies a protocol for the packing process. There are a number of tests that allow one to set up a protocol. Recommended tests to evaluate the packing properties of silicas are

- Particle size analysis (volume and number distribution)
- Observation of sedimentation behavior (wetting of particles, formation of aggregates, settling velocity, density of sediment)
- Shear force experiments using the suspensions
- Rheological properties (thixotropic and dilatant behavior)

Analysis of the particle size distribution (PSD) is recommended for the following reasons: The user has an indication of the amount of fines that might lead to a high back pressure or a total clogging of the column outlet frit. Fines with an average particle diameter $<1\,\mu m$ should be removed by subjecting the packing to sedimentation. It is advisable to look at both the number and the volume averaged PSD: the PSD based on the number average overweighs the small particles, the PSD based on the volume average overweighs the larger particles (see also Section 3.2.6.1).

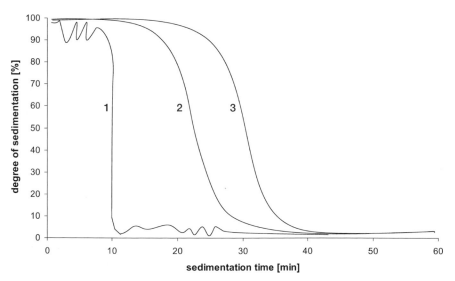

Fig. 3.29 Sedimentation kinetics of a 10 w/w suspension of LiChrospher Si 100, 7 μm in three different solvent mixtures using a Laser photosedimentometer (Hallmann, 1992). Solvents: (1) n-heptane, (2) dioxane–cyclohexane (50/50 v/v), (3) dioxane–cyclohexane–n-octanol (45:45:10 v/v/v).

Table 3.13 Plate number as the function of the packing pressure. Conditions: column 50 × 171 mm, Kromasil 100, d_p = 13 μm (EKA Nobel AB, Bohus, Sweden), linear flow velocity v = 0.3 mm s^{-1}, injection volume 100 μl, solvent: dichloromethane HPLC grade, solute Ceres yellow R (dye).

Packing pressure (bar)	Number of theoretical plate per m of column length
20	8 000
40	16 000
60	19 000
80	23 000
100	24 500

To choose an appropriate slurry liquid the wetting behavior can be viewed under a light microscope and the sedimentation velocity of the particle in the suspension can be measured in a glass cylinder. The slurry liquid should wet the particles and not lead to agglomeration. The settling velocity of the particle at a given suspension should be low to maintain the stability of the suspension. Figure 3.29 shows the sedimentation behavior (sedimentation of particles in dependence of time) of a silica suspension in three different solvents (Hallmann, 1992). The ternary mixture gave the best results, i.e. the most stable suspension.

Thixotropic and dilatant behavior of the suspension can be monitored by viscosimetric measurements as a function of shear rate and the duration of the experiments. Studies with a 25 wt.% suspension of LiChrospher 100, 7 μm in a mixture of dioxane–cyclohexane–n-octanol (45/45/10, v/v/v) have shown that the suspension has rather low thixotropic and pseudo-elastic behavior (Hallmann, 1992). The packing pressure of the axial dynamic column packing technique has a significant influence on the column performance (Tab. 3.13).

Table 3.13 clearly indicates an increase of column performance with packing pressure, having a saturation tendency between 80 and 100 bar. A piston pressure of 10 to 20 bar should be maintained during column operation.

3.3.1
Characterization of the Column Bed Structure

The column bed structure, in the context of column performance and column stability, has been subject to intensive studies over the years, mostly by Cherrak et al. (2002). Attempts have also been made to model the column filling mechanisms and to explain the specific feature of the column bed structure at analytical and preparative/process columns. Table 3.14 lists the studies characterizing the column bed structure.

The results of examining the bed structure of a d_c = 50 mm inner diameter silica column with Kromasil and LiChrospher silicas were (Marme, 1991):

- Interstitial porosities between 0.4 and 0.5
- Regimes of different porosities and packing densities along the column were monitored
- Bed structure was less dense at the wall regimes than at the core of the column
- Fines at the column top were enriched after column operation.

Table 3.14 Methods to characterize the column bed structure.

Method	Reference
Pulsed-field-gradient NMR spectroscopy	Kärger et al. (1988)
NMR-imaging	Bayer et al. (1989)
Hydrodynamic chromatography	Kraak et al. (1989)
NMR imaging	Miller (1991)
Visual monitoring of elution bands in glass columns	Hallmann (1995)

3.3.2
Assessment of Column Performance

Column performance is characterized by the theoretical plate height or the plate number per column length under defined test condition with solutes spanning a retention coefficient range between zero (dead volume marker) and five. Testing is conducted under analytical non-overload conditions. Usually, the performance parameters are a function of the flow-rate or, more precisely, linear velocity of the eluent. These dependencies are presented as plate height–linear velocity curves. Graphically, they show a hyperbolic function with a minimum where the plate height is lowest. The minimum plate height corresponds to 2–5 times of the average particle diameter of the packing. The position of the minimum is inversely proportional

Table 3.15 Column performance of several types of silicas packed into d_c = 50 mm columns by the dynamic axial compression technique (Marme, 1991).

Material	d_p (μm) ($d_{(90)}/d_{p(10)}$ ratio)	Optimum linear velocity u (mm s^{-1})	Number of theoretical plates N	Retention coefficient k' of solute	Reduced plate height h
Kromasil 100	13 (3.2)	0.35	24 300	1.06	3.1
Kromasil 100	11 (5.2)	0.38	28 200	1.26	3.3
Kromasil 100	10 (2.1)	0.39	39 000	1.00	2.5
LiChrospher Si 100	16 (7.1)	0.30	31 100	1.15	2.0
LiChrospher Si 100	15 (3.3)	0.36	24 600	1.26	2.7
LiChrospher Si 100	10 (2.6)	0.39	51 100	1.00	1.8

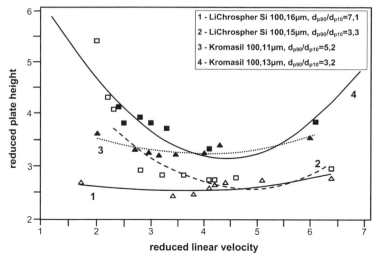

Fig. 3.30 Reduced plate height as a function of reduced linear velocity for four columns (Marme, 1991).

to the average particle diameter. Table 3.15 shows typical performance data of preparative $d_c = 50$ mm columns packed with Kromasil and LiChrospher silicas with an average particle diameter between 10 and 16 µm.

Figure 3.30 shows the corresponding graphs in reduced, i.e. dimensionless, units for a better comparison between the columns.

Column performance studies of $d_c = 50$ mm silica columns revealed:

- Optimum column performance (reduced plate height between 2 and 3) was achieved with silicas of average particle diameters between 10 and 16 µm
- Silica with an average particle diameter of approximately 15 µm and a broad particle size distribution ($d_{p(90)}/d_{p(10)}$ ratio of approximately 7) resulted in an excellent column performance
- PSD width appears to be more critical with silica of an average particle diameter < 7 µm
- Piston pressure should be kept between 10 and 20 bar

3.4
Column Testing

3.4.1
Test Systems

A simple test mixture consisting of a t_0-marker and different phthalic acid esters can be used to check the performance and some basic selectivities of packed columns.

Table 3.16 Test system for normal phase and reversed phase columns.

Normal phase silica	Reversed phase silica
Mixture of	
1 ml dimethyl phthalate	
1 ml diethyl phthalate	
1 ml dibutyl phthalate	
Filled up with	Filled up with
n-heptane–dioxane 90:10	MeOH–H$_2$O (80:20)
to 100 ml	to 100 ml
Addition of 50 µl toluene	Addition of 10 mg uracil
Mobile phase:	Mobile phase:
n-heptane–dioxane (90:10)	MeOH–H$_2$O (80:20)
Injection amount: 0.1 µl per ml V_c (40 µl for a 50 mm × 200 mm column).	

The advantage of phthalic acid esters as test substances is their liquid nature and good miscibility with different mobile phases. In addition they can be used in straight phase as well as reversed phase systems. Table 3.16 gives the compositions of test mixtures for normal and reversed phase columns.

Figure 3.31 shows a typical chromatogram of the test mixture described in Tab. 3.16 on a d_c = 50 mm column packed with LiChrospher Si 60, 12 µm.

The diethyl phthalate peak (third peak of the test mixture in normal phase systems as well as reversed phase ones) can be used to calculate the number of theoretical plates (*N*) per metre of bed height according to Eq. 2.19.

3.4.2
Hydrodynamic Properties and Column Efficiency

For complete characterization of the column performance the efficiency has to be tested over the whole operation range of linear velocities. The injection of a test mixture at different flow-rates can be easily automated with modern HPLC-equipment

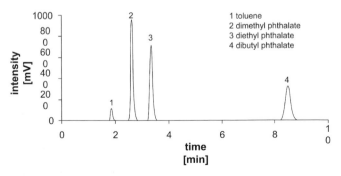

Fig. 3.31 Test chromatogram of a t_0-marker and three phthalic acid esters.

and yields the pressure drop vs. flow curve of the adsorbent as well as its HETP curve. By using marker substances with different molecular weights the influence of the mass transfer resistance (C term) can be investigated. Figure 3.32 shows the HETP curve for two similar adsorbents, which exhibit large differences at high flow-rates.

Two test compounds with molecular weights of 110 and 11000, respectively, were used to obtain the data. The minimum plate height at the lowest velocity was set to 100% and the percentage increase in plate height is plotted against the linear veloc-ity. For both adsorbents, the plate height for the large molecule clearly shows a steeper increase than for the small molecule. At a linear velocity of approximately 8 mm s^{-1} the plate height for the large molecule increases dramatically for one of the adsorbents. This gives a hint of some obstacles within the pore system of that adsor-bent, which are of course more severe for larger molecules. Even if the preparative process are not operated at such high linear velocities, the adsorbent properties should be carefully examined beforehand, to avoid any unnecessary extra contribu-tion to the axial dispersion.

3.4.3
Mass Loadability

For lab-scale purifications as well as a quick adsorbent screening method the deter-mination of the resolution as a function of the load is used. Increasing amounts of the feed mixture are injected onto the column and the corresponding resolution of the compounds of interest is determined. With this methodology the maximum load for 100% recovery and yield can be extrapolated from the graph at that point where the resolution of 1.5 (baseline resolution) cuts the curves for the single adsorbents at a given load. Even if no interactions between the substances in the non-linear range of the isotherm are taken into account, this approach is useful for the near-linear

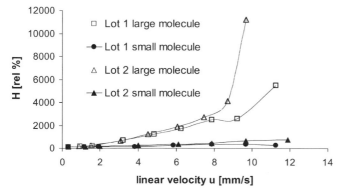

Fig. 3.32 Plate height – linear velocity curve of two different adsorbents obtained with a low molecular weight (MW = 110) and a high molecular weight (MW = 11 000) marker.

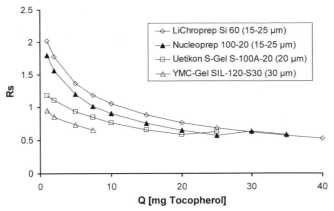

Fig. 3.33 Determination of the resolution factor at different loads for four different silicas.

range with resolution factors down to 1.0–0.8. Figure 3.33 shows the resolution versus load for different adsorbents. All adsorbents are normal phase silica with similar particle sizes. A mixture of two tocopherol isomers from a sunflower oil source was injected and the resolution between the two main peaks calculated. The graph shows that the starting resolution is quite different for each adsorbent, which depends on the selectivity α and the efficiency of the column. As all resolution factors decline with increasing amounts of tocopherol loaded onto the column, the higher the load on the column the smaller the differences between the columns. In that region interactions between the compounds in the feed mixture have to be taken into account, and the complete isotherm should be determined. Nevertheless from Fig. 3.33 a good solution for the isolation of mg amounts of the target compound can be easily obtained.

3.4.4
Comparative Rating of Columns

The rating of columns is performed according to two aspects:
    1. Assessment of column performance data obtained under analytical conditions. This includes:

- Plate height of the column: The plate height should correspond to two to three times the average particle diameter of the packing
- Plate height of the column as a function of the linear velocity of mobile phase: The flow rate should have an optimum where the plate height is at a minimum
- Number of plates per column length: Commonly the plate number is between 5000 and 10 000 to sufficiently separate mixtures
- Retention coefficient of a solute

All these parameters depend on the mass loadability of the column and change significantly when a critical loadability is reached. The critical mass loadability of analytical columns is usually reached at a 10% reduction of the retention coefficient or at 50% decrease of column plate number. At higher values the column is, in chromatographic terms, overloaded.

2. Assessment of parameters under preparative conditions.

Preparative columns are compared with regard to the column productivity, yield, purity and cost. Productivity is defined as the mass of isolate per kg of packing per day at a set purity and yield. It depends, essentially, on the applied technical chromatographic process. For example, the productivity for the separation of racemates into enantiomers by the simulated bed technology has increased from 2 kg per kg packing per day to ca. 10 kg of product per kg packing per day. The yield can be low provided the purity is high. Desired purities are sometimes 99.9% and higher. The overall cost is the only measure that is taken for chromatographic production processes. The main cost contributions come from costs related to the productivity (column size, amount of stationary and mobile phase), yield losses of the product and work-up costs linked to the product dilution (see also Chapter 7.3).

3.5
Column Maintenance and Regeneration

Preparative chromatography in the non-linear range of the isotherm means applying a high mass load of substance onto the adsorbent. Some compounds from the feed mixture may alter the surface of the adsorbent and change its separation activity. Therefore, regular washing procedures and re-activation should be performed to ensure a long column lifetime. Different goals for washing procedures can be distinguished

- Removal of highly adsorptive compounds (Cleaning in place, CIP)
- Re-conditioning of silica surfaces
- Sanitization in place (SIP) of adsorbents
- Column and adsorbent storage conditions

3.5.1
Cleaning in Place (CIP)

Highly adsorptive compounds should always be removed before subjecting the feed onto the main column. This can be achieved by a pre-column, which has to be changed at regular intervals. Sometimes, simple filtration of the feed over a small layer of the adsorbent is sufficient to remove sticking compounds. Nevertheless, it has to be checked that the feed composition is not drastically changed after the pre-purification step so that the simulation of the chromatogram and thus the process parameters are still valid. If compounds stick to the adsorbent (irreversible adsorption) and if the performance of the column in terms of efficiency and loadability decreases, a washing procedure has to be carried out (Cleaning in place, CIP). For this washing procedure two general considerations have to be kept in mind:

- The flow direction should be always reversed so that the sticking compounds are not pushed through the whole column
- As the adsorbent has to be washed with solvents of very different polarity, care has to be taken that all solvents used in a row are miscible. Especially when a solvent with a different viscosity is introduced into the column or the mixture of two solvents results in a higher viscosity than the single solvents, e.g. methanol and water, the pressure drop can rapidly increase and damage the packing. Therefore a maximum pressure should be set in the system control and the flow rate for the washing procedure should be reduced if the resulting pressure drop is unknown.

If the column can be opened at the feed injection side, which is not always possible for technical or regulatory (GMP) reasons, the first one or two centimetres of packing can be removed. The top of the bed is then re-slurried, fresh adsorbent added and the column re-compressed. Up to a column diameter of 200–300 mm this procedure might make it possible to avoid re-packing of a column and ensures the further use of the packed column for some time. Table 3.17 shows some CIP-regimes for silica-based adsorbents with different functionalities.

As a rule of thumb at least 10 to 15 column volumes of each washing solvent should be pumped through the column. No general rules can be given as to whether column washing and regeneration is more economic than replacing the packing material as this strongly depends on the nature of the components to be removed

Table 3.17 CIP procedures for silica-based adsorbents after Majors (2003a) and Rabel (1980).

	Reversed phase	Normal phase	Ion-exchange	Protein removal
	C18, C8, C4, Phenyl, CN	Si, NH$_2$, CN, Diol	SAX, SCX, DEAE, NH$_2$, CM on silica	C18, C8, C4, Phenyl
	Distilled water (up to 55 °C)	Distilled water (up to 55 °C)	Heptane–methylene chloride	Distilled water (up to 55 °C)	Distilled water (up to 55 °C)
	Methanol	Methanol	Methylene chloride	Methanol	0.1 % trifluoroacetic acid
	Acetonitrile	Isopropanol	Isopropanol	Acetonitrile	Isopropanol
	THF	Heptane	Methylene chloride	Methylene chloride	acetonitrile
	Methanol	Isopropanol	Mobile phase	Methanol	Distilled water
	Mobile phase with buffer removed	Mobile phase with buffer removed		Mobile phase with buffer removed	Mobile phase

and the type of the packing material. Naturally, if coarse, cheap packing materials are used the solvent and time consumption for washing and regenerating a column is more expensive than unpacking and repacking with new adsorbent.

3.5.2
Conditioning of Silica Surfaces

One of the drawbacks of chromatography with normal phase silica is the strong and changing interactions the silica surface can undergo with the mobile phase and feed components. In contrast to reversed phase adsorbents, where the interaction takes place with the homogeneously distributed alkyl-chains bonded to the surface, the silica surface has a lot of energetically different active groups such as free, geminal and associated silanol groups, incorporated metal ions in the silica surface and water, which might be physisorbed or chemisorbed (Fig. 4.8 in Chapter 4.2.2). It is of great importance to condition the silica before using it in a preparative separation. Especially, the activity, which is determined by the amount of water, and the surface pH have to be controlled to ensure constant conditions from one column packing to another. The surface pH of a silica adsorbent depends heavily on its production process and is especially influenced by the applied washing procedures (Hansen et al., 1986). The apparent surface pH is measured in a 10% aqueous suspension of the silica. The dramatic influence of different surface pH is illustrated in Fig. 3.34 where the surface pH of a LiChroprep Si adsorbent has been adjusted with sulfuric acid and ammonia respectively and two test compounds (2-aminobenzophenone and 4-nitro-3-aminobenzophenone) have been injected onto the columns. The retention time of

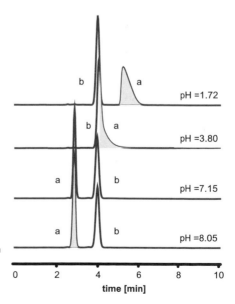

Fig. 3.34 Influence of surface pH on the retention time of two test compounds, 2-aminobenzophenone (b) and 4-nitro-3-aminobenzophenone (a).

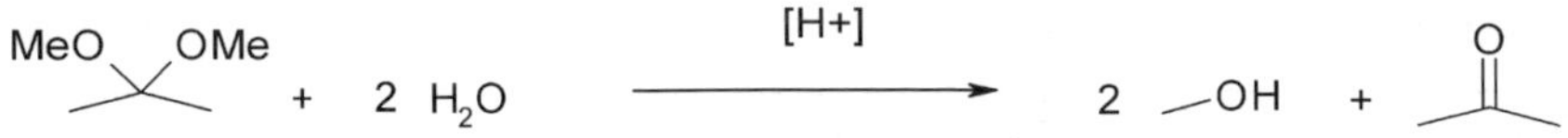

Fig. 3.35 Reaction of dimethoxypropane for chemical removal of water from silica adsorbents.

4-nitro-3-aminobenzophenone is strongly influenced by the surface pH. Even a reversal of the elution order can be achieved at low pH. The adjustment of surface pH is possible even in large batches by washing the silica with diluted acids, removal of the water and afterwards operation of the column with non-polar solvent systems. Under these conditions the apparent surface pH is remarkably stable during operation of the column.

A second adsorbent characteristic that should be controlled is the water content of the silica. The water content can vary from 0 to 10 wt.%. As silica is a well-known desiccant (drying agent) care has to be taken with adsorbent types that are adjusted with low amounts of water. Here, significant uptake of water from the air might influence the chromatographic conditions. Silica should therefore generally be stored in well-closed containers. The adjustment of silica to certain water contents is a necessary standard procedure in large batch sizes for most large-scale silica manufacturers. If water has to be removed from a column a chemical drying procedure can be applied. Dimethoxypropane reacts with water, resulting in methanol and acetone (Fig. 3.35)."

For this procedure the column should be flushed with 10 column volumes of a mixture of dichloromethane, glacial acetic acid and 2,2-dimethoxypropane (68:2:30 v/v/v) Afterwards the column should be subsequently washed with 20 column volumes of dichloromethane and 20 column volumes of n-heptane.

3.5.3
Sanitization in Place (SIP)

For purification of recombinant products from fermentation broth care has to be taken to prevent microbial growth on adsorbents. Therefore, regular sanitization-in-place (SIP) procedures have to be applied. When adsorbents based on organic polymers or sepharose are used (for ion exchange and adsorption chromatography), decontamination is performed at high pH, e.g. by using 0.1 to 1 M NaOH. With silica based adsorbents procedures based on salts, solvents or acids are commonly used due to the chemical instability of silica above pH 8–9. Some recommended washing procedures are listed in Tab. 3.18.

Table 3.18 SIP procedures.

Solvent	Composition
Acetic acid	1% in water
Trifluoroacetic acid	1% in water
Trifluoroacetic acid + isopropanol	(0.1% TFA)/IPA 40:60 (v/v) viscous → reduce flow rate
Triethylamine + isopropanol	40:60 (v/v), adjust 0.25 N phosphoric acid to pH 2.5 with triethylamine before mixing
Aqueous urea or guanidine	5-8 M (adjusted to pH 6–8)
Aqueous sodium chloride, sodium phosphate or sodium sulfate	0.5–1.0 M (sodium phosphate pH 7)
DMSO–water or DMF–water	50:50 (v/v)

3.5.4
Column and Adsorbent Storage

If packed columns or adsorbents removed from columns are to be stored for a period of time the adsorbent should be carefully washed and then equilibrated with a storage solvent. Compounds strongly adsorbed to the adsorbent and, especially, any source of microbial growth has to be carefully removed before the adsorbent is stored. In addition, any buffer substances or other additives, which might precipitate during storage, have to be washed out. For flushing and storage different solvents are recommend (Tab. 3.19).

3.6
Guidelines for Choosing Chromatographic Columns and Stationary Phases

1. Choose a stationary phase that is manufactured reproducibly in large (kg) quantities and is allowed to be used under regulatory issues.
2. Prefer native and pure adsorbents over those with a sophisticated surface chemistry. Biopolymer purifications with affinity type adsorbents are not considered here.
3. Use stationary phases whose retention and selectivity can be adjusted by simple solvents. Solvent removal and solvent recovery can become an expensive step.
4. Follow carefully the recommendations of the packing manufacturer with respect to solvent use, washing and regeneration.
5. Apply pre-columns packed with the same material or with one showing a high adsorptivity, e.g. zeolites, towards undesired impurities or by-products.

Table 3.19 Recommended solvents for storage and flushing.

Adsorbent type	Flushing solvent	Storage solvent
Normal phase	Methanol or other polar solvent	Heptane + 10% polar compound (isopropanol, ethyl acetate)
Reversed phase	Methanol or acetonitrile, THF or dichloromethane for oily samples	>50% organic in water
IEX (silica based)	Methanol	>50% organic in water

4
Selection of Chromatographic Systems

W. Wewers, J. Dingenen, M. Schulte, J. Kinkel

This chapter aims to provide guidelines for the selection of chromatographic systems related to different key issues of the whole process in preparative chromatography. These key issues not only focus on the engineer's point of view. Along with engineering parameters such as productivity, yield and recovery we have to take into account economy, scale of the separation, speed, time pressure, hardware requirements and availability, automation and legal aspects towards documentation, safety and others. Obviously, rules of thumb related to these criteria may not cover all possible practical scenarios, but they may be useful in avoiding pitfalls.

The selection of chromatographic systems is critical for process productivity and thus process economy. While the selection of the chromatographic system must be regarded as holding the biggest potential for optimizing a separation process it is also a source of severe errors. The choice of the chromatographic system includes the selection of the adsorbent and the mobile phase for a given sample. This selection can be based on a systematic optimization of the system by extensive studies of solubility, retention and selectivity, but, sometimes, the use of generic gradient runs with standard systems is sufficient.

One source of severe mistakes in system development is a common misunderstanding about preparative chromatography. The classical differentiation between analytical and preparative chromatography is based on the size of the equipment, the size of adsorbent particles or the amount of sample to be separated. This old view of preparative chromatography leads to a geometric scale up of systems that are dedicated for analytical applications. Scale up from analytical to preparative systems is problematic because the goal and restrictions for the two operation modes are different (Tab. 4.1) and thus the parameter to be optimized are also different. The goal of analytical separations is to generate information whereas preparative applications aim to isolate a certain amount of purified products. For these reasons preparative and analytical chromatography should be differentiated on the basis of the goal and not the geometric size of the system.

For analytical separations, the sample can be processed, handled and modified in any way suitable to generate the required information (including degradation, label-

Table 4.1 Differentiation between analytical and preparative chromatography.

	Analytical	Preparative
Goal	information fast information generation	material optimal product recorvery
Restrictions	all separation modes allowed	recovery of unchanged solutes must be guaranteed
Optimization parameters	• necessary separation sensivity • peak capacity ($1 < k' < 5$) • speed • selectivity	• purity • yield • productivity • economics

ing or otherwise changing the nature of the compounds under investigation). In preparative mode, the product has to be recovered in the same condition that it was in before undergoing the separation (i.e. no degrading elution conditions, no reaction etc.). This determines the chromatographic process and chromatographic system far more than any consideration of process size.

Preparative HPLC separations are used in all stages of process development. The three main stages in process development as well as examples for the tasks and corresponding criteria for process development are described in Fig. 4.1. They differ in time pressure, frequency of the separations and amount of target products, which dominate the economic structure of the process. In laboratory and technical scale the development of the separation is dominated by the time pressure and the separation problem includes only a limited amount of sample and a limited number of repetitions. Process optimization in this small- to semi-scale is not helpful, because the effort for optimization will not be repaid in reduced operational costs.

In the early stages of process development too little sample is available for extensive optimization studies. These studies are not needed and are counterproductive to economic success because separation costs in these early stages are dominated by the time spent developing the separation and not by the operational costs. If purified products are needed for further research (e.g. lead finding) time delays are problematic. Furthermore, the number of different separation problems at this early stage is quite high, while the number of repetitions of one separation is low. In this case generic gradient runs, which can separate a broad range of substances, provide a quick and effective means of separation.

In production-scale processes economic pressure comes to the fore. Next to economic pressure it is very often not possible to separate the huge amounts involved with non-effective chromatographic separations. Although the time pressure is still present, an optimization of the chromatographic system with extensive studies is required. The investment in an optimized system will be repaid though by the reduced operation costs of a subsequent production process.

However, due to the vast number of variables and the restrictions regarding the environment of the preparative separation one may say: „*There is no general chromatographic system to meet all requirements*".

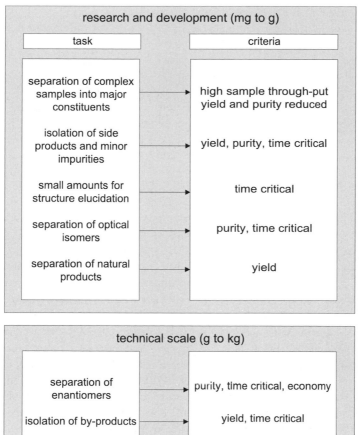

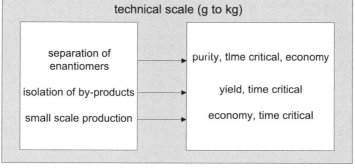

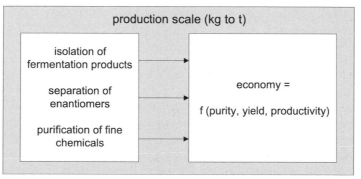

Fig. 4.1   Tasks for preparative chromatography in different development stages.

Comparable to the different scales of processes (Fig. 4.1) the users of preparative separations can also be distinguished by their tasks, and differentiated in the following groups:

There are users, who have to solve many different separation problems in a short time. These users represent the laboratory-scale production of chromatography. The frequency of changes in their tasks and thus the time pressure is high and rules of thumb for a fast selection of suitable chromatographic systems are needed. Other users have to develop the separation for one production-scale process. This group is interested in guidelines for the systematic optimization of the chromatographic system and for the choice of the process concept. Rules of thumb are only used for a first guess of the system.

The development of a chromatographic separation can be divided into three stages, which are discussed here and in Chapter 5. As shown in Fig. 4.2, the first step is to define the task (Section 4.1), meaning nothing else than a reflection and classification of the separation problem. After the task is defined, a suitable chromatographic system must be found. For this purpose the properties of chromatographic systems are discussed in Section 4.2, while criteria for the selection of chromatographic systems are given in Section 4.3. The last step of method development is the choice of the process concept, which depends on the given task and chosen chromatographic system as well as the available equipment. Different process concepts are introduced in the Chapters 5.2 and 5.3, while Chapter 5.4 gives guidelines for the selection of the chromatographic method. The whole development process is exemplified there and thus Chapter 5.4 can be regarded as a conclusion resulting from Chapters 4 and 5.

Numerous strategies to solve these tasks have been presented and there is nothing really new to say. The main areas of progress have been in equipment design, modeling, and simulation tools. Nowadays, it is possible to predict the chromatographic behavior and economy of a separation task. Next to these predictive procedures the operator of a plant is also aided during process operation by advanced process control systems and documentation of the separation, e.g. with respect to GMP regulations.

4.1
Definition of the Task

Any selection of a chromatographic system for preparative isolation of individual components should start with a definition of the task. The starting point can be described as careful reflection of the main criteria influencing the separation.

Good advice for primary actions is to collect all available data describing the nature of the sample. Information exchange is especially important if synthesis and separation of the products are done by different departments. For this purpose a rough characterization of the sample and the preparative task made by following the questionnaire of Tab. 4.2 eases the planning of experimental procedures.

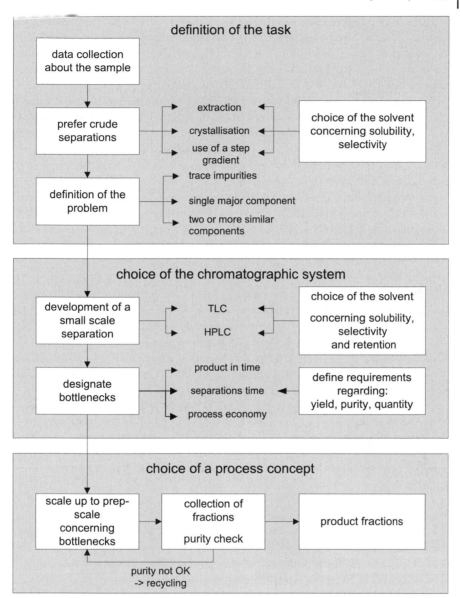

Fig. 4.2 Development of a chromatographic method.

Before starting the selection of any chromatographic system the use of crude separation steps such as extraction, crystallization or flash chromatography should be checked. The separation task can be simplified by the use of these crude separations if impurities or the target product are removed in the first capture steps.

An example of these steps of downstream processing is the isolation of Paclitaxel, an anti cancer drug that is extracted for example from the pacific yew tree. This is a very challenging task due to the low concentration (0.0004–0.08%) in the bark (Pandey, 1997 and Pandey et al., 1998). The target product is extracted from the herbal material by Soxhlet extraction. The concentrated extract contains a large range of impurities, which is shown in the first chromatogram of Fig. 4.3. Many pre-purification steps are necessary before the separation is downsized to a separation problem of paclitaxel and closely related taxanes. Owing to the difficulty of this last separation step, paclitaxel is isolated by preparative chromatography.

The isolation of paclitaxel exemplifies that most preparative separations must be downsized to a level where a limited number of individual compounds are present to ease the final purification steps. This downsizing of the separation problem can be done by crude separations or by a cascade of consecutive chromatographic separation steps. One finally ends up at a point where a multicomponent mixture with a broad concentration range of the different substances has to be fractionated to a series of mixtures. This approach is described in Fig. 4.4. In general, a mixture can be split into three types of fractions, which each represent a specific separation problem. These three fractions exemplify possible separation scenarios that differ with regard to the ratio of target products and impurities.

The first separation scenario is the most comfortable problem. The target product is the major component in the mixture and thus it is easy to reach a yield and purity in the desired range. Thermodynamic effects can be used to displace impurities during the separation process.

Table 4.2 Questionnaire for the collection of the information.

- What is the consistency of the sample (solid, liquid, suspension, etc.)?

- What is the source of the sample (chemical synthesis, fermentation, extraction from natural products, etc.)?

- Which possible solvents for the sample as well as solubilities are known?

- Is the sample a complex mixture of unknown composition?

- Which chemical or physical properties of the sample are available (stability, structures, UV spectra as well as toxic and biohazard data)?

- Is the structure, e.g. physical, or chemical behavior of the target compound(s) known?

- Is the structure, e.g. the physical, or chemical behavior of the impurities known?

- Is any kind of chromatographic data available?

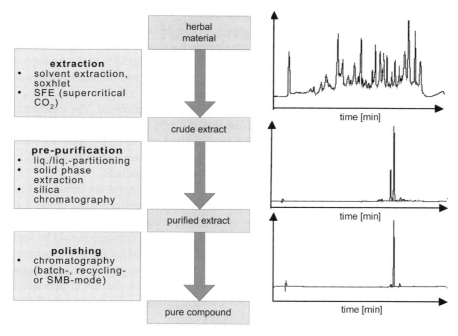

Fig. 4.3 Purification of Paclitaxel.

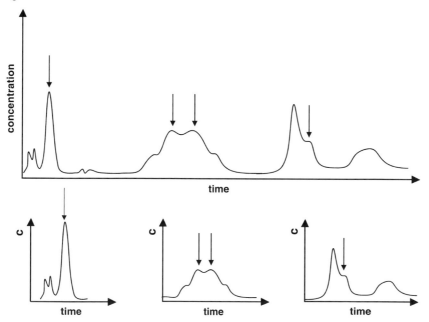

a) target products are the dominant components

b) target products and impurities are of similar amount

c) impurities and by-products dominate the desired products

Fig. 4.4 Main types of preparative chromatographic separation scenarios.

Table 4.3 Properties of solvents for preparative chromatography.

Solvent	Solvent strength	Boiling point (°C)	Viscosity$_{25°C}$ (cP)	RI (–)	UV cut-off (nm)	LD 50 (µg kg^{-1})	ICH-class	Vapor pressure 25°C (bar)	Explosion limit (vol.%)
n-Hexane	~0	69	0.31[a]	1.375	195	28710	2	0.202	1.2-7.8
n-Butyl ether	2.1	142	0.64	1.397	220	7400		–	0.9-8.5
di-Isopropyl ether	2.4	68	0.38	1.365	220	8470		0.200	1-21
Methyl tertiary butyl ether	2.7	55	0.27[a]	1.369	210	3870	3		
di-Ethyl ether	2.8	35	0.24	1.35	218	1213		0.669	1.8-
n-Butanol	3.9	118	2.6	1.397	210	790	3	0.009	1.4-11.3
2-Propanol	3.9	82	2.4[a]	1.377	205	5045	3	0.05	
1-Propanol	4	97	1.9	1.385	240	1870	3	0.028	
Ethanol	4.3	78	1.2[a]	1.361	195	7060	3	0.079	3.4-19
Methanol	5.1	65	0.55[a]	1.328	205	5628	2	0.169	5.4-44
Tetrahydrofuran	4	66	0.55[a]	1.407	212	1650	2	0.216	
Pyridine	5.3	115	0.88	1.507	210	891	2	0.028	
Methoxy ethanol	5.5	125	1.6	1.4	210	2370	2	0.013	
Dimethylformamide	6.4	153	0.8	1.428	268	2800	2	0.005	
Acetic acid	6	118		1.447			3	–	4-17
Formamide	9.6	210	3.3	1.447		5800	2	–	
Dichloromethane	3.1	40	0.44[a]	1.424	233	1600	2	0.573	–
1,2-Dichloroethane	3.5	83	0.78	1.442	228	670	2	0.105	
Ethyl acetate	4.4	77	0.45[a]	1.372	256	5620	3	0.126	2-12
Methyl ethyl ketone	4.7	80	0.38	1.376	329	2600	3	–	
1,4-Dioxane	4.8	101	1.2	1.42	215	5200	2	0.050	
Acetone	5.1	56	0.36[a]	1.359	330	5800	3	0.306	2.2-13
Acetonitrile	5.8	82	0.38[a]	1.344	190	2460	2	0.118	3-16
Toluene	2.4	111	0.59[a]	1.497	284	636	2	0.038	
Benzene	2.7	80	0.6	1.498	280	930		–	1.4-8
Chloroform	4.1	61	0.53	1.443	245	908	2	0.259	
Water	10.2	100	1[a]	1.333	<190			0.032	

[a]Value for 20°C

4.2.1.1 Stability

The stability of all components of the chromatographic system must be assured for the complete operation time. The solvent must be chemically inert to all kinds of reaction. Neither an instable solvent, which for example tends to form peroxides, nor a solvent that reacts with the sample or the adsorbent is suitable for an economically successful solution of the separation problem. Of course, corrosion of the HPLC unit must be prevented, too.

4.2.1.2 Safety Concerns

The safety of preparative processes depends on the flammability and toxicity of the solvent. Owing to the huge amounts of solvents handled in preparative chromatography, low flammability should be preferred. Flammability is described by the vapor pressure, the explosion limits or the temperature class of the solvent. Generally, the use of flammable solvents can not be avoided and thus the risk must be minimized by good ventilation and other precautions in the laboratory.

The toxicity of the solvent is important with reference to safety at work and product safety. Toxicity is classified, for example, by the LD_{50} value. If no alternative solvent is available any danger of toxification of the employees during the production process has to be avoided by precautions in the laboratory. The safety of the product is another concern linked to the choice of certain eluents. A contamination of the product might occur through the inclusion of solvent in the solid product due to inadequate drying of the product. The residual amount of solvents in the product is regulated by the different pharmacopoeia as well as the International Guidelines for Harmonization (ICH-Guideline Q3C Impurities: Residual Solvents [www.fda.gov/cber]). Class 1 solvents should be avoided by any means, while the use of Class 2 solvents can not be avoided and, therefore, has to be carefully optimized. If possible, Class 3 solvents with low toxic potential should be used. The ICH classification as well as the classification limit of the single solvents in pharmaceutical products is given in Tab. 4.3.

4.2.1.3 Operating Conditions

For operational reasons it is important to look at detection properties, purity, recycling ability and viscosity of the mobile phase. If solvent mixtures are used as mobile phase, the miscibility of the solvents is an absolute condition for their use, which has to be guaranteed for the whole concentration range of the solute as well (Fig. 4.6). For example, acetonitrile and water are miscible, but when sugars are added at high concentrations (e.g. fructose) the system demixes.

The detection system is another boundary condition for the choice of the mobile phase. In most cases ultraviolet (UV) absorbance or refractive index (RI) detectors are used. An imprecise detection leads to insufficient recognition of the target substances as well as the impurities, thus causing purity problems. To avoid these problems the UV cut-off value and the refractive index (RI) are given in Tab. 4.3 for different pure solvents.

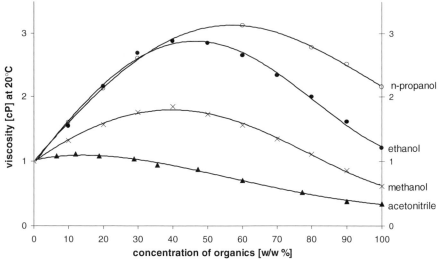

Fig. 4.7 Viscosity of different mixtures of water and organic solvents.

Especially with strongly associating solvent mixtures (alcohol–water, acetonitrile--water) the viscosity shows anomalous variations with composition. The viscosity of the mixture is very often larger than of the pure solvents (Fig. 4.7). For example, the maximum viscosity of an ethanol–water mixture is more than two-fold higher compared with the pure solvents. Next to the composition, the viscosity is strongly influenced by temperature. At higher temperatures the viscosity reduces dramatically. However, chromatographic processes are very often operated at ambient temperature, because it is difficult and expensive to guarantee a constant elevated process temperature. Further information for a few solvent mixtures is given by Gant et al. (1979).

4.2.2
Adsorbent and Phase System

The previous chapter discussed the solvent and its interaction with the solute. To complete the chromatographic system the adsorbent has to be selected. As mentioned in Chapter 3.2.1 one has to distinguish between enantioselective and non-enantioselective adsorbents. Both groups of adsorbents are classified into polar, semi-polar and nonpolar adsorbents (see Tab. 4.4). This classification is based on the surface chemistry of the packing material. Interaction between mobile phase and adsorbent characterizes the phase system, which is distinguished between normal phase (NP) chromatography and reversed phase (RP) chromatography. This differentiation is historic and appointed by the ratio of the polarity of the adsorbent and the mobile phase.

Table 4.4 Survey of adsorbents and operational modes.

Chemistry	Non-enantioselective			Enantioselective (CSP)		
	Polar	Semi-polar	Non-polar	Polar	Semi-polar	Non-polar
Material	Silica, aluminium oxide	Silica with bonded cyano; amino and diol groups	Silica with bonded n-alkylsilyl groups; polymer coatings; cross-linked organic polymers; porous carbon	Cellulose/amylose ester; cellulose/amylose carbamate;	Cyclodextrines; chemically modified silicas (Pirkle phases)	
Mode	NP	NP and RP	RP	Operation either at NP or RP mode		

4.2.2.1 Normal Phase System

Normal phase systems consist of a polar adsorbent and a less polar mobile phase. Because these were the first available chromatographic systems they were named normal phase systems. They are for instance silica gels or other oxides in conjunction with a non-polar solvent such as heptane, hexane or some slightly polar solvents like dioxane. Semi-polar adsorbents such as cyano or diol phases can be operated in the normal phase mode as well. The combination of water and silica is not recommended due to the strong interaction between water and adsorbent. Furthermore silica is slightly soluble in water, which results in a shortened lifetime of the adsorbent. Normal phase systems are limited to organic solvents and thus the solutes have to be soluble in these solvents.

The interaction mechanism in NP chromatography is adsorption on the polar surface of the adsorbent. Because silica is the most widespread NP adsorbent the following focuses on the use of silica. As mentioned in Chapter 3.2 active centers for adsorption are statistically spread over the surface of the silica. Generally these are siloxane groups and different kind of silanol groups (Fig. 4.8). The silanol groups appear as single (free) or as pairs. Single silanol groups have the strongest polarity while the polarity of silanol pairs is reduced. Impurities such as metal atoms (Ca, Al or Fe) produce a disturbed surface chemistry and the formation of polarity hot spots, resulting in unwanted strong adsorption on these sites.

Obviously, due to the polar surface, only polar parts of solutes interact with the adsorbent, while non-polar groups like alkyls are not adsorbed. This behavior determines the range of applications for NP chromatography. For instance, it is not suitable for the separation of homologues that only differ in the length of an alkyl group. Conversely, solutes that differ in the number, nature or configuration of their substituents are mainly separated by NP chromatography. This includes the separation of stereoisomers, if they are not enantiomeric. This stereoselectivity can be explained by the isotropic nature of the adsorption on silica. Silanol groups are rigidly fixed on

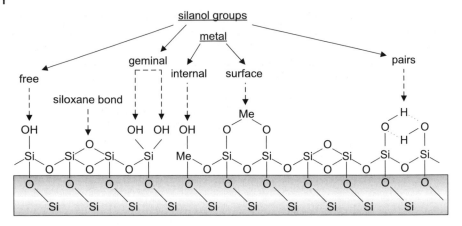

Fig. 4.8 Surface groups on silica.

the surface and, due to the short bond length of the silanol groups, the interaction between solute and surface is very isotropic.

Aqueous mobile phases are not used because the interaction between water and the polar stationary phase would be too strong. However, a minimal amount of water is needed to saturate the above-mentioned hot spots of polarity due to metal atoms. For these two reasons the water content of the solvents should be controlled very carefully (Section 4.3.2.2). Changes in the range of a few ppm, resulting for example from air humidity, can cause significant changes in retention time and peak shape. As a result of this sensitivity to changes in water content the equilibration time for the system is very high. A constant baseline and system behavior is reached after about 50 column volumes, while Poole (2003) remarked that column equilibration for mobile phases of different hydration level takes at least 20 column volumes.

4.2.2.2 Reversed Phase Chromatography

The demand for separation of water-soluble solutes that differ in their non-polar part led to the development of adsorbents for reversed phase (RP) chromatography. In RP systems the polarities of the adsorbent and mobile phase are inverted compared with NP systems. Non-polar or weak polar adsorbents are used together with polar solvents. The transition to polar mobile phases affords the use of water. A typical solvent for RP systems consists of a mixture of water and a miscible organic solvent, such as methanol or tetrahydrofuran (THF). Due to the aqueous solvents, precise control of the water content is not necessary. The non-polar packing material consists, for instance, of a porous silica support coated with a monolayer of alkyl groups (C_2–C_{18}), aryl groups or hydrophobic polymers. Alternatives to these reversed phase silica are hydrophobic cross-linked organic polymers or porous graphitized carbons. One important property is the hydrophobicity of this surface. It is mainly afforded by the carbon content and the endcapping of the silica (see also Chapter 3.2.3.1).

Adsorption–desorption mechanisms of RP systems are a turnaround from NP systems. In this case the non-polar, or hydrophobic, portion of solute molecules adsorbs to the surface of the stationary phase, while the polar part of the molecule is solvated by the mobile phase solvent. The result is a reversed elution order – polar before less polar solutes (Fig. 4.9).

Reversed phase silicas are called brush type, because of their structure. Numerous alkyl groups point from the surface into the mobile phase and thus the surface is similar to a brush, with bristles of alkyl groups. To describe the adsorption process a simple two layer model of the system has been proposed (Galushko, 1991), which means that the surface layer of the alkyl groups is assumed to be quasi-liquid. The retained solutes penetrate into the surface layer and retention can be regarded as partitioning between hydrophobic stationary phase and mobile phase, similar to liquid–liquid chromatography.

RP chromatography is mainly used to separate substances that differ in their lipophilic part or in the number of C atoms, e.g. the separation of homologues. However, due to the anisotropic nature of the interaction between surface layer and solute, RP phase systems reach lower selectivities for the separation of stereoisomers than NP phase systems.

The elution order in NP and RP systems is described in Fig. 4.9. The retention time and thus the elution order for two solutes of different polarity is exemplified for

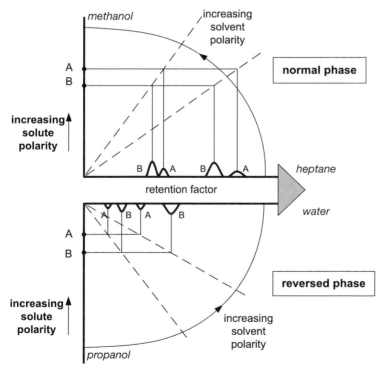

Fig. 4.9 Retention behavior in both normal and reversed phase chromatography.

RP in the lower part of the Fig. 4.9 and for NP in the upper part. Component A is more polar than B, which results in A eluting before B in RP systems and an inverted elution order for NP systems. The influence of solvent polarity is shown by the two diagonal lines from the origin. Each line illustrates a solvent with a certain polarity. Retention times are given by the point of intersection with the horizontal line, which describes the solute polarity.

Most analytical separations are performed by reversed phase systems while most preparative separations are performed using an NP system. For example, 84% of the analytical separations at Schering AG, a mid-sized pharmaceutical company, are done in RP mode while 85% of the preparative separations are performed on normal phase silica gels (Brandt and Kueppers, 2002).

RP systems dominate analytical applications because of their robustness and reduced equilibration time. This fast column equilibration makes RP phases applicable for gradient operation. For preparative chromatography normal phase systems are preferred because of some disadvantages of the RP systems. Although water is a very cheap eluent it is not an ideal mobile phase for chromatography because it evaporates at higher temperatures, the enthalpy for evaporation is quite high and the viscosity of aqueous mixtures is much higher than the viscosity of mixtures of organic solvents.

Since the polarity range of adsorbents is bordered by silica on the polar side and RP-18 or hydrophobic polymeric phases on the non-polar side it is easy to assign these absorbents to normal or reversed phase systems. Medium polar packings (Chapter 3.2.1) possess polar properties because of the functional group as well as hydrophobic properties contributed by the spacer (Unger and Weber, 1999). Owing to this mixed nature these phases can not be directly assigned to a certain type of phase system.

4.3
Criteria for Choice of Chromatographic Systems

The last two sections characterized the elements of a chromatographic system and discuss different separation tasks. This section provides guidelines to develop the chromatographic system for a given separation problem. The classical recommendation for preparative separations performed by elution chromatography is as follows:

Find a mobile phase with high solubility, a compatible stationary phase and then optimize selectivity, capacity and efficiency followed by a systematic increase in column load.

This is certainly true for the technical and production scale domain, but for high-throughput separations or other projects, where time dominates productivity, other solutions have to be applied based on limited or even not optimized methods.

These two main fields of application for chromatography result of course in a dissimilar approach to selecting the chromatographic system. If time pressure minimizes the possible number of experiments the use of generic gradient systems (Section 4.3.1.3) is recommended. In such cases solubility problems are also important, but they should be solved quickly (Section 4.3.1.1) and not by extensive solubility studies.

For production-scale processes economic pressure arises and thus a successive optimization of the chromatographic system will pay off by a reduction in operating costs. Therefore, the development of the chromatographic system has to start with the search for a suitable mobile phase with high solubility for the given solute (Section 4.3.1) and a compatible adsorbent. Dependent on the nature of the phase system the separation is optimized by adjusting the mobile phase composition (Sections 4.3.2–4.3.4).

4.3.1
Choice of Phase System Dependent on Solubility

The most important step in developing a chromatographic system is the choice of mobile phase. Although the mobile phase influences the separation process in many ways, the main decision parameters for the choice of the mobile phase are the maximum solubilities c_s of the target components and the selectivity of the separation. Solubility is essential for the productivity of the chromatographic separation process, because it is a precondition for exploiting the whole range of column loadability.

An old alchemist maxim, „similia similibus solvuntur" („like dissolves like"), is the oldest rule for selecting suitable solvents, meaning that the nature of the solute determines the nature of the solvent. Due to the classification of phase systems in Section 4.2.2, organic and aqueous solvents are distinguished. Non-polar to medium-polar substances show best solubility in typical organic solvents while medium-polar to polar substances show best solubility in aqueous solvents.

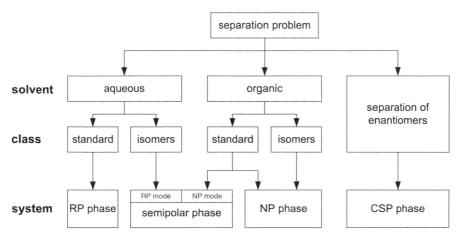

Fig. 4.10 Selection of phase systems dependent on the separation problem.

elutes unretained with the dead time of the column. The strong injection solvent displaces the sample until the sample and the solvent are sufficiently diluted with the mobile phase solvent. At high injection volumes, methanol and sample are no longer diluted and, thus, a front of both migrates through the bed. This behavior is exemplified in the left-hand part of Fig. 4.13.

These problems are intensified if the sample is dissolved in a solvent completely dissimilar to the mobile phase solvents. Many samples delivered for preparative separations are dissolved in common universal organic solvents such as dimethyl sulfoxide, dimethylformamide or other polar organic solvents (Neue, Mazza et al., 2003). Guiochon and Jandera have examined the influence of sample solvent and advise avoiding injections with a solvent of higher elution strength than the mobile phase, especially at high sample loads (Jandera and Guiochon, 1991). Injection of a large volume of the saturated mobile phase is preferred.

Peak distortion and precipitation problems due to injection of solvents of high solubility can be defused by changing the injection technique. The right-hand part of Fig. 4.13 shows the principle of the at-column-dilution (ACD) technique. Sample is dissolved in a strong solvent but, before it is injected into the column, it is mixed with the weak solvent to adjust the solvent strength to the mobile phase. The solubility of the sample in the mixture is of course decreased but, due to the short path between the mixing point and the packed bed, the growth of particles large enough to clog the frits can be excluded, especially since precipitate formation is preceded by a supersaturated solution (Neue, Mazza et al. 2003). ACD significantly increases the amount of sample that can be purified relative to conventional injection methods. A desirable peak shape is obtained for the ACD technique although the separation is performed at mass loadings more than ten times greater than conventional loading where the conventional chromatographic separation already fails (Szanya, Argyelan et al. 2001).

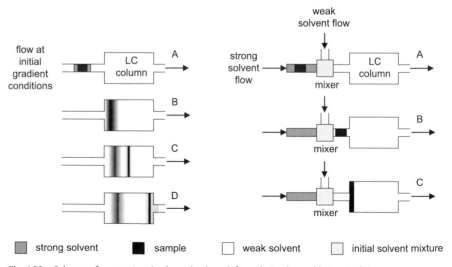

Fig. 4.13 Scheme of conventional column loading (left) and at-column-dilution (right).

Another method for separating low-solubility samples, which is well known in the downstream processing of biopolymers, starts with a large volume of sample dissolved in a chromatographically weak solvent. This solution is fed to the column. As the solute has a relatively strong affinity to the stationary phase, e.g. silica, it is adsorbed and concentrated at the column inlet. This has the effect of introducing the sample as a concentrated plug.

When enough sample solution has been injected, the solvent is changed to a chromatographically stronger mobile phase. The sample is thus eluted with a high concentration in a step-gradient-type procedure. When this technique is used, coarse filtration of the original sample solution is advisable to prevent solid material entering the system.

4.3.1.2 Dependency of Solubility on Sample Purity

During scale-up of a multistep chemical synthesis the chemical purity of samples is often very variable. This is a major issue for process optimization of chromatographic processes as chemical impurities can strongly influence a sample's solubility. The solubility of the sample can decrease dramatically when it becomes purer.

Figure 4.14 shows the quantitative analysis of the racemic pharmaceutical intermediate HG290 to determine its maximum solubility. Two different samples, with 90 and 100% purity, have been available. For both samples a big excess of the solid compound were dissolved under slight agitation for 24 h in the solvent. It has to be assured that a residuum remains at the bottom of the sample vial to achieve a saturated solution. After 24 h a sample of the clear supernatant is taken and analyzed by HPLC. The peak areas in Fig. 4.14 show that the solubility is about 2 times higher for the sample with 90% purity.

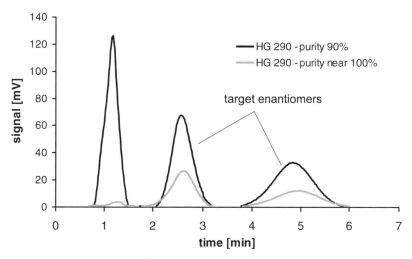

Fig. 4.14 Maximum solubility of a sample for two different purities; sample volume 20 μl. (Reproduced with permission of H. Gillandt SiChem GmbH, Bremen.)

In production-scale processes the purity of the samples also depends on the quality of the raw material (e.g. educts for chemical reactions) as well as on the equipment used. To avoid problems in the subsequent separation process the influence of purity on solubility must be checked after the first amounts of the actual sample are available.

4.3.1.3 Generic Gradients for Fast Separations

For components to be isolated in the mg to g range, tedious method development is not advisable as these components have to be delivered under time pressure and not with maximum economy. Specialized laboratory units that perform high-throughput purification are often operated within the pharmaceutical industry.

A fast solution for these separation problems is the use of generic gradients. Table 4.5 lists typical solvents and conditions for gradient runs for RP and NP systems.

Gradient separations always start with a mobile phase composition of 90–100% of the solvent with low elution strength (weak solvent). This leads to the highest possible retention at the beginning. For instance, RP systems start with 100% water, if the stationary phase can tolerate this high amount of water without any collapse of the alkyl chains. Otherwise, water has to be mixed with a small amount of an organic solvent. Section 4.3.3.2 describes the development of a typical RP gradient. The sample should be injected in a solvent composition with weak elution strength. Otherwise the sample breaks through with the front of the injection solvent. Solvent strength is now increased linearly by increasing the volume fraction of the solvent with high elution strength. The mobile phase composition is linearly changed to 100% of the strong solvent during the so-called gradient time, t_g. This gradient time should be adjusted to 10- to 15-fold of the dead time of the column.

4.3.2
Criteria for Choice of NP Systems

If selecting a chromatographic phase system with regard to Fig. 4.10 results in a normal phase system, neat silica gels are often applied. They have excellent separating power characteristics for samples with low to moderate polarity and intermediate molecular weight (<1000). As long as the sample to be separated does not contain extremely polar or dissociating functional groups, it is the method of choice for preparative chromatographic applications. In general, small organic molecules have bet-

Table 4.5 Typical gradient runs for reversed and normal phase systems.

Phase system	Adsorbent	Weak solvent	Strong solvent
Reversed phase	RP-18 , RP-8	Water	Acetonitrile
Reversed phase	RP-18 , RP-8	Water	Methanol
Reversed phase	RP-18 , RP-8	Dichloromethane	Acetonitrile
Normal phase	Silica	Heptane	Ethyl acetate
Normal phase	Silica	Dichloromethane	Methanol

ter solubility in organic solvents than in the water-based mobile phases used in reversed phase chromatography.

In addition, more product can be loaded on a silica gel than on reversed phase materials. Furthermore the separated products are readily recovered by simple evaporation procedures, while the isolation of products from an RP separation process often require additional work-up steps. These steps include the removal of salts or other aqueous phase additives and extraction of the target product after evaporation of the organic phase (Section 4.3.3.5).

Besides neat silica gel, it is also worth exploring the separation characteristics of polar modified silica gels (amino, diol, cyano) under NP conditions, because these phases often enable unexpected selectivities. This development can be performed by HPLC runs and by thin-layer chromatography (TLC).

4.3.2.1 Pilot Technique Thin-layer Chromatography

Thin-layer chromatography is a simple, fast technique that often provides complete separation of fairly complex mixtures. Any sample type that has been successfully separated by thin-layer chromatography on silica or alumina plates can be separated by means of liquid–solid chromatography.

Mobile phase optimization using HPLC can easily be carried out using fully automated instruments, but this approach is extremely time-consuming. Conversely, modern thin-layer chromatographic techniques allow several mobile phases to be tested in parallel within a very short period of time. Furthermore, different methods are available to visualize the sample components, and the thin-layer chromatogram immediately provides information about the presence of products in the sample that remain at the point of application.

Taking some basic rules into account, the mobile phase composition established by means of TLC can be used directly for preparative chromatographic applications. A pragmatic approach using TLC as a pilot technique for normal phase preparative chromatographic separations is elucidated in this chapter.

Similar to the retention factor in HPLC the retardation factor R_f describes retention behavior during TLC experiments. It represents the ratio of the distance migrated by the sample to the distance traveled by the solvent front. Boundaries are $0 < R_f < 1$. For $R_f = 0$, the product does not migrate from the origin and for $R_f = 1$, the product is not retained. R_f values are calculated to two decimal places, while some authors prefer to tabulate values as whole numbers, as hR_f values (equivalent to $100R_f$) (Poole, 2003).

4.3.2.2 Retention in NP Systems

Different models have been developed to describe the retention of substances in adsorption chromatography. The Snyder model (Fig. 4.15) assumes that in liquid--solid chromatography the whole adsorbent surface is covered with a monolayer of solvent molecules and the adsorbent together with the adsorbed monolayer has to be considered as the stationary phase (Snyder, 1968; Snyder and Kirkland, 1979; Snyder et al. 1997). Adsorption of the sample occurs by displacement of a certain volume of solvent molecules from the monolayer by an approximately equal volume of sample,

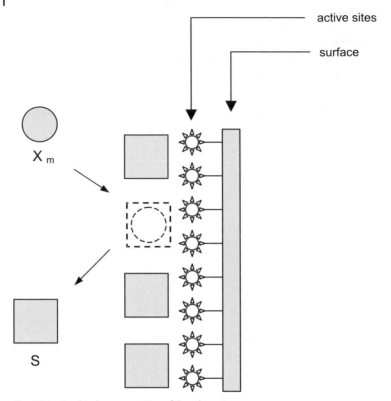

Fig. 4.15 Graphical representation of the adsorption process.

by which these molecules can adsorb on the adsorbent and become part of the mono-layer. In this model, product retention is always caused by displacement of solvent molecules from the monolayer.

Snyder's model provides a good understanding of separations on alumina as the adsorbent and is fairly good at explaining separations on silica gel using weak solvents. An almost similar model developed by Soczewinski (Soczewinski, 1969) is more suitable for separations on silica gel using strong eluents.

In contrast, the Scott-Kucera model considers a solvent system composed of an apolar solvent A and a polar solvent B (Scott and Kucera, 1975). When this mixture is pumped through a column, a monolayer of the most polar solvent B is formed by adsorption of B on the adsorbent. Sample molecules are adsorbed on this monolayer instead of on the adsorbent surface. In other words, there is no displacement of adsorbed solvent molecules, and interaction between the molecules of the monolayer and the sample molecules determines the retention of the component. This theory has been adapted by saying that the model is only valid for medium polar mobile phases and solutes with a polarity lower than the most polar solvent in the eluent. These medium polar solvents are called hydrogen-bonding solvents (esters, ethers, ketones). A monolayer of these solvents behaves as a hydrogen-bonding phase. Interaction between this monolayer and the product molecules takes place mainly by means of hydrogen bonding. When the monolayer of strong solvent B is completely formed, a second layer comes into being. Advancement in the build-up of this second layer determines whether the sample can displace solvent molecules. However, product molecules will never displace solvent molecules from the monolayer as long as the sample molecules are less polar than the strongest solvent in the eluent.

As already described, molecules of the mobile phase and solute compete for the active sites of the adsorbent. The competition is represented by Eq. 4.2 (Snyder, 1968).

$$X_m + nS_{ads} = X_{ads} + nS_m \tag{4.2}$$

X_m and X_{ads} represent the sample molecule in the mobile phase and the adsorbed state, respectively. The corresponding states of the mobile phase molecules are represented by S_m and S_{ads}; n is the number of solvent molecules that have to be displaced to accommodate the sample molecule.

Equation 4.3 gives the equilibrium, which shows that the relative interaction strength of the solvent and the sample molecules for the active sites of the adsorbent, determine the retention of a product.

$$K = \frac{(X_{ads})(S_m)^n}{(X_m)(S_{ads})^n} \tag{4.3}$$

Both silica gel and alumina contain surface hydroxyl groups and possess some Lewis acid–base type interaction possibilities, which determine their adsorption characteristics. The more hydroxyl groups, the stronger a solute molecule will be retained. The number and the topographical arrangement of these groups determine the activity of the adsorbent.

Silica and alumina have the highest surface activity when the adsorbents are free of physisorbed water. Addition of water blocks the most active sites on the surface since water, as a polar adsorptive, is preferentially adsorbed. Other polar compounds such as alcohols can also adsorb irreversibly at the surface. Consecutive adsorption of water deactivates the adsorbent surface and the solute retention will decrease concurrently. For this reason the water content of the mobile phase solvents should be controlled carefully if an apolar mobile phase is used (Unger, 1999). The water content only influences retention when apolar eluents are used, e.g. hexane or heptane. When these solvents are mixed with 10% or more of a moderately polar solvent (e.g. acetone, ethyl acetate) the dependency disappears. The influence can be decreased by the addition of a small amount of acetonitrile to the mobile phase.

The steric orientation of the functional groups in the solute molecule in relation to the spatial arrangement of the hydroxyl groups on the surface enable silica and alumina to separate isomeric components (e.g. cis–trans/positional isomers).

4.3.2.3 Solvent Strength in Liquid–Solid Chromatography

Knowing that the solvent and the solute are in competition for active sites of the adsorbent, it is easily understood that the more the mobile phase interacts with the adsorbent the less a solute molecule is retained. Therefore, the major factor determining product retention in adsorption chromatography is the relative polarity of the solvent compared with the solute.

It is possible to set up a polarity scale by empirically rating solvents in order of their strength of adsorption on a specific adsorbent. A solvent of higher polarity will displace a solvent with a lower rank on the polarity scale. Solvents ranked according to such a polarity scale are called an eluotropic series. Table 4.6 gives an eluotropic series developed for alumina as adsorbent. It can be transferred to silica by a simple factor. The ε^0 value in Table 4.6, called the „solvent strength parameter", is a quantitative representation of the solvent strength and can be used for the calculation of retention factors for different solvent compositions. Assuming that the adsorbent surface is energetically homogeneous and the solute–solvent interaction mechanism in the mobile phase does not influence interactions in the adsorbent phase, Poole (2003) recommends the following empirical equation (Eq. 4.4) for binary mixtures:

$$\log k'_2 = \log k'_1 + a\, A_s(\varepsilon_1^0 - \varepsilon_2^0) \tag{4.4}$$

ε_1^0 and ε_2^0 are the solvent strength parameters for the mixture, while $k_1{}'$ and $k_2{}'$ represent the retention factors for the solute S, which occupies the surface cross section A_s at the stationary phase; a stands for the activity parameter of the adsorbent phase (see below).

Values in the table only serve as a guide because the relative polarity can change, depending on the type of sample and adsorbent.

An eluotropic series is a tool to influence the retention of sample components. If the retention of a product has to be reduced, a solvent of higher ranking in the eluotropic series is chosen. Due to increased competition between product and solvent molecules for the active sites on the adsorbent, the retention factor of the product will automatically decrease. The same holds true for the opposite situation.

According to L.R. Snyder (Snyder, 1968) it is possible to calculate the solvent strength of binary mixtures using Eq. 4.5.

$$\varepsilon_{ab}^0 = \varepsilon_a^0 + \frac{\log\left[N_b 10^{a\, n_b(\varepsilon_b^0 - \varepsilon_a^0)} + 1 - N_b\right]}{a\, n_b} \tag{4.5}$$

n_b describes the molecular cross section occupied on the adsorbent surface by the solvent molecule b in units of 0.085 nm² (which is 1/6 of the area of an adsorbed benzene molecule, corresponding to the effective area of an aromatic carbon atom on the adsorbent surface), while the parameter a describes the activity parameter of the adsorbent surface. The adsorbent activity parameter is a measure of the adsorbent's ability to interact with adjacent molecules of solute or solvent and is constant for a given adsorbent. For silica gel a value of 0.57 is given by Poole (2003). The molar solvent ratio is calculated by Eq. 4.6 with the molar volumes of the molecules (Eq. 4.7).

$$N_b = \text{molar solvent ratio} = \frac{\text{Vol.\%B}\left(\dfrac{1}{V_b}\right)}{\text{Vol.\%B}\left(\dfrac{1}{V_b}\right) + \text{Vol.\%A}\left(\dfrac{1}{V_a}\right)} \tag{4.6}$$

Table 4.6 Eluotropic series for alumina.

Solvent	$\varepsilon^{\circ}_{Al_2O_3}$
n-Pentane	0.00
Isooctane	0.01
Cyclohexane	0.04
iso-Propyl ether	0.28
Toluene	0.29
Dichloromethane	0.42
Methyl iso-butyl ketone	0.43
Methyl ethyl ketone	0.51
Triethylamine	0.54
Acetone	0.56
Tetrahydrofuran	0.57
Ethyl acetate	0.58
Methyl acetate	0.60
Diethyl amine	0.63
Acetonitrile	0.65
1-Propanol	0.82
2-Propanol	0.82
Ethanol	0.88
Methanol	0.95
Acetic acid	Large

$$\left(\frac{1}{V_a}\right) \text{ and } \left(\frac{1}{V_b}\right) = \frac{\rho(\text{density})}{MW(\text{molecular weight})} \tag{4.7}$$

In the following example the solvent strength of a mixture of dichloromethane and methanol with 90% (V/V) dichloromethane is calculated.

Dichloromethane   Methanol

$\varepsilon_a^0 = 0{,}42$ and $\varepsilon_b^0 = 0.95$

$n_a = 4.1$ and $n_b = 8$

$a = 0.7$

$$\left(\frac{1}{V_a}\right) = \frac{1.325\,\text{g ml}^{-1}}{84.93\,\text{g mol}^{-1}} = 0.0157\,\text{mol ml}^{-1}$$

and

$$\left(\frac{1}{V_b}\right) = \frac{0.79\,\text{g ml}^{-1}}{32\,\text{g mol}^{-1}} = 0.0249\,\text{mol ml}^{-1}$$

$$N_b = \frac{10 \times 0.0249}{10[0.0249 + (90 \times 0.0157)]} = 0.1489$$

$$\varepsilon_{ab}^0 = 0.42 + \frac{\log\left(0.1498 \times 10^{0.7 \times 8(0.95-0.42)} + 1 - 0.1498\right)}{0.7 \times 8}$$

$$= 0.42 + 0.38325 = 0.80325$$

4.3.2.4 **Selectivity in NP Systems**

Separations by means of liquid–solid chromatography are carried out on polar adsorbents. The primary factor that determines the adsorption of a product is the functional groups present in the sample. Relative adsorption increases as the polarity and number of these functional groups increase, because the total interaction between the solute and the polar adsorbent surface is increased.

Uniqueness of retention and selectivity in NP-LC arise from two characteristic phenomena:

- Competition between sample and solvent molecules for the active sites on the adsorbent surface
- Multiple interactions between functional groups of the sample and the rigidly fixed active sites on the adsorbent surface

The broad selectivity potential of different solvent systems in NP-LC results from the adjustable concentration of the polar components that are adsorbed into the monolayer. This leads to an enormous variety of possible stationary phases (adsorbed solvent plus adsorbent) and a corresponding broad range of potential selectivities. Due to differences in the content of the polar modifier, different stationary phases are created.

For both TLC and HPLC many mobile phase optimization procedures and criteria have been described in the literature. Mainly, two strategies are followed:

- Keeping the solvent mixture simple by considering binary mixtures only.
- Tuning selectivity by using more solvents, which can, of course, result in very complex mixtures.

For binary mixtures a limited number of experiments are required and this approach is, naturally, the preferred method for preparative chromatographic applications.

4.3.2.5 **Mobile Phase Optimization by TLC Following the PRISMA Model**

Before starting extensive experiments, a procedure recommended by Kaiser and Oelrich (1981) to rule out adsorbents by fast experiments should be employed. Each elution experiment takes about 20 s. For this purpose samples are applied on a 50 × 50 mm TLC plate at 9 points, which are exactly 10 mm apart. Five microlitres of methanol are drawn into a micro-capillary with a Pt–Ir point. By applying the point of the filled capillary on one of the sample points on the plate, methanol is introduced onto the plate. A miniature radial chromatogram of ca. 7 mm diameter is produced. If the sample components remain at the point of application, the use of this adsorbent type is ruled out for HPLC usage. To make sure, the procedure is repeated with 5 µl of acetonitrile and tetrahydrofuran, respectively. If the products still remain at the point of application, the situation will not be changed by using any other mobile phase that is suitable for preparative chromatography work.

However, if the entire sample migrated to the outer ring with methanol, the test is repeated with n-heptane. If the entire sample still migrates to the outer ring, then the used adsorbent is again ruled out as stationary phase. If, with n-heptane, the entire sample or part of it remains at the starting point, the unlimited possibilities of combining solvents can be exploited to further optimize the method (Kaiser and Oelrich, 1981).

The PRISMA model developed by Nyiredy and co-workers (Nyiredy et al., 1985; Dallenbach-Tölke et al., 1986; Nyiredy and Fater, 1995; Nyiredy, 2002) for use in Over Pressured Layer Chromatography is a three-dimensional model that correlates solvent strength and the selectivity of different mobile phases. Silica gel is used as the stationary phase and solvent selection is performed according to Snyder's solvent classification (Tab. 4.7).

The PRISMA model is a structured trial-and-error method that covers solvent combinations for the separation of compounds from low to high polarity. Initial experiments are done with neat solvents, covering the eight groups of the Snyder solvent classification triangle.

The PRISMA model (Fig. 4.16a) has three parts, an irregular truncated top prism, a regular middle part with congruent base and top surfaces, and a platform symbolizing the modifier. The model neglects the contribution of the modifier to the solvent strength, because modifiers are usually only present in a low and constant concentration. In NP chromatography, the upper frustum is used in optimizing mobile phases

Table 4.7 Classification of the solvent properties of common liquids (Snyder and Kirkland, 1979).

Solvent	S_t	x_e	x_d	Group
n-Hexane	0	–	–	
n-Butyl ether	2.1	0.44	0.18	I
DIPE	2.4	0.48	0.14	I
MTBE	2.7	0.49	0.14	I
Diethyl ether	2.8	0.53	0.13	I
iso-Pentanol	3.7	0.56	0.19	II
n-Butanol	3.9	0.56	0.19	II
2-Propanol	3.9	0.55	0.19	II
n-Propanol	4	0.54	0.19	II
Ethanol	4.3	0.52	0.19	II
Methanol	5.1	0.48	0.22	II
Tetrahydrofuran	4	0.38	0.2	III
Pyridine	5.3	0.41	0.22	III
Methoxyethanol	5.5	0.38	0.24	III
Methylformamide	6	0.41	0.23	III
Dimethylformamide	6.4	0.39	0.21	III
Dimethyl sulfoxide	7.2	0.39	0.23	III
Acetic acid	6	0.39	0.31	IV
Formamide	9.6	0.36	0.23	IV
Dichloromethane	3.1	0.29	0.18	V
Benzylalcohol	5.7	0.4	0.3	V
Ethyl acetate	4.4	0.34	0.23	VI
Methyl ethyl ketone	4.7	0.35	0.22	VI
Dioxane	4.8	0.36	0.24	VI
Acetone	5.1	0.35	0.23	VI

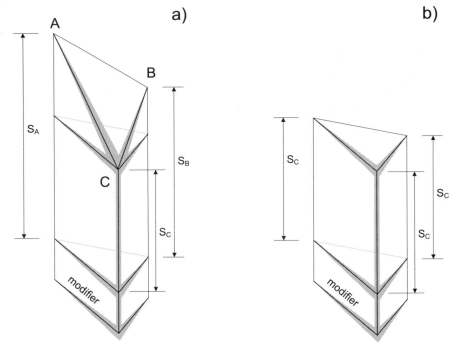

Fig. 4.16 Graphic representation of the PRISMA model: (a) complete model (b) regular part of the model.

for polar and/or semi-polar substances. The regular centre part of the prism is used in solvent optimization for apolar and semi-polar components.

The three top corners of the model represent the selected undiluted neat solvents. The corner to the longest edge of the prism (A) represents the solvent with the highest strength, while the solvent with the lowest strength (C) corresponds to the corner of the shortest edge. Because the three selected solvents have unequal solvent strengths, the length of the edges of the prism are unequal and the top plane of the prism will not be parallel and congruent with its base. If the prism is intersected at the height of the shortest edge and the upper frustum is removed, a regular prism is obtained (Fig. 4.16b). The height of this prism corresponds to the solvent strength of the weakest solvent. All the points of the equilateral triangle formed by the cover plate of the prism have equal solvent strength. One of the corners (C) represents the neat solvent with the lowest strength. The composition of the two other corners can be obtained by diluting these solvents with a solvent of zero strength (n-hexane for NP and water for RP applications)

In the regular prism, horizontal intersections parallel to the base can be prepared by further diluting the selected solvents with solvent of zero strength (represented by the line from P_s^* to P_s^{**} in Fig. 4.17). All points on the obtained triangles represent the same solvent strength, while all points on a vertical straight line correspond to the same selectivity points (P_s).

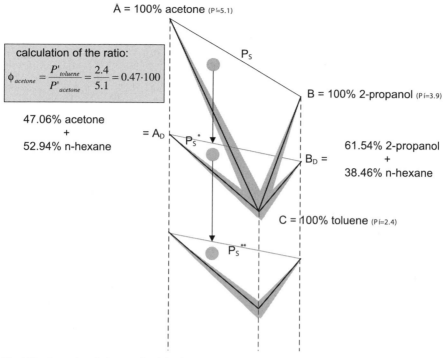

Fig. 4.17 Example calculation of mobile phase composition.

Points along the edges of the irregular part of the model, which represent mixtures of two solvents (AB, BC and AC), are shown in the right-hand part of Fig. 4.18, while the left-hand part describes the selectivity points of mixtures of the three solvents.

Each selectivity point can be characterized by a coordinate, defined by the volume ratio of the three solvents. The ratio is written in the order of decreasing solvent strength of the undiluted solvent (A, B, C). 100% solvent A with the highest solvent strength is represented by $P_s = 10$–0–0 and the solvent with the lowest strength (C) by $P_s = 0$–0–10.

From these data, all other basic selectivity points can be defined by a three digit number. For example, the selectivity point with coordinates 325 means 30 vol.% A, 20 vol.% B and 50 vol.% C (Fig. 4.19). Finer adjustment of the volume fractions can be expressed by three two-digit- numbers ($P_s = 55$-25-20)

The whole strategy of solvent optimization via the PRISMA model includes the following steps:

- Start – selection of neat solvents
- Step 1 – solvent strength adjustment
- Step 2 – optimization of selectivity
- Step 3 – final optimization of solvent strength
- Step 4 – determination of optimal mobile phase composition

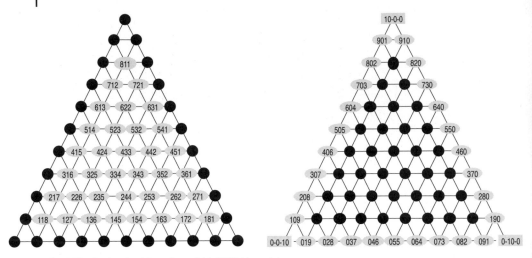

Fig. 4.18 Basic selectivity points of the PRISMA model.

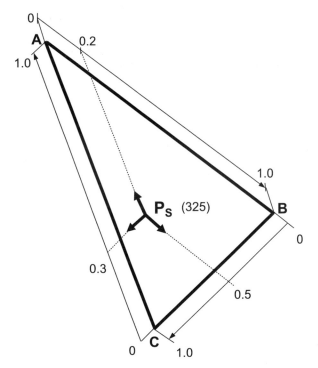

Fig. 4.19 Combination of three neat solvents in the irregular part of the model.

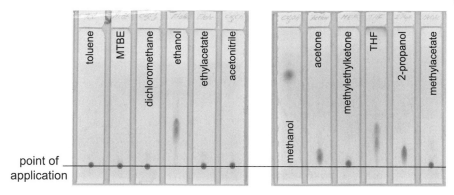

Fig. 4.20 TLC experiments for the separation of two isomers using neat solvents.

The optimization strategy will be exemplified by a rather difficult isomer separation. The product to be separated was a relatively strong base and first experiments using neat solvents on different types of stationary phases (silica gel, amino-, cyano-, and diol-modified silica gel) revealed that it would be very difficult to separate the two isomers on a larger scale. Therefore, synthetic chemists were contacted to modify one of the secondary base functions with an easily removable BOC (tert-butoxycarbonyl) group.

Preliminary TLC experiments in unsaturated chromatographic chambers were performed using ten neat solvents. Solvents listed in Tab. 4.8 represent the most common starting solvents from each group of the Snyder model. For the example of the isomers this list was modified slightly to fit the experience of previous experiments. Figure 4.20 shows the results of the solvent selection experiments.

Based on experience (R_f values, selectivity and spot shape), one to three solvents were selected. Preferentially, solvents are chosen that demonstrate small, well-defined spot shapes for the title components. Binary mixtures or single solvents that result in round finite spots have to be selected for the first set of experiments.

Figure 4.20 clearly demonstrates that only ethanol, acetone and tetrahydrofuran showed some selectivity between the two isomers, while only methanol gave a nice round spot. This example shows the practical usefulness of screening with TLC experiments. Because the experiment with each plate took roughly 20 min, the use of 12 different solvents could be examined in 40 min.

Table 4.8 Solvents for preliminary TLC experiments.

Solvent	Group	Solvent	Group
Diethyl ether	I	Dichloromethane	V
2-Propanol	II	Ethyl acetate	VI
Ethanol	II	Dioxane	VI
Tetrahydrofuran	III	Benzene	VII
Acetic acid	IV	Chloroform	VIII

As a standard approach for preparative chromatography, binary solvent combinations are always investigated before considering ternary mixtures. Figure 4.21 shows the results of experiments with binary mixtures where selectivity was obtained. Other binary combinations were tested, but none resulted in a good spot shape and an acceptable separation between the isomers. Therefore, some additional tests were performed with ternary solvent combinations.

Step (1) Solvent Strength Adjustment

Once the different solvents have been selected, a first experiment is done using the mobile phase composition corresponding to the centre of the cover triangle ($1/3$ solvent A, $1/3$ solvent B and $1/3$ solvent **C**). If the observed R_f values are too high, additional experiments are performed to adjust the solvent strength by the addition of zero strength solvent.

Step (2) Optimization of Selectivity

In the triangle with the desired solvent strength, three selectivity points near the edges of the triangle, [**811**(80% A–10% B–10% C)] – [**181** (10% A–80% B–10% C)] – and [**118** (10% A–10% B–80% C)], are the mobile phase compositions for the following three experiments (Fig. 4.22).

For binary mixtures the selectivity points (75% A–25% B–0% C), (25% A–75% B–0% C), (75% A–0% B–25% C), (25% A–0% B–75% C), (0% A – 75% B – 25% C), (0% A – 25% B – 75% C) were investigated.

The obtained chromatograms and chromatograms of step 1 serve as guidelines for the next experiments. If necessary, selectivity is further optimized by choosing new points in the triangle near the coordinate or between coordinates that gave the best resolution. Generally, a limited number of additional experiments are required to

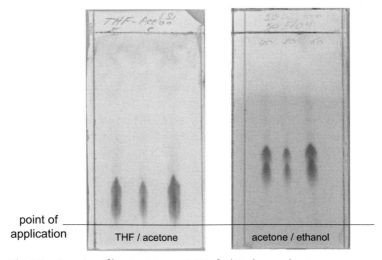

point of
application   THF / acetone   acetone / ethanol

Fig. 4.21 Screening of binary mixtures (50:50) of selected neat solvents.

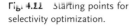

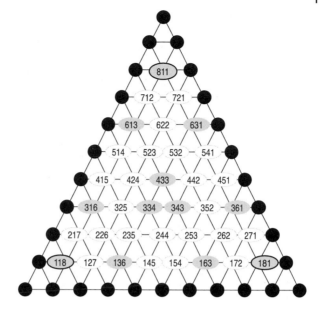

Fig. 4.22 Starting points for selectivity optimization.

obtain optimum conditions for separation. The resulting selectivity point (P_s) is used in the final optimization step. To separate the isomers the following solvent combinations were chosen for steps 1 and 2:

- Acetonitrile, methanol and ethanol
- Acetone, methanol and ethanol
- Acetone, ethyl acetate and ethanol
- Acetone, methyl acetate and methanol
- Acetonitrile, methyl acetate and methanol

First experiments were done with equal volumes of the three neat solvents as described in step 1. Based on this data some additional experiments were performed using different volume ratios of the most promising solvent mixtures.

Figure 4.23 shows the results of this optimization. The R_f values are in a promising range and thus a zero strength solvent is not needed. The TLC chromatograms clearly illustrate that different ternary solvent combinations separate the isomers well. Both spot shape and resolution are better than for the binary solvent mixtures (Fig. 4.21).

Step (3) Final Optimization of the Solvent Strength
This step is used for the final optimization of solvent strength. It corresponds to a vertical shift in the regular part of the prism, starting from the optimal selectivity point (P_s) established in step 2. If all products of interest are sufficiently separated, the solvent strength can be increased or decreased to reach the desired goal.

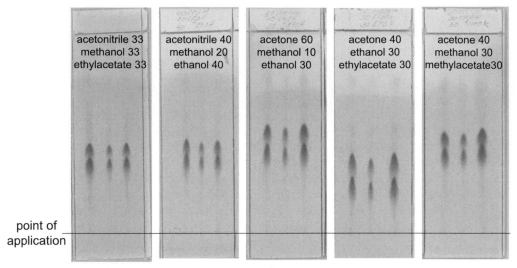

point of
application

Fig. 4.23 Screening of ternary solvent mixtures.

Step (4) Determination of the Optimum Mobile Phase Composition

After the basic tests, were suitable solvents are selected and the desired solvent strength is established, the effect of eluent changes at a constant solvent strength are further investigated to find the optimum mobile phase composition.

To obtain reliable results the TLC measurements have to be carried out in saturated chambers, because in unsaturated chambers the reproducibility of R_f may be poor and the formation of secondary and tertiary solvent fronts can affect the interpretation of the generated results. The most versatile TLC chamber in which to perform such experiments is the Vario-KS chamber developed by Geiss and Schlitt (1965). This chamber is suitable for evaluating the effect of different solvents, solvent vapors and relative humidity. In this type of chamber the chromatoplate is placed face down over a conditioning tray containing several compartments to hold the required conditioning solvents. The design of the chamber ensures that saturating and developing solvents are completely separated. The chamber's major advantage is that, for the purpose of solvent optimization, up to ten activity and/or saturating conditions can be compared on the same plate using the same eluent.

To formulate a mathematical model for the dependence of hR_f on mobile phase composition, the obtained hR_f are displayed against the selectivity points, which symbolize the composition of the eluent.

Correlations were tested along the axes of the triangle and from the three basic selectivity points (118–811–181) across the middle point to the opposite side of the triangle.

A minimum of 12 experiments are required to determine a local optimum. By measuring hR_f for the selectivity points 118–316–613–811–631–361–181–163–136–433–334 and 343 (Fig. 4.22), hR_f for the selectivity points 217–415–514–712–721–

511 451 271–172–154–145–127–235–253–325–352, 523 and 532 can be calculated using the mathematical functions obtained from the measured values.

Functions along the axes:

- 811–613–316–118 (solvent A) (calculation of 217–415–514–712)
- 181–361–631–811 (solvent B) (calculation of 271–451–541–721)
- 118–136–163–181 (solvent C) (calculation of 127–145–154–172)

In general the correlation between hR_f and the selectivity points at a constant solvent strength level can be expressed by Eq. 4.8.

$$hR_f = a(P_s)^2 + b(P_s) + c \tag{4.8}$$

$$\ln hR_f = a(S_t) + b \tag{4.9}$$

To obtain a global solvent optimum, the vertical relationship between the solvent strength and hR_f values has also to be investigated and, therefore, some additional tests are required. Using quaternary solvents, the correlation was found to be described by Eq. 4.9. Because a linear mathematical function requires a minimum of three measured data points, the vertical relationship between solvent strength and hR_f values has to be tested at three different solvent strength levels. To collect accurate data, the chosen solvent strength levels have to differ individually by 5 to 10%.

To determine accurately the vertical relationship function, the following selectivity points were investigated:

- Solvent strength level 1:   811–631–118–343–334–181
- Solvent strength level 2:   811–433–118–316–361–181
- Solvent strength level 3:   811–613–118–334–163–181

4.3.2.6 Strategy for an Industrial Preparative Chromatography Laboratory

The effort required to develop a separation method strongly depends on several factors:

1. Recurrence of a separation problem (repeatedly or only once)
2. Amount of product to be separated
3. Do we have to isolate a single target component or all relevant peaks in the mixture?
4. Quality requirements
5. Purpose of the purification process
   - Purification of key intermediates or separation of a final product
   - Preparation of reference standards for analytical purposes
   - Isolation and purification of impurities in the 0.05–0.1 mg ml^{-1} concentration range
   - cGMP related investigations

Obviously, when separating small quantities of many different products, standard separation methods using gradient elution are preferred. Conversely, for larger scale separations it is important to design a robust separation process that can run fully automated 24 h a day. Clearly, for a production-scale process that has to be performed under cGMP conditions many experiments are required to be able to design a reproducible, robust process that can be transferred from one production location to another without the need for substantial modifications. For this reason two examples, covering both separations are exemplified.

Example 1: Separation of Large Numbers of Different Products in the Range 1–10 g

As mentioned before, for this type of work, standard procedures using gradient elution are the method of choice and there are two possible approaches:

- Normal phase chromatography on silica gel
- Reversed phase chromatography (Section 4.3.3.2)

For such products, HPLC chromatograms using a standardized method and mass spectrometry data are usually available and, therefore, it is very tempting to perform a preparative chromatographic purification using these experimental conditions.

The major advantage of this approach is that for mixtures containing several peaks of approximately equal size, the analytical HPLC-MS data allow immediate recognition of the desired product in the preparative run and, therefore, no additional structural analyses are required to identify the isolated product. However, the major disadvantage of reversed phase applications is often the limited solubility of the samples in eluents having high water content. Furthermore, due to the relatively high viscosity of the eluents used in reversed phase chromatography, the flow rate that can be pumped through a preparative chromatography column is generally lower than for the pure organic solvent mixtures used in normal phase chromatography. Another point that certainly has to be considered is the loading capacity of a C18 reversed phase material, which is generally only $1/2$–$1/3$ of the loading capacity of bare silica gel.

However, is it possible to run, reproducibly, solvent gradients on bare silica? Many people have their doubts, especially concerning the necessity of very long equilibration times when returning to the initial solvent composition. Furthermore, on the first generation of silica gels it was impossible to perform gradient elutions using very polar solvents, e.g. methanol, because some silica gel was dissolved, which certainly does not help to convince these people. Nonetheless, new spherical materials prepared from ultrapure silica no longer demonstrate this pronounced solubility problem. On these materials, gradient runs ending up with 100% of very polar organic solvent are easily performed. The problem of activity changes when going from an intermediate polarity to a relatively polar solvent is also non-existent when, for two or more injections in succession, exactly the same procedure is followed, i.e. same gradient profile, same rinsing time with polar solvent, same time to go back to the start solvent composition, and, very important, same time at the initial composition of the gradient to start the next injection. Only when, for the next series of samples, the gradient has to start with zero strength solvent (n-heptane) equilibration problems can be expected, and then some measures may have to be taken to reactivate the stationary phase.

Standard Gradient Elution Method on Silica Gel

To select suitable gradient conditions, TLC experiments are performed using an eluotropic series based on dichloromethane–methanol mixtures. By means of a Camag Vario chamber (Fig. 4.24), the eluent combinations following Tab. 4.9 have been tested.

The mixture dichloromethane–methanol (50:50%) has been chosen on the basis of experience that products remaining in the point of application with solvents of lower strength often can be moved more easily to higher R_f values with this mixture than with pure methanol as the eluent. Probably, this phenomenon is due to the very good solubility characteristics of this solvent mixture for a broad variety of products.

Based on the observed chromatographic profile, the start and end conditions for a preparative gradient run are selected. The solvent composition that corresponds to the solvent mixture where the product of interest just leaves the point of application is used as the starting composition for the gradient. The solvent mixture moving the product of interest closely to the solvent front is the final composition of the gradient.

Simplified Procedure

Instead of testing the whole dichloromethane–methanol eluotropic series, only one TLC run is performed, using a 90:10 volume ratio mixture of dichloromethane––methanol. The location of the product of interest on this plate is then used to establish the preparative gradient conditions. This procedure is shown in Fig. 4.25.

The type of silica gel selected to perform this type of preparative gradient elutions is a 5 to 10 μm good quality spherical material with excellent mechanical stability (Kromasil®, Akzo – Nobel). About 200–250 g of this stationary phase is packed at 100–150 bar in a 50 mm internal diameter dynamic axial compression column. Chromatography is executed at a flow rate of 110–125 ml min^{-1} and the average sample amount applied is about 3 g. This gradient approach on silica gel as stationary phase has proven to be a very versatile and robust method as long as the sample and eluent are carefully filtered prior to their use, thus avoiding partial blockage of the porous metal plates in the top flange and the piston of the column.

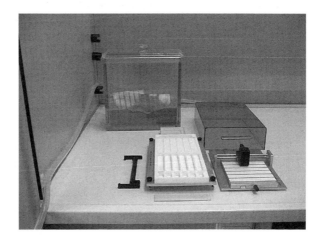

Fig. 4.24 Vario chamber with accessories.

Table 4.9 Solvent series based on dichloromethane–methanol mixtures.

Dichloromethane–n-heptane (50–50 v/v) ($S_T = 1.55$)
Dichloromethane ($S_T = 3.1$)
Dichloromethane–methanol (98–2 v/v) ($S_T = 3.14$)
Dichloromethane–methanol (95–5 v/v) ($S_T = 3.195$)
Dichloromethane–methanol (90–10 v/v) ($S_T = 3.3$)
Dichloromethane–methanol (50–50 v/v) ($S_T = 4.1$)

Example 2: Pilot Chemistry Laboratories and Pilot Plant Scale Separations (A few Hundred Grams to Tenth's of Kilograms Scale)

For this type of problem, time constraints generally dominate. Very often, the scale-up of a chemical reaction without extensive optimization work results in low yields and bad qualities. The necessity to deliver a certain amount of product within a very narrow time frame, in general does not allow preparative chromatography workers to thoroughly optimize the separation method. Therefore, a standard approach is required.

At first the polarity of the product is investigated by means of a TLC analysis using an eluotropic series based on dichloromethane–methanol mixtures such as described in Tab. 4.9.

Very often, one of the tested mixtures can be used to perform the separation, or a few additional experiments using the same solvent combination have to be performed to fine-tune the chromatography process.

Simultaneously, two additional silica TLC plates are made to test 12 neat solvents that, from a safety, health, environmental and practical point of view, are acceptable for use on a larger scale.

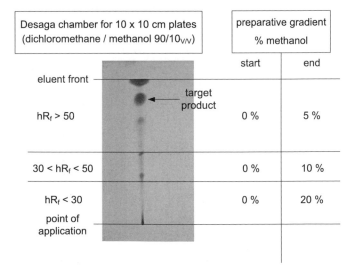

Fig. 4.25 Simplified procedure to establish the start and end composition for a preparative gradient elution on silica gel.

From the series of solvents proposed in Tab. 4.8, diethyl ether, dioxane, benzene, and chloroform are excluded. Instead, the following solvents are used:

Plate 1:

Toluene–TBME–dichloromethane–ethanol–ethyl acetate–acetonitrile

Plate 2:

methanol–acetone–methyl ethyl ketone–tetrahydrofuran–2-propanol–acetic acid

Toluene, acetone and methyl ethyl ketone are, due to their cut-off values, not the first choice if UV detection is used to control the process. Tetrahydrofuran has to be avoided because for larger scale use it has to be stabilized with an anti-oxidant (butylated hydroxytoluene) that will finally end up in the product, often making an additional purification step necessary.

However, for a broad variety of substances, some solvent mixtures based on toluene or acetone or THF have proven to be universally applicable.

- Toluene–2-propanol
- Toluene–ethyl acetate–ethanol
- Toluene–2-propanol–25 % ammonia in water
- n-Heptane–methyl ethyl ketone (or acetone)
- Acetone–ethanol
- Acetone–ethanol–ethyl acetate
- Dichloromethane (or acetonitrile)–THF

Based on the chromatographic profiles observed on both plates, suitable binary solvent mixtures are chosen. Using another TLC plate, an eluotropic series of these solvents is prepared and tested. The start and end composition of this eluotropic series depends on the R_f values measured for the strongest solvent of the mixture.

If on silica gel no satisfactory results are obtained with the neat solvents, the two solvent series (except acetic acid) are, respectively, tested on an aminocyano and diol modified TLC plate.

Most problems can be solved on silica gel as the stationary phase. If either some specific selectivities or the addition of a basic modifier is required, amino modified silica gel (90:10 v/v) often brings the solution.

Clearly, this solvent selection procedure will not always result in optimal preparative chromatographic conditions, but it offers a method that can be applied immediately.

Direct application of a TLC method for preparative chromatography requires consideration of some basic differences between the techniques:

1. Driving force for solvent transport:
   - Capillary forces (TLC)
   - External force (pump) in HPLC
2. Specific surface area of silica gel used to prepare a TLC plate can differ from the packing material in the column

3. Organic or inorganic binders are used to fix the silica layer on the surface of a glass plate. Furthermore, the layer generally contains an inorganic fluorescence indicator
4. TLC starts from a dry layer and we have to deal with a mobile phase gradient over the whole migration distance

Preferential adsorption of the most polar component of the solvent mixture in TLC means that the use of the same mobile phase composition in an HPLC experiment will always result in shorter retention times. (In HPLC the solvent mixture is continuously pumped through the column and, after some time, the active surface spots are occupied with polar solvent molecules.)

Therefore, it is advisable to perform a test injection of the sample on the HPLC column using a mobile phase containing a somewhat smaller amount of the polar component.

4.3.3
Criteria for Choosing RP Systems

In reversed phase liquid chromatography (RPLC) silylated silicas are preferred. The surface of these silicas is covered with chemically bonded non-polar groups such as alkyl chains or polymeric layers (Chapter 3.2.3). Silica modified with medium polar groups such as cyano, diol or amino might be used in NP as well as RP mode. Alternatively, cross-linked polymers such as hydrophobic styrene divinyl benzene-copolymers can be used (Chapter 3.2.4). Polymer packings show stability in a pH range 2–14 while silica based packings show limited stability for pH > 7.

4.3.3.1 Retention and Selectivity in RP Systems
Reversed phase liquid chromatography is used in separating polar to medium polar components. Their separation is based on the interaction of the lipophilic part of the solutes with the non-polar surface groups. Retention depends on the nature of the active groups bonded on the silica surface as well as the functional groups of the solute. The hydrophobicity of reversed phase packings differs in the relative sequence:

Porous graphitized carbon > polymers made from cross-linked styrene/divinylbenzene > n-octadecyl (C18) bonded silicas > n-octyl (C8) bonded silicas > phenyl-bonded silicas > n-butyl (C4) bonded silicas > n-propylcyano-bonded silicas > diol-bonded silicas

The retention time in RP chromatography increases with the surface hydrophobicity. Therefore, for instance, C18 groups cause higher retention times than C8 groups at constant ligand density – as demonstrated in Fig. 4.26, which compares the retention on RP-8 and RP-18 phases. As well as a higher retention time, RP-18 silicas also exhibit different selectivities.

In addition to the surface groups of the adsorbent the nature of the functional groups of the solutes determines the interaction and, thus, the retention time as well. The following sequence shows the elution order for solutes of similar size with different functional groups (Unger and Weber, 1999):

Amine/alcohols/phenols < acids < esters < ethers < aldehydes/ketones < aromatic hydrocarbons < aliphatic hydrocarbons

Figure 4.26 presents the influence of functional groups on the retention time. Benzene and some aromatic derivates are separated by RP chromatography. Compared with benzene the retention time is decreased for derivates with polar aldehyde or hydroxyl groups while it is increased for the more bulky apolar esters, e.g. benzoic or terephthalic methyl esters.

In RP chromatography, mixtures of organic solvents such as methanol, acetonitrile and tetrahydrofuran with water are used as mobile phases. The organic (more hydrophobic) solvent also interacts with the non-polar surface groups of the packing and thus is in competition with the solutes for the non-polar adsorption sites on the adsorbent. Consequently, the retention time of the solutes decreases with increasing fraction of the organic solvent in the mobile phase. The dependency of the retention factor on the volume fraction ϕ of the organic solvent is described by Eq. 4.10 for medium values of ϕ. In this equation $k'_{0,S}$ is the retention factor for pure water and S is the elution strength of the organic modifier (Snyder and Kirkland,1979; Snyder et al., 1997; Snyder and Dolan, 1998).

$$\ln k' = \ln k'_{0,S} - S\phi \tag{4.10}$$

Owing to the exothermic behavior of the adsorption, the retention time increases at lower temperatures. An Arrhenius-type equation describes the retention factor's dependence on temperature (Eq. 4.11)

$$\ln k' = \ln k'_{0,T} - \frac{b}{T} \tag{4.11}$$

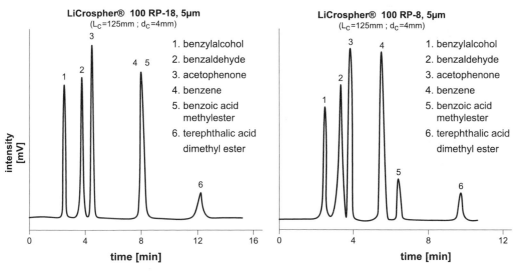

Fig. 4.26 Retention of aromatic components on RP-8 and RP-18 columns with water–methanol (50:50 v/v); flow rate 1 ml min⁻¹. (Reproduced from Unger and Weber, 1999.)

The impact of the mobile phase composition and the temperature is individual for each component and thus the selectivity can be optimized by these parameters.

4.3.3.2 Gradient Elution for Small amounts of Product on RP Materials

RP chromatography is widely used to isolate small amounts of target molecules in automated high-throughput systems. Due to good reproducibility, stability and a broad application range, RP instead of NP chromatography is preferred for this task. For reversed phase separations, TLC is seldom used as a pilot technique, mainly due to the large difference in the degree of derivatization between RP materials used in column chromatography and the materials that can be used for TLC applications. With the now ready availability of mass spectrometry (MS) detection, and since MS is coupled to columns with diameters of up 25 mm by splitting systems, compound isolation is often directly combined with its identification. The structure of the molecules to be separated largely determines the type of mobile phases that can be used in RP chromatography. For some product types, mixtures of water and organic solvents can be used as the mobile phase. However, in this case one has to deal with basic or acidic molecules and so, in general, the aqueous phase has to be buffered to obtain acceptable peak shapes. It is common practice for HPLC-MS identification work to add an ammonium salt to the aqueous phase.

Figure 4.27 gives experimental conditions that can be used as a standard HPLC-MS analysis procedure. After the column has been equilibrated the operation starts with a 10 min gradient separation (from 70 % A, 15 % B, 15 % C to 50 % B, 50 %C) followed by 5 min washing and 5 min re-equilibration.

The obtained analytical results can be directly transferred to a preparative chromatography column. For the standard method depicted in Fig. 4.27 a combination of acetonitrile and methanol is used as the organic modifier. Therefore, some addi-

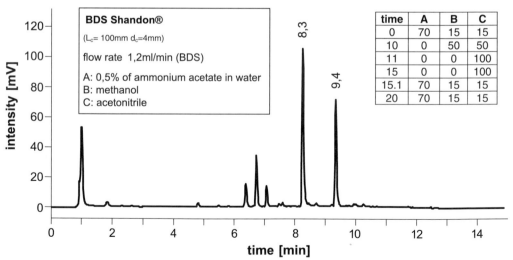

Fig. 4.27 Standard HPLC-MS reversed phase analysis procedure.

tional experiments, investigating the effect of each type of organic modifier individually, are advisable because, very often, large differences in selectivity can be observed.

Owing to microbiological growth, a solution of ammonium acetate in water cannot be stored for a long period. In other words, when for preparative chromatographic applications larger volumes of an ammonium acetate solution have to be stored, some measures have to be taken to avoid this microbiological growth. Different remedies are possible, one of which is to add 5–10 vol.% acetonitrile to the mixture of ammonium acetate and water. Obviously, the presence of this amount of organic modifier has to be taken into account when transferring data from an analytical to a preparative column.

Therefore, the standard gradient for HPLC-MS analysis is modified by using an aqueous ammonium acetate solution containing 10 vol.% acetonitrile. Care has to be taken that the polarity at the start of the gradient remains approximately equal. Chromatograms in Fig. 4.28 depict the separation using this modified weak solvent.

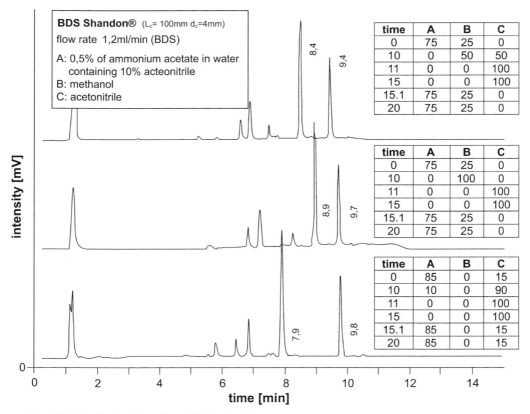

Fig. 4.28   Standard gradient with modified weak solvent.

4.3.3.3 Rigorous Optimization for Isocratic Runs

After developing a first gradient method, the optimization of an isocratic method should be considered for process-scale separations. Gradient runs help to find optimal conditions for isocratic operation, which should be preferred for preparative purposes as the advantages for large-scale separations are obvious:

- Easier equipment can be used (only one pump, lower number of eluent tanks)
- No re-equilibration time necessary (touching band-methods as well as closed-loop recycling and SMB-chromatography can be applied)
- Easier eluent recycling due to constant eluent composition

In the following the development of a separation method is explained in detail by the separation of a pharmaceutical sample consisting of three intermediates.

The general tasks during optimization are:

- Minimize total elution time
- Minimize cycle time (= time between first and last peak)
- Optimize selectivity
- Use isocratic conditions

Minimization of the total elution and cycle time results in higher productivity and lower eluent consumption. Optimizing the selectivity can be contradictory to these parameters. However, if the cycle time is minimized the productivity for touching band operation increases.

Figure 4.29 presents a procedure for optimizing RP chromatography by isocratic experiments. It usually begins with gathering solubility data for the sample, using water-miscible solvents. The solvent with highest solubility for the sample is the first mobile phase. A test run is made on an alkylsilica and the retention factor k' as well as the selectivity are determined. In most cases, this solvent (e.g. methanol) will not separate the solutes and they elute unretained within the dead time of the column. Solvent strength is then decreased by successively adding water to the mobile phase. As the polarity of the mobile phase solvent is increased, the retention factors also increase.

When the selectivity α is inadequate for preparative purposes, the stronger solvent, in this case methanol, is substituted by another water-miscible organic solvent, such as THF or acetonitrile. The optimization of k' is then repeated by adjusting the water:organic ratio. This approach works well for most polar to moderately polar compounds that remain non-ionic in the solution.

As an example, Fig. 4.30 shows the separation of three pharmaceutical intermediates on LiChrospher® RP-18 with a mobile phase composition of acetonitrile–water (80:20). Intermediate 1 elutes at 2.28 min while intermediates 3 and 2 co-elute at 3.16 min. Obviously, the separation problem is not solved and, thus, the mobile phase composition must be adjusted in subsequent batch runs with higher water fractions.

Table 4.10 gives the selectivities and retention factors for increased water fractions, while Figure 4.31 shows the plot of $\ln k'$ versus ϕ. Corresponding to Eq. 4.10, the

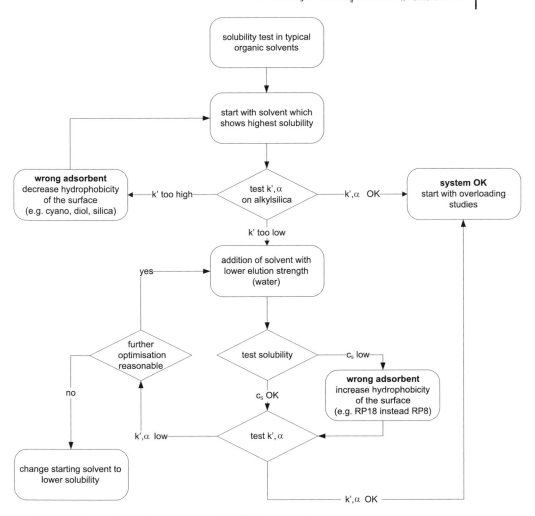

Fig. 4.29 Choice of optimal solvent mixture for RP chromatography.

influence of ϕ on the retention factor of the intermediates is described by the slopes of these straight lines.

The parameters of Eq. 4.10 are determined from Fig. 4.31, and thus the selectivity can be calculated for the whole range of mobile phase composition by Eq. 4.12. The influence of ϕ on selectivity is mainly determined by the difference between the two slopes S_1 and S_2. If the difference is positive ($S_2 < S_1$) the selectivity increases for higher fractions of the organic solvent.

$$\alpha_{21} = \frac{k'_2}{k'_1} = \frac{e^{(\ln k_{0,S2} - S_2 \phi)}}{e^{(\ln k_{0,S1} - S_1 \phi)}} = \frac{k_{0,S2}}{k_{0,S1}} e^{[(S_1 - S_2)\phi]} \tag{4.12}$$

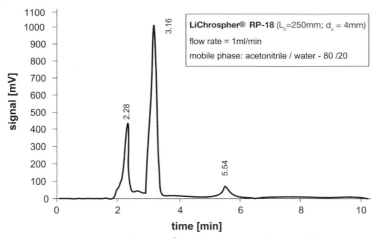

Fig. 4.30 Separation with LiChrospher® RP-18 with acetonitrile–water (80:20).

The mobile phase composition can now be adjusted to the needs of the separation. For example, if intermediate 1 must be isolated with high purity its retention factor should not be too small ($k' > 1$; $\ln k' > 0$), and thus the volume fraction should not exceed 43% acetonitrile.

The selectivities $\alpha_{2,1}$ and $\alpha_{3,2}$ show a contrary behavior: $\alpha_{2,1}$ increases with ϕ while $\alpha_{3,2}$ decreases at higher volume fractions of acetonitrile. This is due to the differences in slopes in the plot of Fig. 4.31; $S_1 > S_2$ and thus $S_1 - S_2$ is positive, while $S_3 > S_2$ and thus $S_2 - S_3$ is negative.

Selection of the optimal mobile phase composition has to take into account the desired product components. If there are more than two product components in the

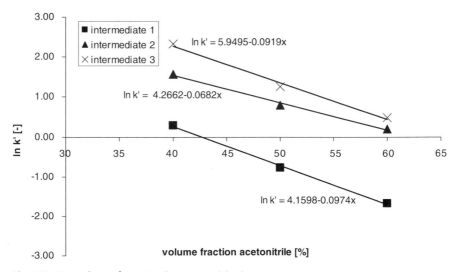

Fig. 4.31 Dependency of retention factor on mobile phase composition.

Table 4.10 Optimisation of an RP separation with LiChrospher® RP 18 and different volume fractions ϕ of acetonitrile.

ϕ (%)	k'_1	k'_2	k'_3	$\alpha_{2,1}$	$\alpha_{3,2}$
40	1.34	4.80	10.23	3.58	2.13
50	0.46	2.22	3.51	4.78	1.58
60	0.19	1.23	1.63	6.42	1.33
80	0.04	0.45	0.45	12.13	1.00

sample a crucial separation must be defined. In the example case a volume fraction of 60% acetonitrile shows best results for the three-component separation because a good selectivity $\alpha_{2,1}$ is observed at low retention times and the reduced k'_1 is still acceptable.

4.3.3.4 Rigorous Optimization for Gradient Runs

The optimization of reversed phase separation by gradient runs is sometimes unavoidable even in process scale. Mobile phase gradients can reduce the development time dramatically as the solvent strength is varied during the chromatographic

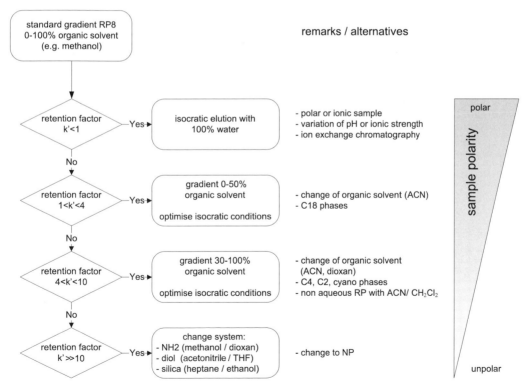

Fig. 4.32 Optimization of mobile phase composition by gradient operation.

separation by changing the mobile phase composition. Figure 4.32 shows the development scheme for optimizing mobile phase composition by gradient operation.

The optimization by gradient operation starts with a linear gradient over the whole range of 100% water to 100% organic. If the solubility in different solvents is unknown, methanol or acetonitrile can be taken as starting solvents. Gradient time is adjusted to 10–15 fold the dead time of the column. If the target solutes elute at retention factors <1 the solutes are either polar or they ionize in aqueous solvents. Here, the interaction of polar solutes with the non-polar adsorbent is very low and a polar solvent with low interaction (e.g. 100% water) is recommended. Ionization must be suitably suppressed (Section 4.3.3.5) or the solutes should be separated by ion-exchange chromatography.

If the retention time is much higher than the time of the initial gradient (retention factor >> 10) the substances seem to be too lipophilic for RP separation on the initial adsorbent. The polarity of the adsorbent should be increased by the use of amino or diol phases. If the adsorbent is already polar the phase system must be changed to NP-chromatography (Section 4.3.2).

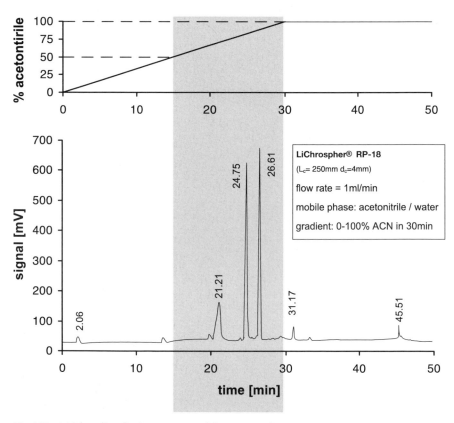

Fig. 4.33 Initial gradient for the separation of three intermediates.

If the retention times are in a suitable range ($1 < k' < 10$–15), solvent composition is adjusted by further gradient runs. Conditions for subsequent gradient runs are derived from the retention times of the solutes during the previous run. Mobile phase composition at the elution point of the first target solute is a good starting point for the mobile phase composition of the following gradient run, while the mobile phase composition at the elution point of the last peak can be taken as the final composition of the mobile phase.

The following example for the optimization of mobile phase composition by gradient runs refers to a separation of the same pharmaceutical mixture as specified in Fig. 4.30. LiChrospher® RP-18 ($L_c = 250$; $d_c = 4$ mm) is used as adsorbent and the mobile phase is a mixture of acetonitrile and water. Figure 4.33 shows the initial gradient from 0 to 100% acetonitrile over 30 min and a flow rate of 1 ml min^{-1}. The three target intermediates elute at 21.21, 24.75 and 26.61 min. The range of the elution times of the three intermediates is marked by the grey box in the chromatogram. It exceeds the above recommended starting and ending points by a few minutes. Corresponding to this time range, the gradient conditions for further optimization are determined according to the upper part of Fig. 4.33.

The first mobile phase composition for the subsequent gradient run corresponds to ϕ at the beginning of the grey marked time range (50%), while the final composition is equal to ϕ at the end of this range (100%). By this strategy the gradient (and thus the mobile phase) is optimized step by step. Table 4.11 shows the gradient optimization. The initial gradient is followed by a second gradient from 50–100% acetonitrile. A third gradient from 70–100% has been examined, but the retention times are too low and, therefore, the components were not separated.

To check the selectivity with different chromatographic systems the separation with LiChrospher® RP-18 is compared with the gradient separation with two different adsorbents. For LiChrospher® RP-8 the separation with methanol–water is shown in

Table 4.11 Optimization of RP separation of three intermediates by gradient operation and resulting isocratic conditions.

Adsorbent	Organic solvent	Gradient	t_G (min)	Cycle time (min)	k'_1 (–)	k'_2 (–)	k'_3 (–)	$\alpha_{2/1}$	$\alpha_{3/2}$
RP-18	Acetonitrile	0–100	30	5.40	8.64	10.26	11.10	1.19	1.08
RP-18	Acetonitrile	50–100	30	5.99	0.46	2.00	3.18	4.35	1.59
RP-18	Acetonitrile	70–100	30	1.24	0.07	0.07	0.64	1.00	8.71
RP-18	Acetonitrile	60	isocratic	3.16	0.19	1.23	1.63	6.42	1.33
RP-8	Methanol	0–100	30	2.29	11.54[a]	12.35[a]	12.58	1.07	1.02
RP-8	Methanol	50–100	30	6.27	4.17[a]	6.73[a]	7.02	1.62	1.04
RP-8	Methanol	70–100	30	0.99	0.81[a]	1.12[a]	1.26	1.39	1.12
RP-8	Acetonitrile	0–100	30	5.02	8.52	9.78	10.80	1.15	1.10
CN	Acetonitrile	0–100	30	7.58	5.83	8.44	9.28	1.45	1.10
CN	Acetonitrile	30–100	30	8.49	0.56	3.45	4.42	6.22	1.28
CN	Acetonitrile	60	isocratic	1.77	0.06	0.63	0.87	9.80	1.38

[a] Elution order between intermediates 1 and 2 inverted due to the use of methanol.

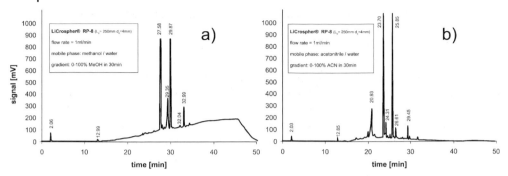

Fig. 4.34 Gradient runs for two different organic solvents with LiChrospher® RP-8: (a) methanol and (b) acetonitrile.

Fig. 4.34a while Fig. 4.34b shows the separation with acetonitrile–water. With methanol the elution order of the first two peaks was inverted compared with the separation with acetonitrile. However, this change of the elution order results in decreased selectivities and thus only acetonitrile is further tested as organic solvent. The third tested adsorbent is LiChrospher® CN. Retention factors and selectivities are shown in Tab. 4.11. For the RP-8 phase no suitable system was found while the selectivity for the RP-18 and CN phase was satisfying and appropriate gradient conditions could be identified. Based on the good selectivities on these two adsorbents isocratic conditions in the medium polarity region of the gradient composition were also tested and gave good results. The two isocratic separations on RP-18 and CN-silica would be preferred for large-scale separations of these feed mixtures as they show short cycle-times and still have good separation factors. Isocratic mobile phase compositions for both phases are also presented in Tab. 4.11.

4.3.3.5 Practical Recommendations
Although RP systems are quite robust, different sources of practical problems during the operation exist:

- Solutes might ionize in aqueous mobile phases
- Effects of residual silanol groups
- Operational problems due to high fractions of water in the mobile phase

Since retention in RP systems depends on the interaction between hydrophobic groups of the solutes and the adsorbent surface, ionization of the solutes can result in severe peak distortion. Ions are very hydrophilic and are thus poorly retained in RP systems even if the non-ionized solute shows good retention. Because ionization reactions (e.g. dissociation of organic acids) are determined by their chemical equilibrium, solutes are present in both forms. This leads to broad peaks with poor retention. The ionization reaction must be suppressed by buffering the system. Thus, the stability of the adsorbent for the pH range must be checked. Especially for alkaline mobile phases, silica-based materials are problematic and the use of polymeric phases should be checked. If ionization of the solutes is suppressed by adding salts to the aqueous mobile phases a second effect occurs since the salt increases the polarity of the mobile phase, thereby increasing retention times.

Residual silanol groups on the adsorbent surface are a source of secondary interaction different to the bonded groups. Next to hydrogen bonding and dipole interactions, silanol groups can be regarded as weak ion exchangers (Arangio, 1998). Especially for basic solutes, the ion-exchange effect dominates. Secondary interactions result in significant changes in retention time and, very often, peak tailing. The ion-exchange mechanism can be suppressed by protonation of the silanol groups at low pH. At pH 3 roughly all residual silanols are protonated and no ionic interactions are observed. At pH > 8 nearly all residual silanol groups are present in ionic form. Along with suppression of ionization by adding buffer, chromatographic efficiency is increased by reducing the number of residual silanol groups by endcapping.

Endcapping of adsorbents is very important if high volume fractions (even 100%) of water are used as mobile phase. Due to the hydrophobicity of the adsorbent surface, problems with moistening of the adsorbent can occur and, as a result, the surface groups can collapse. Standard RP-18 alkyl-chains are not wetted and are thus not stable at 100% water. The alkylchains, which should point into the mobile phases like bristles from a brush, will collapse and stick together on the surface. As a result, solutes in the mobile phase will not migrate into the surface of the adsorbent and the retention time will change dramatically. This behavior depends strongly on the endcapping of the surface (Chapter 3.2.3.1). If non-polar groups are used for endcapping, the hydrophobicity of the alkylsilica increases and thus the stability for 100% water mobile phases gets worse. For this reason special adsorbents with polar groups for the endcapping or embedded polar groups in the alkyl chain have been designed for the use in 100% water. Nevertheless, wherever possible non-endcapped adsorbents should be used in process-scale applications, because they are cheaper and less sophisticated.

Owing to the practical problems with RP chromatography, very often aqueous solvents with buffers are used as mobile phase. Because of the high boiling point of water and the presence of salts in the mobile phase, recovery with high purity of the target solutes from the fractions can be simplified with one more adsorption step. A short RP column with a large diameter (pancake column) can be used for a solid-phase extraction step. The target fraction is diluted with water to decrease the solvent strength of the mobile phase. The diluted fraction is then pumped onto the column and the sample is strongly adsorbed at the column inlet. After the complete fraction is collected in the column it is washed with a mobile phase of low solvent strength to remove the salts. In the last step the sample is eluted with a low volume of a solvent with high elution strength (e.g. methanol). The concentration of the sample is now much higher, the salt is removed and the more volatile strong solvent is easier to evaporate.

4.3.4
Criteria for Choosing CSP Systems

In contrast to normal and reversed phase separations, one critical difference with chromatographic enantioseparations on chiral stationary phases (CSP) is that rational development of selectivities on one given stationary phase is nearly impossible. Optimization of the chromatographic parameters for enantioseparation is much more difficult than for non-chiral molecules and involves the screening of several mobile–stationary phase combinations. As a general guide to successful enantioseparation the process should be developed using the following steps (Fig. 4.35):

- Suitability of preparative CSP
- Development of enantioselectivity
- Optimization of separation conditions

4.3.4.1 Suitability of Preparative CSP
Before starting to screen for a suitable CSP some general considerations have to be taken into account to optimize the separation in the best direction right from the beginning. The required purity and amount and time frame should be known as well as some limitations with regard to solvents or other process conditions, e.g. temperature range. In addition it should be known if only one or both enantiomers have to be delivered in purified form.

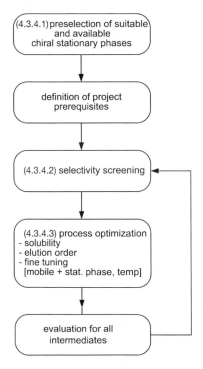

Fig. 4.35 Screening strategy for chiral separations.

Over 100 chiral stationary phases are now commercially available (Chapter 3.2.5). Not all of them are designed and suited for preparative purposes. Certainly, every CSP might be used to isolate some mg amounts of an enantiomer in case of urgent need, but to develop a production process the CSP should fulfill some critical requirements. The main parameters are:

- Selectivity (range)
- Reproducibility
- Stability
- Availability
- Price

A certain selectivity range should be given for the individual CSP. The selectivity range is of great importance for laboratories where moderate amounts (max. in the gram-range) of a multitude of different racemates are separated in a short time. Chiral phases with the broadest selectivity range are the cellulose and amylose derivatives. With the four chiral phases Chiralcel® OD and OJ and Chiralpak® AD and AS, selectivities for ca. 70% of all screened racemates are found (Daicel Application Guide 2003). For production purposes a broad selectivity range is not necessary if only the desired racemate is well separated. Even tailor-made adsorbents could be used, if the production is well controlled and the regulatory and reproducibility issues of the phase are well addressed. The main parameters for overall suitability are related to the mechanical and chemical stability of the adsorbents. Chemical stability is mainly linked to the range of solvents that can be used. As solubility is often an issue for pharmaceutical compounds, the chosen CSP should have a good stability against solvents with different polarities so that the whole spectrum of retention adjustment can be used. For production processes the availability of the CSP in bulk quantities at a reasonable price should also be taken into account. Different CSP groups are characterized according to these parameters in Tab. 3.10.

4.3.4.2 **Development of Enantioselectivity**

For chiral stationary phases the mobile phase selection has to aim first for enantioselectivity α (d, l). The molecules to be separated show identical chemical (same functional groups) and chromatographic behavior (same retention on the bare stationary phase; chiral recognition of enantiomers only by the bonded chiral selector). Therefore, the separation is based on very small differences in complexation energies of the transient diastereomer complexes formed, and similar strategies as for the selections of solvents for liquid–solid chromatography are helpful. Take a good solvent, in which the sample readily dissolves (e.g. ethanol for cellulose or amylose based packings) and increase retention until enantioseparation may be observed.

As pointed out above, no selectivity guarantee can be given for the separation on a certain chiral stationary phase. Therefore, screening routines have to be followed to obtain the appropriate adsorbent. The first possibility is the knowledge-based approach. Some CSP exhibit a certain group-specificity, e.g. the quinine-based CSPs for N-derivatized amino acids or the poly-(N-acryloyl amide derivatives) for five-membered heterocycles. The most advanced statistical and knowledge-based approach is the chiral separation database Chirbase®, developed by Christian Roussel at the University of Marseille (http://chirbase.u-3mrs.fr/). Based on more than 20 000 separations integrated into the database, software has been developed by Chiral Technologies in cooperation with the University of Marseille to suggest the most successful starting conditions for a racemate based on its functional groups (ChiralTOOL, Chiral Technologies Application Guide 2003, www.chiral.fr). More general predictions have been made by the Marseille group based on correspondence analysis of the graphical representation of single CSPs and fifteen empirical molecular descriptors of the racemates included in the database (Roussel et al. 1997).

If no rational starting point for the separation of a racemate can be found, a random screening procedure has to be applied. In most cases 250 × 4 or 4.6 mm columns are used for this purpose. Modern HPLC systems often include a column switching device with up to 12 columns. Most systems have programmable software options for the set of different mobile phase compositions. Combination of UV and polarimetric detection allows even very small selectivities to be found and used as a starting point for further optimization. Commercial systems with deconvolution and optimization options are available, together with a given set of 12 different chiral stationary phases (www.pdr-chiral.com). After an initial detection of selectivity on a CSP, the mobile phase composition and additives as well as temperature have to be varied either manually or by means of an automated system.

For large-scale separations even the design of a new chiral stationary phase might be economically advisable. The most prominent CSP design approach is the reciprocal approach developed by Pirkle and Däppen (1987). One pure enantiomer of the racemate to be separated is bound via a spacer to silica gel. Onto this chiral stationary phase various racemates are subjected whose pure enantiomers are readily available in sufficient amounts. One enantiomer of the racemate with the highest selectivity is chosen as the chiral selector for the CSP. After manufacturing this reciprocal CSP the production of the desired enantiomer can take place. This approach has been used to obtain a very good stationary phase for the separation of non-steroidal anti-inflammatory drugs, e.g. ibuprofen and naproxen (Pirkle and Welch, 1992).

Now that combinatorial and parallel syntheses are available, the number of possible selectors for chiral stationary phases can be drastically increased. Selector synthesis on solid phases and the testing in 96-well plate format have been used to make the CSP-screening process more efficient (Welch, 1999; Murer, 1999; Wang, 2000; Bluhm, 2000; Svec 2001).

The history of columns used for screening racemates for large-scale production purposes should be carefully documented. Cellulose and amylose derivative CSPs, especially, sometimes show strong reactions on small amounts of acids and bases. These compounds might not only be mobile phase additives but also racemates bearing acidic or basic functions that have been previously injected. For large-scale projects it is advisable to use new columns to avoid disappointments if selectivities cannot be reproduced on other or larger columns.

4.3.4.3 Optimization of Separation Conditions

Determination of Racemate Solubility

The solubility of the racemate in the mobile phase is the first parameter to be optimized. Several solvents should be screened for solubility and selectivity at the chiral stationary phase that showed the highest selectivity. Notably, once again, not all chiral stationary phases can be operated with all solvents. The instability of the cellulose- and amylose-derivative CSPs against medium polar solvents such as dichloromethane, ethyl acetate and acetone is widely known. Improved phases with chiral selectors covalently fixed onto the support have been developed and made available in bulk quantities very recently (Francotte and Huynh, 2002). Conversely, some CSP tend to low retention factors if polar solvents are used. Especially, tartar-diamide based CSPs (Kromasil® CHI) need a certain amount of apolar alkane solvent to achieve sufficient retention. For this type of stationary phase the use of supercritical carbon dioxide in combination with a polar modifier might be an alternative option. As mentioned in Section 4.3.1.1, productivity can be increased by changing the injection solvent if the solubility is not sufficient in the mobile phase (Dingenen, 1994). This can be realized only in non-continuous modes, especially in closed-loop recycling chromatography (Chapter 5.2.3).

Selection of Elution Order

With chromatographic production processes the elution order of the enantiomers is of importance. In SMB processes the raffinate enantiomer can often be obtained with better economics as it is recovered at higher purities and concentrations. If the CSP offers the possibility of choosing one of the two optically active forms of the selector, the adsorbent on which the desired enantiomer elutes first should be chosen. This option can be used especially with the brush-type phases with monomolecular chiral selectors. Even if the CSP is not available in both forms, the elution order should be checked carefully as the elution order might be reversed on two very similar adsorbents or with two similar mobile phase combinations. Okamoto (1991) and Dingenen (1994) have shown that by changing only from 1-propanol to 2-propanol, respectively with 1-butanol, the elution order on a cellulose-based CSP might reverse.

Optimization of Mobile/Stationary Phase Composition, including Temperature

It should be taken into account that the highest enantioselectivity is observed at the lowest degree of non-chiral interactions, i.e. at the level of a nearly non-retained first enantiomer. Moreover, enantioselectivity increases with lower temperature according to Eq. 4.13.

$$\alpha = \frac{1}{e^{\left(\frac{\Delta\Delta G_{DL}}{RT}\right)}} \qquad (4.13)$$

This effect on resolution may be counterbalanced by increased viscosity, leading to lower efficiency of the system. Therefore, fine tuning of mobile phase composition and temperature should be carefully taken into account for production-scale systems as some economic benefits have to be considered against a higher complexity of the separation system, e.g. in terms of controlling temperature and small amounts of modifier.

Determination of Optimum Separation Step

Notably, in the synthetic route towards the final enantiomerically pure compound all intermediates should be taken into account for a chromatographic separation step. After introducing the chiral centre the intermediates might differ substantially in solubility and selectivity. No general guidelines can be given as to whether separation on an early or final stage is better economically. Good project coordination involving synthetic chemists and chromatography specialists is the best way to ensure that the optimum separation stage is found.

4.3.4.4 Practical Recommendations

When optimizing the temperature of a chiral separation it should be noted that some chiral compounds tend to racemize at elevated temperatures. On-column racemization might be seen as a typical plateau between the two peaks of single enantiomers. (Trapp and Schurig, 2002). Acidic or basic mobile phase additives might even catalyze such racemization. Conversely, on-column racemization can be used to optimize the yield of a chiral SMB separation. If only one enantiomer is of interest and the other is considered as „isomeric ballast", on-column racemization could be used to enhance the separation yield by transferring the undesired enantiomer into the racemate again. This might be done in an on- or off-line chromatographic reactor (such reactors are discussed in Chapter 8).

In chiral chromatography, especially in SMB production systems, the quite sensitive chiral stationary phases are subjected to a high mass load of compounds to be separated – which might cause some conformational changes of the chiral selector. Once the column is rinsed with pure solvent and checked by a pulse injection, severe changes might be observed. Figure 4.36a shows enantioseparation on a column continuously loaded with feed mixture for several hours. Afterwards, the column was washed with pure eluent and checked with a pulse injection (Figure 4.36b).

Clearly, some severe changes have occurred during the 139 h of operation. The retention time of both peaks has shifted and, especially, the tailing of the second peak has increased tremendously with time. After washing the chiral phase with a

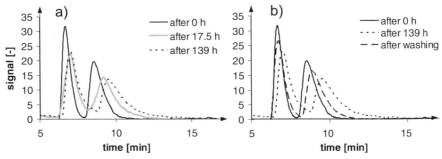

Fig. 4.36 Effect of high feed loading on a chiral stationary phase.

solvent of high polarity the changes are only partly reversible. This behavior is shown in Fig. 4.36b, where the chromatograms at the beginning 0 h and after 139 h of separation are compared with that after washing. Obviously, after the washing step the column does not reach the original condition at the beginning of the separation. For SMB chromatography the increased tailing had to be adjusted by changing the flow rates. However, finally, the separation had to be stopped because no stable conditions were found under which the change in stationary phase behavior could be compensated by small flow-rate changes or short washing procedures.

Chiral stationary phases are also used for the separation of positional isomers. Isomers are normally well separated on straight phase silica or medium polar silica phases, e.g. cyano-modified silica. However, if the two or more chiral centers are in a certain distance apart, they might behave on the columns as if they were pairs of enantiomers. Therefore it is also worth testing CSPs if difficult isomer separations have to be performed. Figure 4.37a–c shows optimized separations of a pair of diastereomers on a straight phase silica (LiChrospher® Si60), a reversed phase silica (Purospher RP-18e) and a chiral stationary phase (Chiralpak® AD), respectively. Clearly, the separation with the CSP is much better and simplifies the preparative separation so that the higher costs of the stationary phase are compensated by improved overall process economics.

Derivatization of racemates is widely used for analytical purposes to enhance detection or reduce the interference of the target compound with the matrix (Schulte,

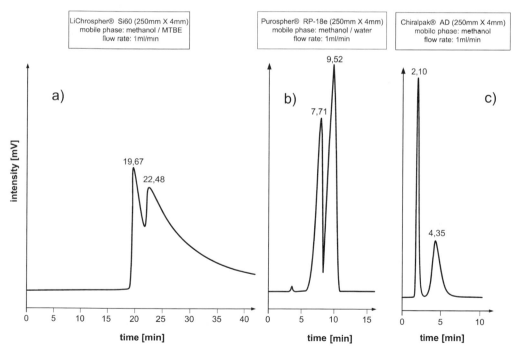

Fig. 4.37 Separation of a pair of diastereomers on different non-chiral and chiral stationary phases (a–c).

2001), but derivatization might also be used for preparative purposes. Francotte (Francotte, 1998) showed that the reaction of different chiral alcohols with a benzoyl ester functionality results in a series of racemates that are well separated on benzoylcellulose-type CSPs. With difficult enantioseparations for a given target molecule it can be worthwhile derivatizing the racemate with different achiral side groups and testing the series of racemates for optimum selectivity. Table 4.12 shows the results of this approach for a series of derivatives of a chiral C3-building block. This building block is widely used in medicinal chemistry. As the side chain is cleaved off during the following synthesis step the derivatization group might be chosen from a wide range of available components. Table 4.12 shows that substantial differences in enantioselectivity can be observed for the single compounds. Nevertheless, it should be taken into account that the derivatization group is ballast in terms of the mass balance of the total synthetic route. The molecular mass of the group should be kept as low as possible to keep the total amount of components to be separated small. This strategy has, for example, been successfully implemented by Dingenen (1994) for α-(2,4-dichlorophenyl)-1H-imidazole-1-ethanol-derivatives.

4.3.5
Conflicts During Optimization of Chromatographic Systems

Up to now many different criteria for the choice of an optimized chromatographic system have been given. Many of these parameters are not independent. For example, a high solubility should always be preferred, but the retention time is very often too low with solvents that provide high solubilities. The most important parameter is high throughput at the desired purity and yield. The resolution gives useful hints for the optimization of a separation, because it merges different effects into one parameter.

$$R_S = \underbrace{\left(\frac{\alpha - 1}{\alpha}\right)}_{\substack{\text{influence} \\ \text{of the} \\ \text{selectivity}}} \cdot \underbrace{\left(\frac{k'_2}{1 + k'_2}\right)}_{\substack{\text{dependent} \\ \text{on the} \\ \text{retention time}}} \cdot \underbrace{\frac{\sqrt{N_2}}{4}}_{\substack{\text{efficiency} \\ \text{of the column}}} \tag{4.14}$$

Resolution can be increased if one of the three terms in Eq. 4.14 is increased. As mentioned in Chapter 2.4.3 the first term describes the influence of selectivity. This term should be maximized by maximizing α in a selectivity screening with different adsorbents and mobile phases. The second term should be kept in a certain range and not be maximized, because the maximum value of 1 is reached for an infinite retention factor. At infinite retention the productivity would decrease due to the high cycle time. The last term of Eq. 4.14 describes the efficiency of the column in terms of the number of plates. Resolution can be increased by selecting efficient adsorbents with small particle size and appropriate narrow particle size distribution. For these adsorbents fluid dynamic and mass transfer resistances are minimized. Con-

Table 4.12 Derivatization of a chiral C3-building block and corresponding chromatographic results.

C3 moiety

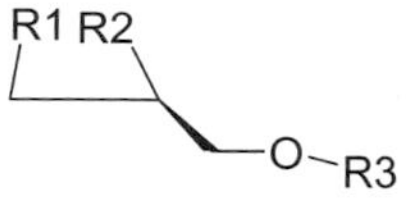

R1 R2 ...—O—R3

R1	R2	R3	CSP	k'_1	α
O		-H	Several	n.a.	1.0
O		(ethylphenyl)	Several	n.a.	1.0
O		SO₂-(p-tolyl)	Chiralpak® AS	0.52	1.08
O		(methoxyphenyl)	Whelk-O-1®	1.19	1.08
O		(methyl-methoxyphenyl)	Whelk-O-1®	1.39	1.11
O		(methylnaphthyl)	Chiralcel® OJ	1.26	1.21
O		(diphenylethyl)	Whelk-O-1®	0.69	1.05
O		(triphenylmethyl)	Whelk-O-1®	2.27	1.07
O		(ethylbiphenyl)	Whelk-O-1®	1.17	1.24
O		(fluorenyl)	Exp. Poly-(N-acryloyl amide)	1.26	1.11
OH	OH	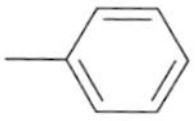 (phenyl)	Chiralcel® OD	3.69	1.25

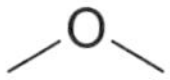

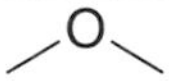

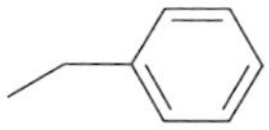

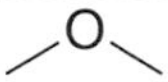

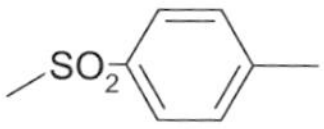

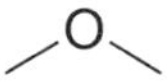

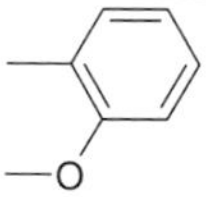

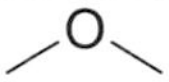

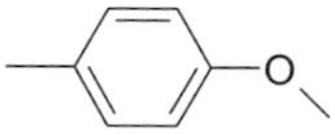

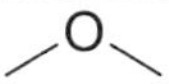

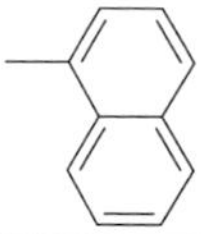

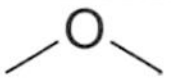

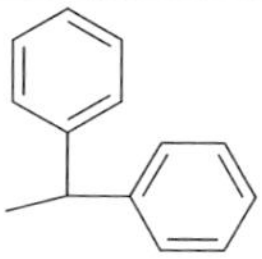

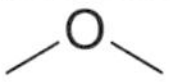

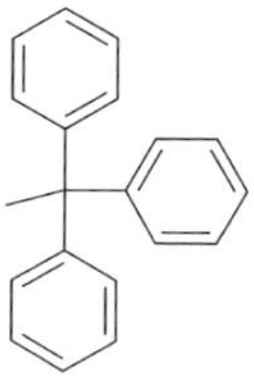

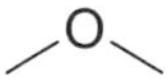

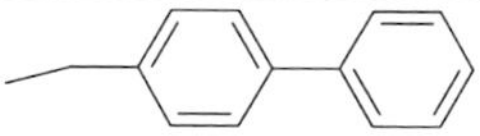

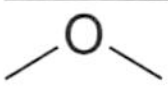

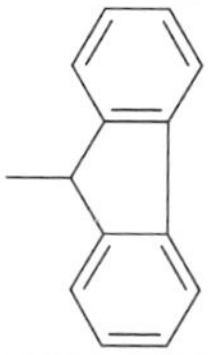

versely, the back-pressure of the column and the stability of the packing must also be taken into account for process-scale separations. Very small particles with a narrow particle size distribution result in high back-pressure and may show reduced efficiency due to packing imperfections (Chapter 3.3).

Next to these practical problems of high efficient column packings, a high number of plates is very often not needed if peak broadening results from thermodynamic effects. Seidel-Morgenstern (1995) has discussed the effects of high column loading (Chapter 2.6) and showed that separation efficiency is not dependent on the number of plates, if isothermal effects come to the fore. A typical isothermal effect is the decreasing retention time for high concentration in the Langmuir range of the isotherm. Highly efficient adsorbents (with high number of plates) are not useful for separations with dominant isothermal effects. Only for separations with low selectivities at low loadings must the number of plates be high to reach a sufficient resolution (Fig. 2.12, Chapter 2.4.3).

This behavior is demonstrated by the different working areas defined in Fig. 4.38. Assuming that a certain selectivity can be reached, the influence of the retention factor and the solubility is shown. The working range of process chromatography should have medium retention factors and medium to high solubilities for the feed compounds. For low feed solubilities the influences of thermodynamics and fluid dynamics dominates. In this range adsorbents with high numbers of plates per metre should be used. At high solubilities high column loading is possible and thermodynamics effects come to the fore. In this range, less efficient adsorbents for high flow rates can be used.

In preparative chromatography the systems are mainly operated at high feed solubilities and thus thermodynamic effects dominate. If separations with low feed solubilities have to be operated, high-efficiency adsorbents should be used and the ACD injection technique could be applied (Section 4.3.1.1).

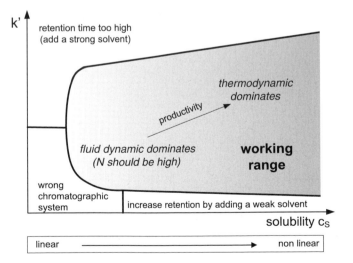

Fig. 4.38 Working ranges of chromatographic separations.

Thermodynamic effects are the main source of increasing productivity in preparative chromatography. The most important parameter is the selectivity, which describes the ratio of the initial slopes H of the adsorption isotherms for both components (Eq. 2.46).

Especially for feed mixtures with different ratios of the single components, the elution order must be considered. The major component should elute as the second peak, because in this case the displacement effect can be used to ease the separation (Chapter 2.6.2). If the minor component elutes as the second peak the tag-along effect reduces the purity and loadability of the system.

If the components to be separated are present in a similar amount (e.g. as for racemic or diastereomeric mixtures) the target component should elute first, because the first peak is normally obtained at higher productivity and concentration.

Very often the cycle time of isocratic batch chromatography is high due to extensive peak tailing of the second peak, which can be explained by a Langmuir-type isotherm (Chapter 2.6). The black line in Fig. 4.39 shows a typical chromatogram of an enantioseparation with a polysaccharide adsorbent with n-hexane–ethanol as the mobile phase. Here ethanol is the solvent with higher elution strength. The second peak shows tailing due to the Langmuir shape of the isotherm and high concentrations. The cycle time (time between the starting point of the first peak and the ending point of the last peak) for the isocratic elution mode is 8.2 min.

The grey line shows the separation when a forced elution step is used. After the first peak has eluted from the column, a second injection is made with the strong solvent (in this case ethanol). If the volume of this second injection is large enough the strong solvent displaces the second peak. Owing to this displacement, the second peak elutes much faster and at higher concentrations. The cycle time is decreased

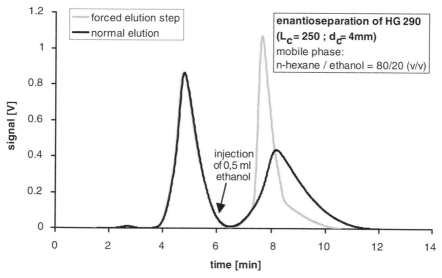

Fig. 4.39 Increase in productivity due to forced elution step (injection of a strong eluent) during enantioseparation of HG 290. (Reproduced with permission of H. Gillandt, SiChem GmbH Bremen.)

from 8.2 to 6.7 min and, thus, productivity is increased by 22%. Eluent consumption is also decreased as well, but recycling of the solvent becomes slightly more problematic because the eluent composition is no longer constant over the whole cycle time. When using the forced elution step-method it should be ensured that the column is in equilibrium at the end of the cycle time. If only small amounts of eluent with high eluting strength are used for this method, isocratic process concepts can still be applied, e.g. touching–band elution.

5
Process Concepts

M. Schulte, K. Wekenborg and W. Wewers

The basis for every preparative chromatographic separation is the proper choice of
the chromatographic system, as described in the previous chapter. Important aspects
in this context are selectivity and solubility, which are influenced and optimised by
the deliberate selection of stationary and mobile phase.

Beside the selection of the chromatographic system its implementation in a preparative process concept plays an important role in serving the different needs in
terms of, e.g., system flexibility and production amount. Depending on the operating
mode, several features distinguish chromatographic process concepts:

- Batch-wise or continuous feed introduction
- Operation in a single- or multi-column mode
- Elution under isocratic or gradient conditions
- Co-, cross- or counter-current flow of mobile and stationary phase
- Withdrawal of two or a multitude of fractions

Starting with the description of the main components of a chromatographic unit,
this chapter gives an overview of process concepts available for preparative chromatography. The introduction is followed by some guidelines to enable a reasonable
choice of the concept, depending on the production rate, separation complexity and
other aspects.

5.1
Design and Operation of Equipment

Before injecting the first sample some general considerations have to be taken into
account with regard to the design of the preparative chromatographic plant. All preparative HPLC plants consist, basically, of the same components (Fig. 5.1): a solvent
and sample delivery system, the preparative column and a detection and fraction collection system.

Preparative Chromatography. Edited by H. Schmidt-Traub
Copyright © 2005 WILEY-VCH Verlag GmbH & Co. KGaA, Weinheim
ISBN: 3-527-30643-9

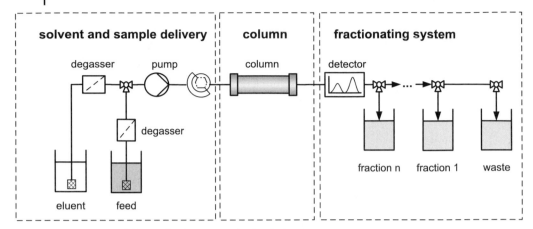

Fig. 5.1   General design of a chromatographic batch plant.

Capillaries, tubes or, in large systems, pipes connect the different parts of the chromatographic plant. The tubes are made of stainless steel, Teflon or polyetheretherketone (PEEK). Selection of the right material depends on the mobile phase used for the separation. The choice of the internal diameter of these tubes should be a compromise between small dead volume and low backpressure. Small diameters result in high pressure drops in the system and higher diameters lead to higher dead volume, causing back mixing and, thereby, peak distortion. More important than the absolute tube length is the smooth connection of tubes via connectors, reducers and ferrules to different system parts. Care has to be taken that only connectors of the same measurement system (metric or foot–inch system) are used.

Figure 5.2 shows the influence of dead volume and connection design on the peak distortion and dead time of the HPLC unit. Three runs were performed with the same flow rate and sample size (injection time: 2 s). The straight black line describes the elution profile of a tracer in an optimal system with small dead volume and optimal connections. In the next example one of the tube connections was insufficient because, within the connector, the tube endings did not touch and a small dead volume is formed. The dashed line shows the new elution profile. A poor connection results in higher backmixing and, thus, peak distortion. This effect should be kept in mind for systems with many connections.

The third experiment, represented by the grey line, shows the elution profile of the same tracer in the same unit, but in this case one tube with a small internal diameter was replaced by a bigger one. Dead volume and peak distortion are dramatically increased. This increase in dead volume results in a loss of base line separation. For preparative separations the higher backmixing in the unit leads to decreased productivity and economy of the process.

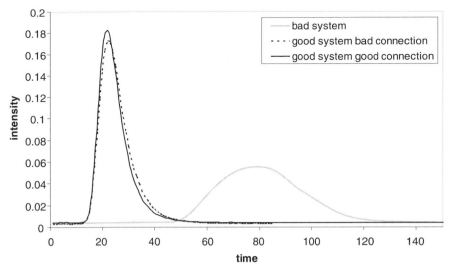

Fig. 5.2 Influence of system design on dead time and peak distortion.

5.1.1
Solvent and Sample Delivery System

5.1.1.1 HPLC Pumps
The flow rate of the eluent and the sample is delivered to the column by a constant displacement pump. The demands on chromatographic pumps are very high, because they must guarantee an exact flow rate even against high back pressure generated by the column. To prevent mechanical stress to the fixed bed in the column the solvent flow must be free from pulsation. This can be achieved by using piston pumps with more than one pump head (in most cases two or three) or a pressure module subsequent to the pump. In the start up period of pumping the adjusted flow rate should be reached with a gradient of the flow rate so as to prevent damage to the packed bed.

The working range of the pump, concerning the maximum pressure difference and flow rate, must be adjusted to the column. An inadequate system causes lower productivities of the separation and thus wastes time and money. Table 5.1 gives typical flow rates and tube diameters for different column geometries.

5.1.1.2 Gradient Formation
The formation of solvent gradients can be realised by two different concepts, mixing at the low-pressure side of one pump (Fig. 5.3a) or mixing the outlets of two pumps at high pressure (Fig 5.3b). The low-pressure option works with proportioning valves, which open for a time interval that is proportional to the solvent concentration at a given time. The two fluids are then mixed in a mixing chamber. This system is easy to realize as only one pump is needed. However, it is not highly accurate,

Table 5.1 Typical flow rates and tube diameters for different column geometries.

d_c (mm)	4	25	50	100	200	450	600	1000
$\dot{V}_{min}$ (ml min^{-1})	0.5	20	80	300	1200	6000	10×10^3	30×10^3
$\dot{V}_{max}$ (ml min^{-1})	2	80	300	1200	5000	25×10^3	45×10^3	120×10^3
Outer tube diameter, (mm)	1.56	1.56–3.12	3.12	3–6	6–8	8	16	25
(inch)	1/16	1/16–1/8	1/8	1/8–1/4	1/4–3/8	3/8	5/8	1
Inner tube diameter, (mm)	0.25	0.75–1.6	1.6	1.6–4.0	4.0–5.0	5.0	13.0	21.0
Sample loop volume (ml)	0.01–0.2	1–2	2–5	10–20	Feed introduction via feed pump			
Max. unit dead volume (ml)	0.5	3–5	5–10					

especially when solvents with great differences in viscosity are used. Better accuracy is achieved by using two independent pumps. The outlet streams of the pumps are mixed at the high-pressure side of the system. The use of two pumps makes the system more complex and thus expensive.

5.1.1.3 Eluent Degassing

The major operating problem of piston pumps is dissolved air and the formation of bubbles in the eluent. Bubbles in the pump heads cause pulsation of volume flow and pressure pulsation. Bubble formation and cavitation problems are promoted at the inlet check valve because the minimum pressure in the system is reached here. For this reason the eluent must be suitably degassed. This can be done online by a membrane degasser, by pearling helium offline through the eluent or by the use of an ultrasonic bath.

5.1.1.4 Eluent Reservoir

Cavitation is a general operation problem, which can be solved by the use of tubings with a higher internal diameter or by increasing the pressure at the suction side of the pump. The easiest way to increase the pressure is to lift the eluent reservoir to a higher level than the pump inlet. If this is not sufficient the pressure in the solvent

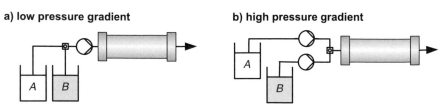

a) low pressure gradient   b) high pressure gradient

Fig. 5.3 Gradient formation.

reservoir must be increased. For this reason the solvent reservoir should be airtight and isolated from the environment. Isolation of the eluent has the positive side effect that the composition and water content can be held constant during operation. Especially in case of normal phase chromatography, it is important to control the water content in the range of a few ppm, because of the hygroscopic character of the adsorbent. In case of small eluent quantities the addition of molecular sieves is a good way of avoiding water adsorption on the stationary phase.

5.1.1.5 Sample Injection

The sample can be injected ahead of the column, either on the suction side or on the pressure side of the pump. Injection on the suction side is performed by a three-way magnetic valve, which switches between the eluent and the feed. Some pumps have an additional pump head for the injection of the feed. If the sample is injected on the pressure side a switch valve is used. The sample is injected into a sample loop, which is connected to the six-port valve. The volume of the sample loop is adapted to the maximum injection volume of the process (see Tab. 5.1).

5.1.2
Chromatographic Column

The heart of every preparative chromatographic system is the column packed with the adsorbent. If all other parts of the equipment are well designed with regard to minimum hold-up volume, the column is responsible for the axial dispersion of the chromatogram. The column has, therefore, to be designed in an optimal way (Chapter 3.1). Notably, the type of adsorbent determines the design of all equipment in the unit. As a result of the significant different pressure drops of particular beds or monolithic columns the working range of the whole system must be adapted to the column.

5.1.3
Detection and Separation System

For a successful separation and production of the target components the detection of the different components must be guaranteed. The assembly of the different detectors is always very similar. The product stream flows continuously through the detector cell, where a physical or chemical property is measured online. The detected value is transformed into an electric signal and transferred to a PC or integrator to record the chromatogram.

The signal should be linear to the concentration of the target component over a wide range – as in Fig. 5.4 – because this simplifies the calibration of the detector and the signal represents the real shape of the peaks. The pure eluent signal is recorded as the baseline of the chromatogram. Fluctuations of the baseline are called the noise of the detection system. To qualitatively identify peaks the signal-to-noise

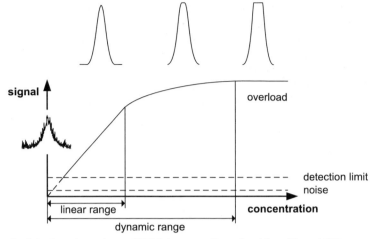

Fig. 5.4 Detection regimes of HPLC detectors. (Reproduced from Meyer, 1999).

ratio (Fig. 5.5) of the chromatogram should be higher than 3, and for quantitative analysis this ratio should be at least 10 (Meyer, 1999). The detection limit is defined as amount of a certain substance that gives a signal two times higher than the noise. A detector's dynamic range is the solute concentration range over which the detector will provide a concentration-dependent response. The minimum is bordered by the detection limit and the maximum is given at the saturation concentration where the

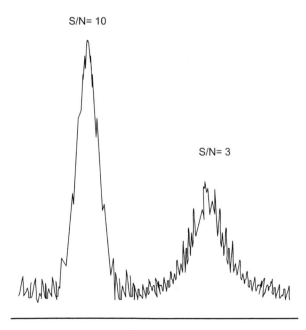

Fig. 5.5 Signal-to-noise ratios for two peaks.

detector output fails to increase with an increase in concentration. The dynamic range is usually quoted in orders of magnitude (Scott, 1986).

An ideal detector for preparative chromatography has the following properties:

- Good stability and reproducibility, even in the production environment
- High reliability and ease of use in routine operation
- Sufficient sensitivity (need not to be too sensitive, as no trace detection is the goal in preparative chromatography, but the sensitivity should be high enough to keep the flow cell volume, which contributes to the axial dispersion, small)
- Short reaction time, independent of flow rate
- High linear and dynamic range

Table 5.2 lists the different detectors and their operation principles that are used in (analytical as well as preparative) chromatography. The optimal detection system only detects the target components and all impurities but is not affected by the solvent. Detectors can be divided into two groups: (1) those that detect any change in composition of the solvent and can, therefore, obviously only be used under isocratic

Table 5.2 Chromatographic detection systems and their detection principles.

Detection principle	Detection sensitivity (g)	Stability (sensitive to)	Dynamic range	Full flow or split flow (max flow rate in ml min⁻¹)	Detection sensibility
Solvent sensitive					
Refractive Index	10^{-6}	Temp shift, solvent composition	10^4	Full (100)	All analytes
Density	10^{-6}	Solvent composition	10^3	Full	All analytes
Conductivity	10^{-9}		10^6	Full	All ionic analytes
Analyte sensitive					
Light scattering ELSD	10^{-7}	Volatile analytes	10^3	Split (4.0)	All analytes
Electrochemical	10^{-12}	Temperature shift, flow rate and mobile phase purity	10^8	Split	All analytes
UV/VIS (DAD)	10^{-9}	UV adsorption of solvents	10^5	Full (10 000)	Compound selective (DAD)
Fluorescence	10^{-12}	Fluorescent substances in eluent	10^3	Split	Compound selective
Mass spectrometer	10^{-10}	Inorganic salt in eluent	Wide	Split (0.1)	Compound selective
Polarimetry	10^{-9}	Temperature shift	10^3	Full (5000)	Compound selective (chirals)
Element selective (nitrogen, sulfur)	10^{-8}		10^2	Split (0.4)	Compound selective

conditions; (2) analyte specific detectors – this group can be subdivided into detectors with a similar sensibility to all analytes and those that can detect certain components selectively. All the detection principles listed in Tab. 5.2 may be applied in preparative chromatography, but some of the detectors have limitations with regard to their applicable flow rate range.

If the maximum allowable flow rate of the detector is too low compared with the flow rate of the chosen column diameter, the total solvent stream has to be split into a detector stream and the eluent stream. The easiest splitting principle is the integration of a T-connector with a certain diameter for the two different streams. Integration of a connector into the eluent stream increases the system pressure drop and the back mixing of the system. This is a severe drawback of all passive splitting systems. In addition, care has to be taken that the time delay between detection and fractionation is carefully determined and taken into account for exact fractionation. The pressure drop problem is solved by active split principles. These are motor valves where, in short intervals, a portion of the eluent stream is captured and injected via an injection pump into the detector. The need for a second pump is one of the drawbacks of active splitting systems. In addition, the valve seal has to be carefully maintained and changed at regular intervals.

UV absorption of the single compounds is commonest detection principle in HPLC. In its easiest set-up a UV detector consists of a mercury lamp and a set of filters for different wavelengths. The main wavelengths used are 220 nm, 254 nm for aromatic compounds and 280 nm for proteins. More advanced UV detectors can be adjusted to all wavelengths between 190 and 370 (740) nm. Extension into the visible wavelength region even allows detection of adsorption bands up to 1020 nm. In new systems, which are especially well suited for preparative chromatography, the flow cell is connected via a fiber optic to the detector. This makes it possible to put the flow cell in a hazardous or ex-proof area or control the temperature of the flow cell more effectively. UV detection has the advantages of high versatility and a large detection range. Its main problems are nonlinear calibration curves, especially in preparative chromatography, the necessity of the solutes to absorb UV light in the range above 200 nm and the UV absorption of most HPLC eluents in the range below 210 nm (Section 4.2.1). Nevertheless in preparative chromatography the UV adsorption of eluents is often tolerable because the information delivered from the detection system is only a rough estimation of the cutting points of the fraction and not the complete identification of all components within the sample. Owing to the high concentrations used in preparative chromatography the absorption of the eluent is a smaller problem than for analytical chromatography. To improve process economics it is advisable to sacrifice detection for feed concentration if the feed components show good solubility in UV-active solvents such as acetone or toluene.

A complete spectrum of the eluting components is obtained if all possible wavelengths are scanned in defined intervals. By means of a diode array detector (DAD) the wavelength scan is possible within less than a second. Measurement of a complete UV spectrum of the eluting solution offers two advantages. Firstly, if a specific calibration function at a certain wavelength can be found for each eluting component, it is possible to calculate the concentration in the detector cell for these substances and thus the purity of the target components. Secondly, the safety of the product purity is increased. An impurity can be seen although it adsorbs the UV light at different wavelengths to the target components. Drawbacks of the DAD system are lower sensitivity and robustness as well as the higher price. Due to the huge amount of data obtained by a DAD system the computing power required must be higher than for other detectors.

A good compromise between a single wavelength UV detector and a DAD system is the use of a multi-wavelength UV detector, which monitors up to four different wavelengths.

If the components to be detected fluoresce, a fluorescence detector can be employed. A mercury or xenon lamp with a monochromator is used as the source for the exitation wavelength. Modern systems use lasers as light source, but such systems are mainly used in trace analysis and not in preparative systems. The main advantage of fluorescent detectors is their high sensitivity. Their reduced robustness and limitation to fluorescent compounds makes them not widely used in preparative chromatography.

Polarimetry detectors are applied to detect optically active components. The emitted linearly polarized light is rotated by optically active components in the eluent stream and the angle of rotation is detected. Since the introduction of these detectors, which use laser light as the light source, the drawback of low sensitivity has been overcome. Similar to the DAD detectors for the UV range, circular dichroism (CD) detectors are available to detect the CD spectrum of substances. Such detectors are, so far, not widely used in preparative chromatography.

5.1.3.1 Solvent-sensitive Detectors

Refractive index detectors (RI) are very versatile because of their universal detection principle, responding to changes in solvent composition. This principle can only be used for isocratic conditions. If the temperature is well controlled they are robust; nevertheless they have a relatively low detection limit.

The other two principles that detect changes in the solvent properties are conductivity and density measurements. Conductivity as a detection principle can only be used for ionic substances. The detection range is quite high, but the detector is sensitive to changes in solvent composition and shows a baseline shift if gradient elution is applied.

For highly concentrated sugar solutions, e.g. in the separation of glucose and fructose, density detectors are often used. Again, these solvent sensitive detectors can only be used under isocratic conditions, as any change in mobile phase composition will also change the detector signal.

Some new specialised detectors are mass spectrometry detectors (MS), evaporating light scattering detectors (ELSC) and atom specific detectors. They all have in common that the mobile phase stream has to be split to secure a safe operation of the fragile detection system. As pointed out earlier, the eluent stream can be split by active and passive principles. For certain applications, mainly in the isolation of impurities with unknown structures in the mg range, the combination of these detectors to preparative equipment is of high value. As soon as the diameter of the preparative column is increased, simple detection principles should be applied or fractions should be taken and analysed off-line by a fast in-process analytical method.

5.1.3.2 Fraction Collection

After the different components of the mixture have been detected, the outlet stream of the detector must be suitably fractionated to collect the components of interest. To collect the fractions, open systems with a fractionating robot (fraction collector) as well as closed systems, e.g. valve blocks, might be used (Fig. 5.1).

Table 5.3 Summary of important system features of chromatographic plants.

System component	Parameter	To be considered
Solvent reservoir	Material	Avoid any incompatibilities
	Temperature	Should be not too different from column temperature. Be careful, if solvent is stored outside the building.
	Stirring	To avoid de-mixing, especially if highly viscous eluents are used
	Inertisation	Over-pressurize with nitrogen
Eluent degasser	Volume	Keep small
Pump	Number of pumps and pump heads	Pumps with 3-pump heads show higher accuracy
	Flow vs. Δp	Adapt to pressure limits of the column
	Precision	Important at higher flow rates
Feed introduction	Feed injection via pump	Check the volume behind the injection point
	Sample loop	Adjust the size, do not use too large loops
	Feed container	Control temperature, stir, over-pressurize with nitrogen
Column	Length and diameter	Adjust to the needs of the separation
	Flow distribution and collection	Secure low contributions to the axial dispersion
Detector	Type	Select robust principle
	Selectivity	Select for the given separation principle
	Linearity	Calibrate with the actual sample
	Max. flow, split flow	Select flow-cell according to maximum flow rate
Fraction collector	Dead volume	Keep small
	Line flushing	Avoid carry-over of samples in the fraction collection tube system
Total system	Total extra-column volume	Keep small
	Connections	Should be smooth, use the same measure (metric or inch)
	Gradient dwell time	Keep short by keeping the system volume before the mixing chamber small

The fraction collector is a robot arm, which distributes the outlet flow into different bottles. This system is very versatile, because the number of possible fractions is large and can be increased by adding more collecting bottles. Its disadvantage is its open nature. The outlet flow is in direct contact with the atmosphere and, therefore, problems with solvent evaporation can occur. In addition, the open handling of fractions is not compatible with GMP requirements.

An alternative way of fraction collection is the use of a valve block subsequent to the detector. The main disadvantage of valve blocks is the need for at least one valve per fraction and the integration into a fixed tube system.

Table 5.3 summarises the important features of the different system components, which should be taken into account when setting up a new preparative chromatographic system.

5.2
Discontinuous Processes

5.2.1
Isocratic Operation

The easiest chromatographic set-up consists of one solvent reservoir, one pump, which can deliver the necessary flow rate against the pressure drop of the packed column, a column and a valve for the collection of pure fractions (Fig. 5.1). If the feed mixture contains no early-eluting impurities or components with high affinity to the adsorbent, the adsorbent can be used very efficiently. With a proper choice of the operating conditions, such as injected amount of substances, flow rate and time between two injections, a touching band situation can be achieved (Fig. 5.6). This means that the peaks of two injections do not overlap at the column outlet (e.g. B_1 and A_2). Complete product recovery requires baseline separation, where the components from one injection do not overlap when leaving the column (e.g. A_1 and B_1). To produce fractions of high purity, baseline separation is not necessarily required. But in this case a waste fraction has to be implemented between the two product fractions. Chapter 7.2 gives a detailed description of design and optimisation strategies for such chromatographic processes in order to maximise productivity, minimise eluent consumption or total separation costs.

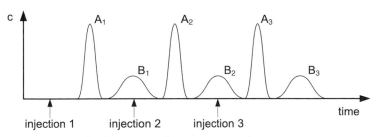

Fig. 5.6 Touching band separation of two components (three injections).

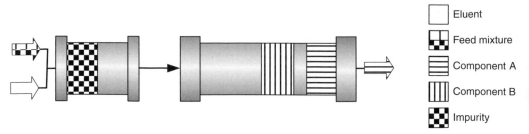

Fig. 5.7 Pre-column for the adsorption of late-eluting impurities.

This process with unchanged conditions of the solvent is called isocratic and exhibits the following benefits:

- Number of pumps and eluent tanks is reduced to the absolute minimum
- Cycle time can be reduced, because no reconditioning of the column is necessary
- Work-up and re-use of the solvent is easier due to its constant composition

Whenever possible the development of a chromatographic separation should start with an isocratic elution mode. Its economic feasibility has to be checked afterwards.

If the feed mixture is contaminated with late-eluting impurities and their adsorption in the main column has to be avoided, the implementation of a pre- or guard-column might be helpful (Fig. 5.7).

This column should be of short bed-length and can be packed with a more coarse material. Its function is only the adsorption of late-eluting impurities. As soon as a break-through of those components can be observed the separation has to be stopped and the pre-column cleaned or emptied and re-packed. Re-packing with a cheap, coarse adsorbent is often more advisable than cleaning the pre-column adsorbent. If the separation should not be stopped for pre-column cleaning, a second pre-column and a switching valve are installed, so that the contaminated pre-column can be cleaned or re-packed offline.

5.2.2
Flip-flop Chromatography

A different concept dealing with highly adsorptive components is the so-called Flip-flop chromatography developed by Colin et al. (1991) (Fig. 5.8). The feed mixture is injected at the one end of the column and the early eluting components are pushed through the column (Fig. 5.8a) and collected at the other end (Fig. 5.8b). When the first components are withdrawn from the column the flow direction is reversed while the late-eluting component is still adsorbed on the column (Fig. 5.8c). After a predetermined time a new portion of feed mixture is injected and the elution is performed in the reversed flow direction (Fig. 5.8d). At the end of the column the late-eluting impurity from injection 1 is now eluted first (Fig. 5.8e) and afterwards the early-

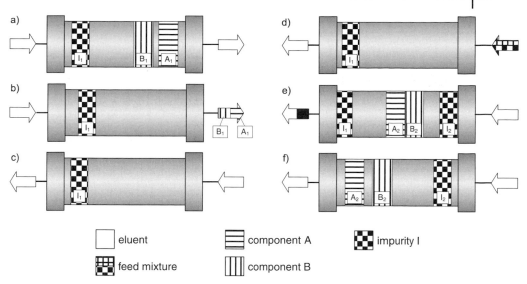

a)

b)

c)

d)

e)

f)

☐ eluent   ☰ component A   ▨ impurity I

▦ feed mixture   ▥ component B

Fig. 5.8 Flip-flop concept.

eluting components from injection 2. When the early-eluting components from injection 2 are withdrawn from the column, the flow direction is reversed again and a new injection is done. The flip-flop concept is an elegant way to deal with feed mixtures consisting of components with very different adsorption behaviour. However, is not very often used, due to the complexity of its operation and the system set-up. If the feed mixture contains components of very different elution behaviour it is more advisable to use the above-mentioned concepts of pre-columns or to divide the whole separation process into two steps: a simple adsorption of the late-eluting impurities and the main separation.

5.2.3
Closed-loop Recycling Chromatography

One of the constraints in preparative chromatography is the inherent dependency between efficiency and pressure drop. If the plate number of a column has to be increased to achieve the desired separation, the pressure drop will increase because a longer column or a smaller particle diameter has to be chosen. One possible way to overcome this situation is to increase the number of plates in a column virtually. When the column outlet is connected to the column inlet and the partly separated mixture is again injected into the column, the components of interest can be separated during the second or following passages through the column. This process concept is called closed-loop recycling chromatography (CLRC).

The critical point in closed-loop recycling chromatography is the increase in axial dispersion due to the additional hold-up volume generated by the pump head and

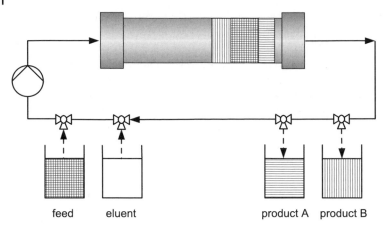

Fig. 5.9 Closed-loop recycling chromatography with peak shaving.

additional piping. Closed-loop recycling chromatography is, therefore, in most cases combined with peak shaving to remove the front and tail ends of the chromatogram (Fig. 5.9).

For this set-up, valves have to be integrated into the cycle. If only one valve is used to withdraw several fractions a flushing line should be integrated to clean the withdrawal line from the first component before the second component is cut from the chromatogram. The main advantage of CLRC is the possibility of isolating even components of low selectivity with good purity and yield, as no fractions with insufficient purities are withdrawn from the system. A second advantage is the reduction of mobile phase consumption. During the closed-loop operation no eluent is consumed

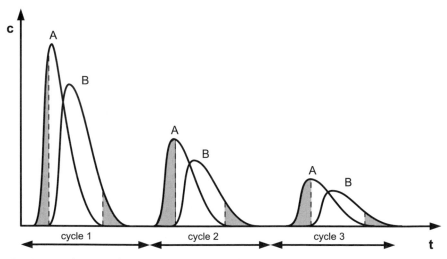

Fig. 5.10 Development of concentrations during closed loop recycling chromatography operation with peak shaving.

as the complete mobile phase is circulated within the system. Fresh eluent has to be introduced only during the feed and fraction collection period.

One drawback of CLRC with peak shaving is the decreasing amount of solutes in the column (Fig. 5.10). After the third or fourth cycle only a small amount of the original feed mixture is left in the column. Therefore, the chromatographic productivity, which indicates the time–space yield, decreases rapidly and reduces process economics, although it is compensated by high purities of the target components as well as low eluent consumption due to internal recycling of the solvent.

5.2.4
Steady State Recycling Chromatography

A set-up to overcome this situation is the CLRC with periodic intra-profile feed injection. The basic set-up of this process is similar to the CLRC concept shown in Fig. 5.9. It is also known as steady-state recycling chromatography (SSRC) or CycloJet-concept (Grill and Miller, 1998; Grill et al., 2004). Figure 5.11 shows a chromatogram at the column outlet after one cycle. When the first component A elutes from the column ($t = 25$ min) the first fraction of pure product can be collected. When the region of non-separated mixture elutes from the column ($t = 26$ min) a new portion of feed is injected into the system for a given time interval. The exact time and duration of the new injection have to be optimised to achieve the desired performance of the process with respect to productivity or any other objective function. After a new injection the chromatogram is again recycled in a closed loop. The desorption front is then withdrawn from the system to separate pure B ($t = 27$ min).

Closed-loop steady state recycling chromatography is an intermediate between simple batch wise or CLRC operation and continuous processes. It combines lower complexity and equipment requirements with higher productivities and lower eluent consumption.

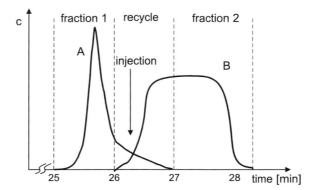

Fig. 5.11 Steady state recycling chromatography (SSRC).

5.2.5
Gradient Chromatography

Under isocratic conditions, the last component might take a very long time to elute. Such late-eluting components are very much diluted, resulting in higher work-up costs. In extreme cases some components do not even elute under the chosen conditions. These drawbacks can be avoided by gradient operation, which changes the elution strength (Chapter 2.6.3).

Process conditions that might be altered during washing and/or elution steps are:

- Mobile phase composition (in terms of organic modifier, buffer amount or strength, salt concentration or pH)
- Mobile phase flow rate
- Temperature
- Pressure (in supercritical fluid chromatography, SFC)

Even enhanced desorption by the introduction of energy might be considered, e.g. microwaves or light to switch the stationary phase orientation.

Figure 5.12 describes various elution modes based on different compositions of the mobile phase. In the diagrams, A represents the composition of the eluent while B indicates the modifier concentration versus time. The simplest gradient elution is the integration of a washing step (Fig. 5.12 b). It is mostly used if the component of interest can be adsorbed onto the stationary phase under non-eluting conditions. After washing off all the impurities the target component is eluted at increased modifier concentration.

If the feed mixture contains components with high affinity to the adsorbent so that they are not eluted with the current mobile phase composition, a second elution step

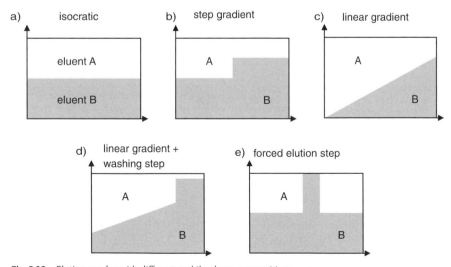

Fig. 5.12 Elution modes with different mobile phase compositions.

with a different composition is necessary. Here, it is also advisable to install a bypass line that allows the flow direction to be reversed. With this set-up sticking components can be withdrawn from the column by flushing them back to the point of injection instead of pushing them through the whole column. The reversed flow makes the washing step much more efficient in terms of time and solvent consumption.

A continuous increase in elution strength of the eluent (Fig. 5.12c) offers the best separation performance for components with very different retention times. If the retention time of the first and the last component differs by more than 30% (Unger, 1994) gradient elution should be applied – or the whole separation should be divided into several process steps. Linear gradients (Fig. 5.13) give good resolution and small peak width, allowing the isolation of pure fractions at higher product concentrations within a shorter cycle time.

Beside the two gradient types, combinations of linear and step gradients might be used. One set-up is the combination of a flat gradient for the part of the chromatogram where the highest separation power is needed, with a step gradient for the quick removal of all unwanted components that elute after the component of interest (Fig. 5.12d).

Gradient elution reduces the overall retention time of the mixed components. However, the cycle time can be longer than for isocratic elution because the mobile phase has to be readjusted to its starting composition by an additional process step.

A more efficient use of a step gradient is the injection of a plug of mobile phase with high elution strength into the part of the chromatogram where the component of interest elutes (Fig. 5.12e). This eluent plug forces the component of interest to desorbed from the column. It thus increases the fraction concentration but not the cycle time due to solvent readjustment. (Chapter 4.3).

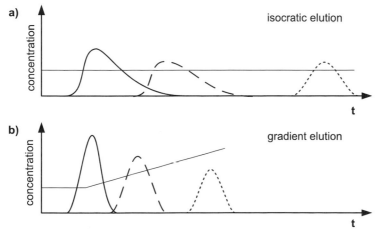

Fig. 5.13 Comparison of isocratic and linear gradient elution.

5.3
Continuous Processes

In the separation processes mentioned so far the introduction of the feed mixture is realised in a batch wise manner on one single column only. Especially for separations in a preparative scale, modes for continuous operation have to be considered to increase productivity, product concentration and to save fresh eluent.

5.3.1
Column Switching Chromatography

The easiest way of transforming a batch separation into a continuous one is the column switching approach, which can be applied for relatively simple adsorption-desorption processes. The feed is pumped through a column until a breakthrough of the target component is observed. At that moment the injection is switched to a second column, while the first one is desorbed by introducing a desorption eluent by a second pump. After desorption is finished the initial eluent conditions have to be re-adjusted before the first column can again be used for adsorption. Obviously, this set-up can be used only for relatively simple separations, where the component of interest can be separated from the impurities by adsorbing it to the stationary phase. This implies that the target component is the last eluting component and that its affinity to the stationary phase is much higher than that of the other components; or vice versa all other impurities stick to the stationary phase and only the component of interest shows no or little adsorption affinity. This set-up is often used as a relatively simple filtration process.

5.3.2
Annular Chromatography

For more complex feed mixtures other approaches for continuous operation of the chromatographic separation have to be considered. One example is annular chromatography with a rotating stationary phase. This concept was developed in the 1950s as a continuous method for paper chromatography by Solms (1955). In annular chromatography the stationary phase is packed between two concentric cylinders and rotates around a central axis (Fig. 5.14).

At the top of the column the feed injection port is situated at a fixed position. Over the whole remaining circumference of the packed bed the eluent is introduced into the column and moves towards the bottom. The annular column rotates at slow velocities. This rotation results in a crosscurrent movement of stationary and mobile phase. The components of the feed mixture are separated into several product streams, which can be withdrawn at the bottom of the column at fixed port positions. A batch chromatogram can easily be transferred to an annular chromatographic system by replacing the retention time by positional degrees. Therefore productivity is

Fig. 5.14 Annular chromatography.

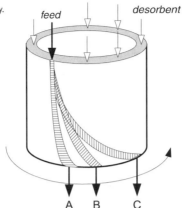

similar to batch systems under the same operational conditions. As in batch chromatography the solvent strength can be altered along the circumference by additional
solvent ports, forming gradients for improved process performance. One drawback
of annular chromatography is the difficulty of packing efficient columns. To operate
an annular chromatograph only rigid adsorbents should be used to obtain a more
equal distribution of the target components over the different withdrawal ports
(Schmidt et al., 2003). The allowable pressure drop of the stationary phase is limited
by the quality of the sealings. Wolfgang and Prior (2002) have given a more detailed
overview of the basic principle, technical aspects and industrial applications.

5.3.3
Multiport Switching Valve Chromatography (ISEP/CSEP)

The concept of an annular chromatographic system can easily be transferred to a
multicolumn set-up where several single columns are mounted within the concentric annulus on a rotating carrousel. The distribution of the different liquid streams
as well as the collection of all outlets is realised by one centre multiport switching
valve (Fig. 5.15). The inlet and outlet lines of the unit are connected to the non-
rotating part of the valve, while the columns are connected to the rotor. In contrast to
the carrousel, which rotates with a constant velocity, the moving head of the switching valve performs this movement in discrete steps, from one position to next at
given switching times.

The introduction of this valve and the fact that the stationary phase is packed into
fixed beds offers a great variety of possible column interconnections. Beside the pure
cross current flow, as it is realised in the annular chromatograph, columns can be
connected very flexibly in series or parallel, allowing a multitude of different process
set-ups (Fig. 5.16). Different sections can be realised to fulfil special tasks within the
process. Common sections are:

port. In section I the solid phase is regenerated with a fresh eluent stream by desorption of the strongly adsorbed component. Finally, in section IV the liquid is regenerated by adsorbing the amount of less retained component not collected in the raffinate. In this way both the solid and the liquid phase can be recycled to sections IV and I respectively. In summary the four sections have to fulfil the following tasks:

Section I   Desorption of the strongly adsorptive component
Section II   Desorption of the less adsorptive component
Section III   Adsorption of the strongly adsorptive component
Section IV   Adsorption of the less adsorptive component

With a proper choice of all individual internal flow rates in sections I to IV and the velocity of the stationary phase, the feed mixture can be completely separated. Complete separation leads to a distribution of the fluid concentrations as displayed in the axial concentration profile in Fig. 5.17. Since the TMB process reaches a steady state, it can be seen from the diagram that pure component B can be withdrawn with the extract stream. Conversely, the raffinate line contains pure component A only.

Unfortunately, the movement of a stationary phase, which in most cases consists of porous particles in the micrometre range, is technically not possible. Therefore, other technical solutions had to be developed. The break-through was achieved with Simulated Moving Bed (SMB)-systems, which were developed by UOP for the petrochemical industry in the 1960s (Broughton and Gerhold, 1961). The stationary phase is packed into single, discrete columns, which are connected to each other in a circle.

Figure 5.18 illustrates the principle of SMB processes. The mobile phase passes the fixed bed columns in one direction. Counter-current flow of both phases is achieved by switching the columns periodically upstream in the opposite direction of the liquid flow. Of course, in a real plant the columns are not shifted but all ports are moved in the direction of the liquid flow by means of valves. The counter-current character of the process becomes more obvious when the relative movement of the packed beds to the inlet and outlet streams during several switching intervals is observed. After a number of switching or shifting intervals equal to the number of columns in the system, one so-called cycle is completed and the initial positions for all external streams are re-established.

The total number of columns and their distribution over the different sections is not fixed to eight with two columns per zone (2/2/2/2) as shown in Fig. 5.18. Another quite common set-up is 1/2/2/1 with only one column in sections I and IV. Naturally, this reduction in total number of columns can reduce investment costs. Conversely, the costs for fresh solvent will increase since regeneration of the solid phase has to take place in a shorter time and, therefore, requires a higher flow rate in section I. For a further decrease of column number, other continuous process concepts like VariCol or ISMB should be implemented (Section 5.3.5).

The SMB concept is, in general, realised for pharmaceutical and fine chemical separation purposes in two different ways. In the first alternative, one centre rotating valve, as introduced in Section 5.3.3, is used to distribute and collect all inlet and out-

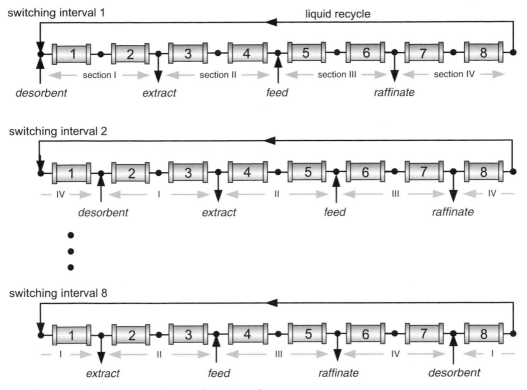

Fig. 5.18 Simulated moving bed (SMB) chromatography.

let streams. The second design concept switches the ports by means of two-way valves between all columns. The advantage in the latter case is that much higher pressures can be realised and the switching of the individual ports can be handled very flexible, which is the main requirement for implementation of the VariCol process (Section 5.3.5).

Another characteristic of the SMB setup is the implementation of a so-called recycle pump to ensure the liquid flow in one direction. In general, four cases can be distinguished (Fig 5.19).

In the first case (Fig. 5.19a) the recycle pump is at a fixed position between two columns. Since all columns are „moving" upstream according to the SMB principle the recycle pump performs the same migration. This means that the flow rate of this pump has to be adjusted depending on the section it is located at present. This design results in a locally increased dead volume because of the recycle line and the pump itself. Such an unequal distribution of the dead volume can be compensated by asynchronous shifting of the external ports, as introduced by Hotier and Nicoud (1995).

The second approach (Fig. 5.19b) is characterised by a „moving" recycle pump that is always located near the desorbent line. This has the advantage that the recycle pump never comes into contact with the products to be separated and the flow rate to be pumped is constant. However, this design needs additional valves.

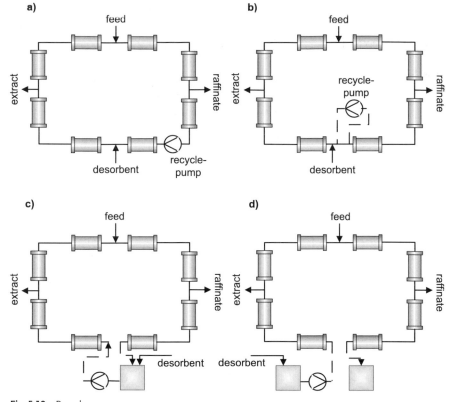

Fig. 5.19 Recycle pumps.

Beside the first alternatives, two more approaches have to be mentioned where no additional recycle pump is implemented. In the third case (Fig. 5.19c) the outlet of section IV is not directly recycled to section I but introduced to the desorbent tank instead. Figure 5.19d shows another possibility where no recycling of the solvent takes place (Ruthven and Ching, 1989). This method is applied when the regeneration of the desorbent turns out to be very difficult and fresh solvent is not too expensive.

When no recycling of the solvent takes place, as described before, the regeneration of the fluid is no longer necessary and, consequently, section IV is no longer required. This leads to a three-section SMB concept without solvent recycle, as depicted in Fig. 5.20.

In addition to three- and four-section SMBs, a five-section system could be useful when a third fraction is required. This would lead to a third product stream beside the extract and raffinate line. With very strongly retained components a second solid regeneration section might be taken into account by implementation of a second desorbent stream with higher solvent strength.

The technology of SMB chromatography has been widely used in the petrochemical (xylene isomer separation) and food industries (glucose–fructose separation) in

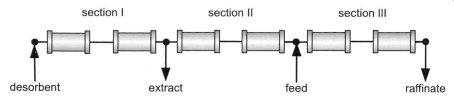

section I   section II   section III

desorbent   extract   feed   raffinate

Fig. 5.20   Three-section SMB concept.

a multi-ton scale. During the 1990s, with the advent of stable bulk stationary phases for chromatographic enantioseparation, the SMB principle was successfully transferred to the pharmaceutical industry. Today it is possible to build SMB systems with column sizes from 4 to 1000 mm inner diameter and to produce pharmaceuticals in small scale as well as up to a 100 ton scale. The advantages of SMB processes are achieved by a higher complexity with respect to layout and operation, which makes an empirical design quite difficult. Therefore modelling and simulation is necessary for process design and optimisation. The modelling part is presented in Chapter 6.7 and aspects regarding model based layout and optimisation are introduced in Chapter 7.3.

5.3.5
SMB Chromatography with Variable Conditions

Continuous counter-current separation by SMB chromatography improves process economics. Nevertheless, more improved processes have been developed recently. They are all based on the standard SMB technology but operated under variable process conditions to reduce the costs of column hardware and stationary phase as well as for fresh eluent and eluent work-up.

5.3.5.1   VariCol
Classical SMB-systems are characterised by the synchronous and constant downstream shift of all inlet and outlet lines after a defined switching period. In that case four defined sections can be distinguished over the complete cycle time (Fig. 5.21a). Within these sections the columns are distributed equally (e.g. 2/2/2/2) or in any other given configuration where the number of columns per section is an integer number (e.g. 1/2/2/1). At periodic steady state, the same process conditions are always reached after one switching interval (t_{shift}) and the number of columns per section is the same.

The VariCol approach, as introduced by Ludemann-Hombourger et al. (2000b), increases the flexibility of the continuous separation system by asynchronous movement of the injection and withdrawal ports. Within a complete process cycle this leads to mean numbers of columns per section that are non-integer. As the minimum number of columns per section in a SMB-system is one, it is possible in VariCol systems to reduce this to virtually any number less than one. Owing to the

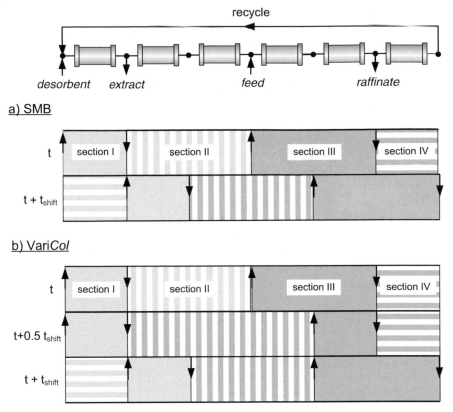

Fig. 5.21 Switching strategies for SMB and VariCol.

asynchronous shift of the inlet and outlet lines in a VariCol process it may even happen that, during a certain interval, there are no columns in a section. In this case the inlet and outlet line of this section coincide in one valve block. By placing the outlet lines upstream from the inlet lines a pollution of the product lines is avoided.

For example, the initial column configuration of a VariCol process at time t might be 1/2/2/1. After a predetermined time interval (e.g. $t + 0.5t_{shift}$) only the feed line is shifted to the next downstream position (Fig 5.21b). Now section II contains three columns for the rest of the time interval while only one column remains in section III. Thus the new configuration is 1/3/1/1. At the end of the switching time ($t + t_{shift}$) all other ports are shifted and the initial set up is re-established. The section configuration can than be calculated according to the number of columns and their residence time within a section. In our case the overall distribution of columns within one process cycle is 1/2.5/1.5/1. These figures demonstrate that the VariCol process offers high flexibility, especially for systems with a low number of columns (five or less columns). The main goal of the VariCol process is, therefore, to decrease the amount of stationary phase and the number of columns. This results in lower investment costs for column hardware and stationary phase.

5.3.5.2 PowerFeed
Another multi-column process based on the SMB principle is the so-called Power-Feed approach proposed by Kloppenburg and Gilles (1999a) and Zhang et al. (2003). In contrast to the asynchronous switching of the external ports in the VariCol process, the switching time remains constant, as it is in a classical SMB plant. The main difference with the PowerFeed approach compared to the different modes of operation described so far is the variation of the external flow rates ($\dot{V}_{feed}$, $\dot{V}_{des}$, $\dot{V}_{ext}$, $\dot{V}_{raf}$) within one switching interval t_{shift}. Consequently, the internal flow rates ($\dot{V}_I$–$\dot{V}_{IV}$) also change within a switching period. In the example illustrated in Fig. 5.22 the flow rate in section IV $\dot{V}_{IV}$ is lowered during the second part of the switching interval for improved adsorption of the less retained component. But, as consequence, as well as the raffinate flow $\dot{V}_{raf}$ the desorbent flow $\dot{V}_{des}$ also has to be increased in order to maintain a constant flow rate in section I. Similar considerations can be made for all other flow rates.

As reported by Zhang the performance of the PowerFeed processes is similar to VariCol with respect to product purities, productivity and eluent consumption and thus better than classical SMB processes. However, this improved performance is achieved with increased complexity of operation and design.

5.3.5.3 Partial feed
For the classical SMB process described above, the composition and flow rate of the feed is constant for the whole switching interval. Following the approach by Zang and Wankat (2002) in the Partial-Feed process, two additional degrees of freedom are introduced, i.e. the feed duration and feed time. Figure 5.23a compares the feed flow

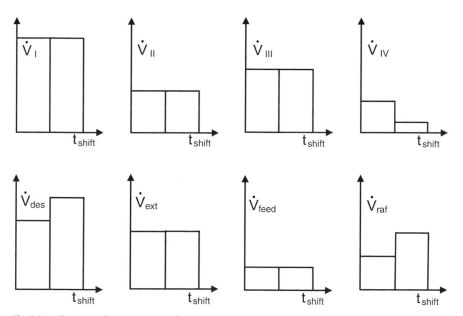

Fig. 5.22 Flow rates during PowerFeed operation.

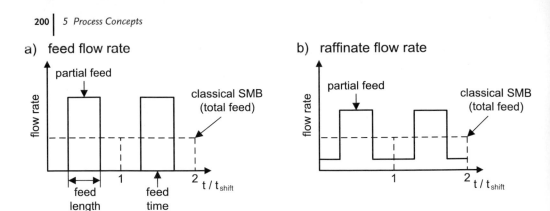

Fig. 5.23 Flow rates during Partial-Feed operation. (Reproduced from Wankat et al., 2002.)

of classical SMB and Partial-Feed processes. To fulfil the mass balance constraints the raffinate flow rate changes according to the changes in the feed flow (Fig. 5.23b); the duration of the feed is shorter, but the introduced flow rate is higher. By this procedure the total amount of feed is kept constant, while productivity and eluent consumption might increase.

5.3.5.4 ISMB

Partial-Feed processes switch the feed stream on and off within one time interval while the other inlet and outlet lines are active all the time. ISMB (Improved SMB) processes partition the switching interval in a different way. In the first part of the period (Fig. 5.24a), all external lines (desorbent and feed inlets as well as extract and raffinate outlets) operate. However, in contrast to a classical SMB unit, the outlet of section IV is not recycled during this part of the switching interval and, consequently, the flow rate in section IV is zero. The first „injection period" is followed by a recirculation period in the second part of the switching interval (Fig. 5.24b). During this time all external ports are closed and recirculation is performed with a constant

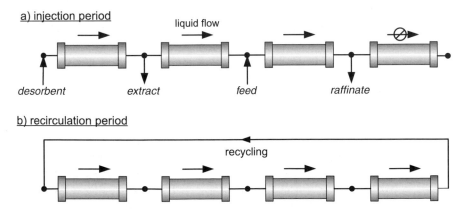

Fig. 5.24 Flow rates during ISMB operation.

Fig. 5.25 Feed concentration during ModiCon operation.

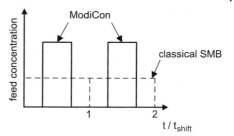

flow rate in all sections of the plant. This operating procedure allows a separation to be carried out with a rather small number of columns, which of course has a positive impact on investment costs. The main application area of the ISMB concept is the sugar industry.

5.3.5.5 ModiCon

The latest modification of the classical SMB process is the so-called ModiCon approach by Schramm et al. (2003). Unlike the Partial-Feed process, where the feed flow rate changes during one switching interval, ModiCon is characterised by a constant flow rate of the inlet but altering feed concentrations. Figure 5.25 illustrates this for the alternation of the feed concentration within one switching interval t_{shift}. At the beginning only pure solvent is fed to the plant while in the second part feed at a rather high concentration is used.

5.3.6
Gradient SMB Chromatography

One restriction of SMB chromatography as described so far is the operation under isocratic conditions. During the selection of the stationary as well as the fluid phase it is therefore one of the main goals to find a separation method where the first component is fairly well adsorbed on the stationary phase while the second one is still eluting under those conditions. This is quiet often a tedious task in which many compromises have to be made. Especially, isocratic elution conditions often lead to a severe tailing of the disperse front of the second component. But, similar to batch chromatography, a gradient can improve the SMB separation if the selectivity of the components is very large or a separation under isocratic conditions is impossible.

5.3.6.1 Solvent-gradient SMB Chromatography

The easiest and most adopted way to produce different solvent compositions inside the plant is a two-step gradient (Fig. 5.26). This is achieved by desorbent entering the plant with high elution strength while the feed stream is introduced with a lower solvent strength. This procedure leads to the formation of a step gradient with a regime of high desorption power in sections I and II and a region of improved adsorption in sections III and IV. According to the functions of the different sections (Section

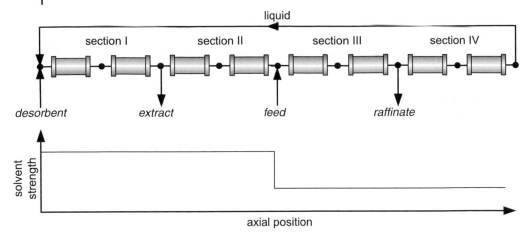

Fig. 5.26 Solvent-gradient operation of a SMB unit.

5.3.4), it becomes obvious that a gradient in a SMB process can improve its performance with respect to productivity, eluent consumption and product concentration.

Improved performance is achieved with a more complex operation and layout of the process, which requires precise process design and control – especially when a recycling of the section IV outlet is applied and a second point of mixing at the desorbent port is implemented. An open-loop operation without eluent recycling between sections I and IV is more robust, especially when regeneration of the adsorbent is necessary.

Some prerequisites for the choice of solvent and stationary phase should be considered at the beginning of the development of the chromatographic system. The solvents used for the gradient should exhibit low viscosities (especially for mixtures), high diffusivity and thus good miscibility. In addition, heat generation through the mixing process and volume contraction should be avoided.

Several research projects have investigated the application of gradient SMB processes (Abel et al., 2002; Abel et al., 2004a). A special focus in this context has been on the separation of bio-products, such as the separation of proteins by ion exchange (Houwing et al., 2002; Wekenborg et al., 2004).

5.3.6.2 Supercritical Fluid SMB Chromatography

An additional degree of freedom is introduced in supercritical fluid chromatography (SFC) with the modulation of the system pressure, which influences the adsorption strength of the solutes. SFC systems are operated above the critical pressure and temperature of the mobile phase systems. In most cases the main component of the mobile phase is carbon dioxide, for which the critical point is reached at 31 °C and 74 bar. In the supercritical region the density and, therefore, the solvating power of the fluid is highly dependent on pressure and temperature, and so is the affinity of a given solute for the supercritical fluid phase itself. With a higher operating pressure, and thus a density increase, the elution strength is improved and smaller retention times can be realised.

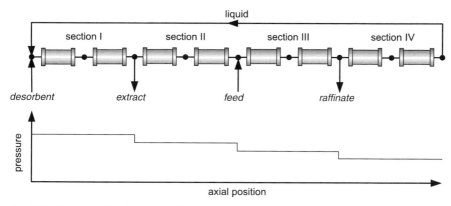

Fig. 5.27 Supercritical fluid operation of a SMB unit.

Supercritical fluids can be seen as intermediates between gases and liquids with liquid-like solvating powers in combination with gas-like diffusion coefficients and viscosities. The main advantages of supercritical conditions are low mobile phase viscosity, high diffusivity, easy recovery of substances from the eluent stream by simple solvent expansion as well as the low cost and no toxicity of the mobile phase („green solvent"). Conversely, the system is more complex and the solubility of most pharmaceutical components in pure CO_2 is rather limited, because drug substances should exhibit a certain hydrophilicity to achieve solubility and resorption in the gastro-intestinal tract. This drawback is often overcome by the introduction of a modifier, in most cases an alcohol or ether. As already pointed out, the adsorption properties of a given separation system can be influenced by adjusting the column pressure. Different adsorption strengths within the four sections of an SMB system are achieved by pressure variation (Fig. 5.27).

In section I desorption of the more adsorptive component has to take place, therefore the elution capability should be the highest and thus the system pressure is also the highest. Through sections II and III towards IV the pressure can be constantly lowered due to adsorption and desorption requirements in the single sections. More detailed information about this process concept can be found in Clavier et al. (1996), Denet et al. (2001), Depta et al. (1999) and Giovanni et al. (2001).

5.3.6.3 Temperature-gradient SMB Chromatography
The basic advantage of any gradient SMB-system is the enhanced desorption, especially of the extract component in section I. Adsorption strength is most readily modulated by variation of solvent composition or column pressure in the case of SFC-SMB; other modes of operation have also been taken into account.

As adsorption depends on the energy of the phase system, the dedicated introduction of energy into section I of the SMB process would lead to enhanced desorption. Recently, a temperature increase in section I has been investigated by Migliorini et al. (2001). One drawback of such a process is the heat capacity of the system and the slow change of temperature. This has to be considered, especially when a column enters another section and its temperature has to be changed. Therefore, a more focused energy introduction can be advantageous in terms of system reaction time.

A sophisticated approach is the change in stationary phase adsorption characteristics by outside stimuli. This effect is related to a conformational change of the three-dimensional structure of the stationary phase on a stimulus such as temperature change or light energy. The most advanced approach is the influence on stationary phases by IPAA selectors (Sun et al., 2004). Another example is the energy supply by microwaves that can be directed towards a single column and more easily stopped after the stationary phase has moved out of the desorption section. Notably, such highly complex systems might only be used if a striking large-scale application can be identified, as their operation is still associated with a lot of uncertainties.

5.4
Guidelines

Every new separation is an individual task that needs the skills of experts to find the optimal chromatographic system and to scale-up the process according to proven routes for successful process development. Developing a new separation based on routines can ensure a trouble-free development but might also risk not finding the optimum process economy and thus endanger the whole project. Therefore, every separation should be developed as open-mindedly as possible.

This section provides some guidelines to assist the decision for an appropriate process concept. The main decision criteria of the approach, as shown in Fig. 5.28, are „scale", „range of k'" and „number of fractions". Processes resulting from this decision tree are explained by examples consisting of an analytical chromatogram representing the early stage of process development and the chromatogram for the final realisation of the preparative process.

5.4.1
Scale

The main criterion to be considered is the „scale" of the project, which distinguishes between large or production scale (kg a^{-1} to t a^{-1}) and small or laboratory scale (mg a^{-1} to g a^{-1}). This implies the question of whether the separation justifies a time-consuming method development and process design to improve process performance in terms of productivity, eluent consumption and yield.

Influencing aspects in this context are for instance:

- Total amount of feed mixture to be separated
- Duration and frequency of the project
- Equipment available in the laboratory or at the production site
- Experience and knowledge about certain process concepts

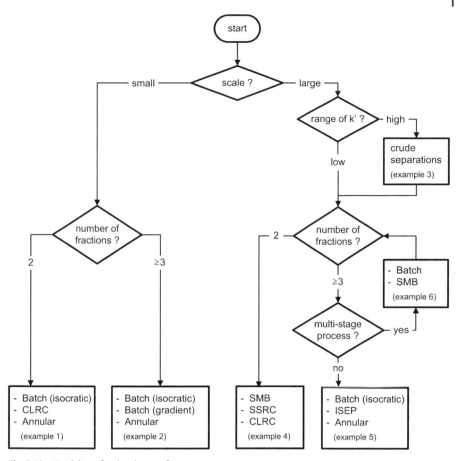

Fig. 5.28 Guidelines for the choice of a process concept.

5.4.2
Range of k'

Chromatographic processes show their best performance when the adsorption behaviour of the components to be separated is not too different. Target components of the feed mixture should elute within a certain window. Under optimised conditions, k' is in the range 2–8. Within that window the selectivity between the target component and the impurities should be optimised (Chapter 4).

If early or late eluting components are present in the feed the front or rear end components can be removed by simple pre-purification steps. These steps need not be chromatographic separations. Alternatives are extraction techniques, as the components normally differ substantially in their polarity and, thus, solubility behaviour.

5.4.3
Number of Fractions

To find a suitable process concept the final differentiation is given by the number of fractions to be collected. Notably, one fraction can contain either one single component (e.g. one target product) or a group of numerous components (e.g. several impurities). Some process concepts, such as SMB, SSRC and CLRC, can only separate a feed mixture into two fractions. In contrast, batch elution chromatography, annular chromatography, ISEP, etc. separate feed streams into three or more fractions.

Multistage processes should be considered for production-scale processes with three or more fractions. An intermediate step by SMB or batch separation reduces the separation problem, finally, to a two-fraction problem that can be performed by applying one of the above-mentioned concepts.

5.4.4
Example 1: Lab Scale; Two Fractions

A racemic mixture has to be separated to produce several grams of a pure enantiomer for further reactions. By optimising the batch conditions a good selectivity but no baseline separation is obtained. Details are given in Tab. 5.4 and Fig. 5.29.

Because only a small amount of the target component was available the yield of the chromatographic separation should be as high as possible. Therefore, closed loop recycling chromatography (CLRC) (Section 5.2.3) with peak shaving was chosen to avoid fractions with insufficient purity, which otherwise had to be reworked. Figure 5.30 depicts the resulting chromatogram of the preparative process with four cycles. Both enantiomers were collected directly during the first cycle. In cycle two the main part of the second eluting component was withdrawn from the system so that in cycle three nearly no second enantiomer is left.

5.4.5
Example 2: Lab Scale; Three or More Fractions

Only a few milligram of a drug metabolite had to be separated from a patient's urine for structure elucidation purposes. Table 5.5 summarizes the conditions of orienting experiments.

Table 5.4 Example 1.

	Analytical scale	Production scale
Sample	Chiral sulfoxide	
Adsorbent	Chiralcel® OD, 20 μm	
Mobile Phase	Heptane–isopropanol (85:15) isocratic	
Column	250×4.6 mm, stainless steel	200×50 mm, stainless steel

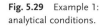

Fig. 5.29 Example 1: analytical conditions.

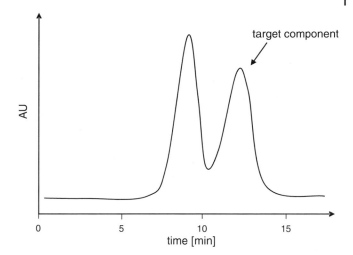

In the resulting chromatogram the target component is surrounded by other components (Fig. 5.31). In addition, all components elute within a wide range of k'. However, since the total amount to be separated is very small, no additional pre-separation steps were implemented. For the same reason no further effort was made to improve the chromatographic system. The elution conditions were linearly transferred to a larger column (Tab. 5.5) and the loading factor increased until the target component eluted immediately behind the earlier eluting impurity (Fig. 5.32). In this case, a smaller adsorbent particle diameter was chosen for the preparative column to achieve the necessary plate number in the column.

As only milligram or grams of product were to be purified in the above examples, relatively little effort was made to optimise the chromatographic system and quite simple concepts were applied. This is completely different to processes with production rates of $100 \, kg \, a^{-1}$ or higher.

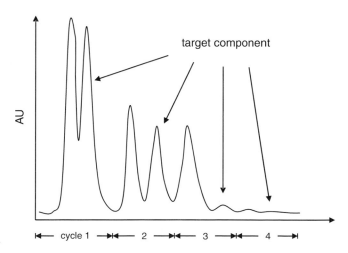

Fig. 5.30 Example 1: preparative conditions.

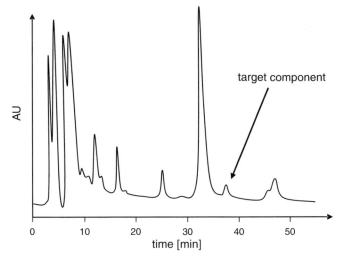

Fig. 5.31 Example 2: analytical conditions.

Table 5.5 Example 2.

	Analytical scale	**Production scale**
Sample	Drug metabolites from urine	
Adsorbent	LiChrospher® 100 RP-18, 5 μm	Superspher® 100 RP-18, 4 μm
Mobile Phase	Acetonitrile–water; gradient	Acetonitrile–water; gradient
Column	250×4 mm, stainless steel	200×50 mm, stainless steel

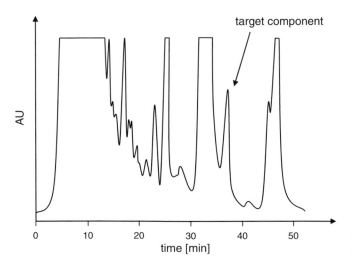

Fig. 5.32 Example 2: preparative conditions.

5.4.6
Example 3: Production Scale; Wide Range of k'

Several flavonoids have to be separated from a plant extract, where the desired production rate is in the range of several tons per year. At the very beginning of the project the complete feed stock has been analysed under the conditions listed in Tab. 5.6.

The resulting chromatogram (Fig. 5.33) reveals a rather wide range of retention times for the multiple components in the mixture, even though a gradient has been applied. This indicates that the components have very different physical properties, and a crude separation at the beginning is sufficient to remove the main impurities.

Adsorption under isocratic conditions was chosen as a crude separation step. Chromatographic parameters were adjusted to a simple water–alcohol solvent on a coarse RP-18 silica (40–63 µm). Figure 5.34 shows the corresponding chromatogram. Using this pre-purification step, the components of interest are obtained at higher concentrations. In subsequent polishing steps single flavonoides can be further purified.

Table 5.6 Example 3.

	Analytical scale	Production scale
Sample	Flavonoids from plant extract	
Adsorbent	LiChrospher® 100 RP-18, 5 µm	LiChroprep® 100 RP-18, 40–63 µm
Mobile Phase	Acetonitrile–water; gradient	Water–ethanol (90:10), isocratic
Column	250×4 mm, stainless steel	125×25 mm, stainless steel

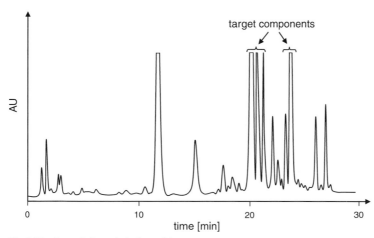

Fig. 5.33 Example 3: analytical conditions.

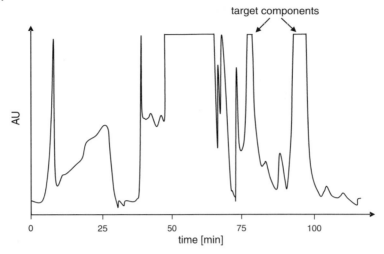

Fig. 5.34 Example 3: preparative conditions.

5.4.7
Example 4: Production Scale; Two Main Fractions

A pharmaceutical component had to be isolated from a multi-component mixture at a production rate of >100 kg a^{-1}. After adjusting conditions for the preparative separations (Tab. 5.7) a chromatogram was obtained on a small column (Fig. 5.35).

This analytical chromatogram indicates that some early eluting impurities are present in the feed mixture as well as two main impurities eluting after the target component. No baseline separation could be obtained between the component of interest and the late-eluting impurities. To avoid intermediate fractions with insufficient purities, which would have to be stored and reworked afterwards closed-loop recycling chromatography (CLRC) was applied. Here the early-eluting impurities are sent to waste during the first cycle, reducing the separation problem to a two-component separation. By only one additional cycle the target component can be obtained in good purity and yield.

Complete conditions for the preparative separation and corresponding chromatogram are given in Tab. 5.7 and Fig. 5.36.

Table 5.7 Example 4.

	Analytical scale	**Production scale**
Sample	Pharmaceutical component	
Adsorbent	LiChrospher® Si 60, 15 µm	
Mobile Phase	n-Heptane/ethylacetate, isocratic	
Column	250×4 mm, stainless steel	250×25 mm, stainless steel

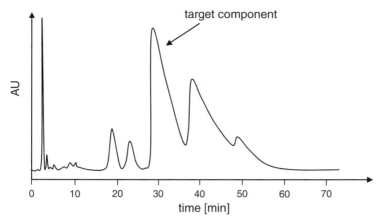

Fig. 5.35 Example 4: analytical conditions.

5.4.8
Example 5: Production Scale; Three Fractions

For the production scale separation of a prostaglandine derivative (ca. $100\,kg\,a^{-1}$), the chromatographic system in terms of stationary and mobile phase was optimised in advance according to the considerations described in Chapter 4 (Tab. 5.8).

Figure 5.37 indicates that, beside the target product, early and late-eluting impurities are also present. Following the considerations made in the beginning of this sec-

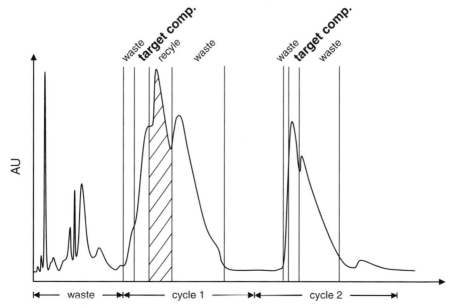

Fig. 5.36 Example 4: preparative conditions.

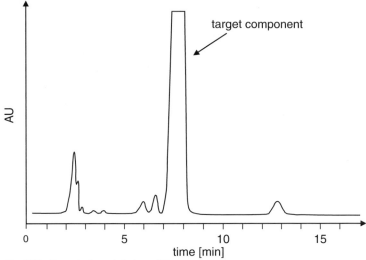

Fig. 5.37   Example 5: analytical conditions.

Table 5.8   Example 5.

	Analytical scale	**Production scale**
Sample	Prostaglandine	
Adsorbent	LiChrosorb® Si 60, 10 μm	LiChroprep® Si 60, 25-40 μm
Mobile Phase	n-Heptane–isopropanol–methanol–THF (96/2.4/1/0.6)	
Column	250×4 mm, stainless steel	600×200 mm, stainless steel

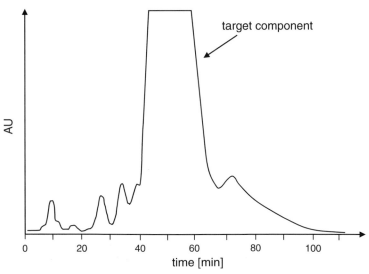

Fig. 5.38   Example 5: preparative conditions.

tion the decision has to be made as to whether the separation should be performed with a process concept, allowing a multi-fraction withdrawal, or a multi-stage process. In the present case, isocratic batch chromatography on an existing process-scale HPLC system was used. The chromatographic system was kept constant with regard to stationary and mobile phase. Only the particle diameter of the silica sorbent was increased. As the stationary phases are manufactured by the same production process, the different particle sizes did not change the adsorbent's properties, especially in terms of selectivity. Figure 5.38 shows the chromatogram under preparative conditions.

Beside the chosen batch separation, preparative annular or ISEP chromatographic processes are alternatives if the required equipment is available.

5.4.9
Example 6: Production Scale; Multi-stage Process

$100\,\mathrm{kg\,a^{-1}}$ of a cyclic peptide had to be separated from a fermentation broth. Analytical experiments (Tab. 5.9) afforded an initial chromatogram (Fig. 5.39).

Table 5.9 Example 6.

	Analytical scale (step 1)	Analytical scale (step 2)
Sample	Cyclic peptide from fermentation broth	
Adsorbent	LiChrospher® 100 RP-18e. 5 µm	LiChrospher® 100 RP-18. 12 µm
Mobile Phase	Acetonitrile–water; gradient	Acetonitrile–water. isocratic
Column	250×4 mm, stainless steel	250×4 mm, stainless steel

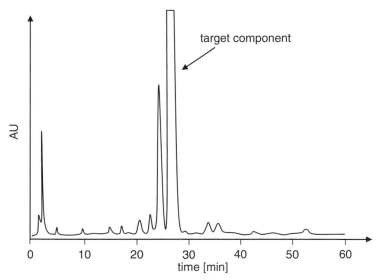

Fig. 5.39 Example 6: analysis of the complete feed stock.

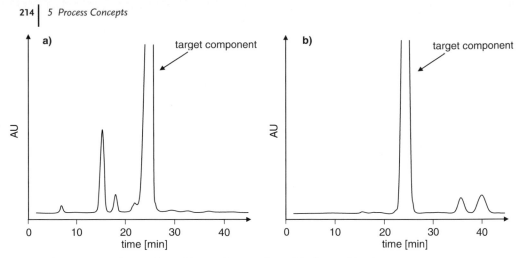

Fig. 5.40 Example 6: analysis of the two fractions obtained by the initial batch separation.

Again the target component is surrounded by both early- and late-eluting impurities. However, in contrast to the foregoing example, here a multistage process consisting of a batch separation followed by two SMB separations was applied. The first batch unit divided the complete feed into two fractions. The first fraction contains early-eluting impurities and approximately half of the target component, while the rest of the main product and late-eluting impurities are collected in the second fraction. Figure 5.40 shows an analysis of these two fractions.

Both fractions obtained from the first separation can be processed in subsequent SMB units where, again, the respective portion is split into two fractions. In this case fraction (a) is fed to a SMB unit and the target product is collected as extract. An SMB separation of fraction (b) leads to a raffinate stream containing the main product and the extract line where all impurities are collected.

Even though three chromatographic steps are now involved instead of just one batch separation, the overall process economy is increased because a higher yield is achieved and no intermediate fractions with lower purities have to be collected and reworked (Voigt et al. 1994).

6
Modeling and Determination of Model Parameters

M. Michel, A. Epping, A. Jupke

The first steps in the design of production-scale chromatographic processes are the selection of the chromatographic systems and the decision for possible processes as described in Chapters 4 and 5. Next is the scale-up from small-scale experiments to an economically optimal HPLC plant. The requirement of high purity under optimal operation conditions in combination with complex nonlinear system behavior makes a sole empirical design too difficult or time consuming. „Virtual experiments" by numerical simulations can considerably reduce time and amount of sample needed for process analysis and optimization. To achieve this, accurate models and precise model parameters for chromatographic columns arc nccdcd. If the models are validated, they can be used predictively for optimal plant design. Other possible fields of application for process simulation include process understanding for research purposes as well as personel training.

This chapter starts with an introduction to modeling of chromatographic separation processes, including discussion of different models for the column and plant peripherals. After a short explanation of numerical solution methods, the next main part is devoted to the consistent determination of the parameters for a suitable model, especially those for the isotherms. These are key issues towards achieving accurate simulation results. Methods of different complexity and experimental effort are presented that allow a variation of the desired accuracy on the one hand and the time needed on the other hand. Appropriate models are shown to simulate experimental data within the accuracy of measurement, which permits its use for further process design (Chapter 7). Finally, it is shown how this approach can be used to successfully simulate even complex chromatographic operation modes.

6.1
Introduction

The basic physical phenomena occurring during a chromatographic separation are described in Chapter 2. A quantitative description of these effects is possible using

Preparative Chromatography. Edited by H. Schmidt-Traub
Copyright © 2005 WILEY-VCH Verlag GmbH & Co. KGaA, Weinheim
ISBN: 3-527-30643-9

mathematical models, which are generally based on material, energy and momentum balances in addition to equations for the thermodynamic equilibrium between the solutes and the different phases. A model has to take into account all requirements of the process as well as the actual problems to be tackle. Therefore, it has to be as detailed as necessary but also as simple as possible. For steady state, time-independent processes, which are very often encountered in the production of bulk chemicals, a steady state model is sufficient. However, if the system variables change with time, the process has to be described by means of dynamic models. Another way to classify different models is based upon the nature of the control volume. A macroscopic balance can be applied if all physical quantities are assumed to be constant throughout the control volume, whereas microscopic balances are necessary for processes with spatial changes of the variables of state.

Since chromatography like every adsorption process is time and space dependent, dynamic and microscopic balances are employed to describe the process behavior.

Most processes consist of several, rather than one, operation units that are each described by different models. Then a flow-sheeting system should be used, which links all models by streams representing the material flow and solves the individual and overall heat and mass balances.

6.2
Models for Single Chromatographic Columns

6.2.1
Classes of Chromatographic Models

Different kinds of modeling approaches, including many analytical solutions, are comprehensively summarized in monographs by Guiochon et al. (1994b), Guiochon and Lin (2003), Ruthven (1984) and Seidel-Morgenstern (1995) as well as in review articles by Bellot and Condoret (1991) and Klatt (1999). Most of these models take into account two or more of the following effects:

- Convection
- Dispersion
- Mass transfer from the bulk phase into the boundary layer of the adsorbent particle
- Diffusion inside the pores of the particle (pore diffusion)
- Diffusion along the surface of the solid phase („surface diffusion")
- Adsorption equilibrium or adsorption kinetics

Figure 6.1 gives an example for a one-dimensional model. Mass balances for the fluid mobile phase as well as the stationary adsorbent phase are derived on the basis of differential volume elements. Section 6.2.2 gives further information regarding the derivation of the material balance.

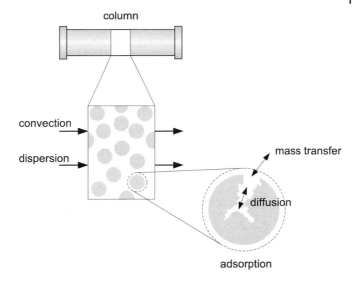

Fig. 6.1 Principle of differential mass balances for a chromatographic column.

Figure 6.2 shows the classification of the various models based on the number and type of effects considered.

In modeling chromatographic processes, the following assumptions are made:

- The adsorbent bed is homogeneous and packed with spherical particles of constant diameter
- Fluid density and viscosity are constant
- Radial distributions are negligible
- The process is isothermal
- The eluent is inert. Its influence on the adsorption process is taken into account implicitly by the parameters of the adsorption isotherm
- There is no convection inside the particles. The liquid phase inside the pores is assumed to be stationary and is not affected by the movement of the mobile phase
- No size-exclusion effects are taken into account – thus all solutes are assumed to penetrate the whole particle pore space if nothing else is specified

Consequently, models for liquid chromatography consist of one-dimensional mass balances for constant fluid velocities. The pressure drop can be calculated by Eq. 3.7. As will be shown in Section 6.6, the models can describe preparative chromatography within the accuracy of measurement.

6.2.2
Derivation of the Mass Balance Equations

Figure 6.3 visualizes differential volume elements and all in- and outgoing streams as well as source and sink terms for a general model. Elements include the mobile

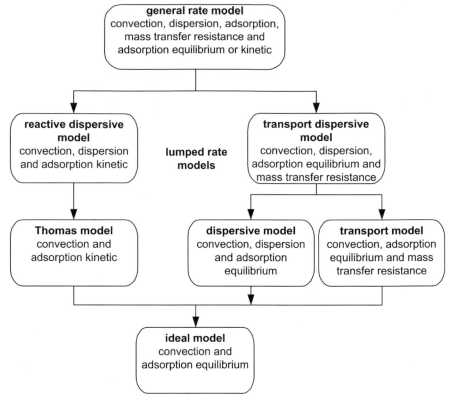

Fig. 6.2 Classification of different model approaches for a chromatographic column. (Reproduced from Klatt, 1999, and Guiochon et al., 1994b.)

fluid and the stationary adsorbent phase, which have to be accounted for separately. Additionally, the adsorbent is split into the stagnant fluid inside the particle pores and the actual solid structure of the particles.

In the following, the different terms of mass transport are derived separately. Subsequently, it will be shown which of these terms are included in the different models shown in Fig. 6.2.

6.2.2.1 Mass Balance Equations

According to the assumptions in Section 6.2.1 the general mass balance for one component i in one differential volume dV_c of the mobile phase is (Fig. 6.3)

$$\frac{\partial}{\partial t}\left[m_{\text{acc},i}(x, t)\right] = \dot{m}^x_{\text{conv},i}(x, t) - \dot{m}^{x+dx}_{\text{conv},i}(x, t) + \dot{m}^x_{\text{disp},i}(x, t)$$
$$- \dot{m}^{x+dx}_{\text{disp},i}(x, t) - \dot{m}_{\text{mt},i}(x, t)$$

(6.1)

Note that all terms given here, and also the variables derived below, are functions of time and space, but for brevity will not be indicated in the following.

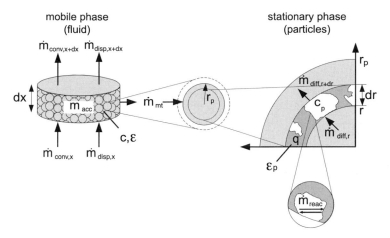

Fig. 6.3 Differential model elements for a chromatographic column.

Equation 6.1 includes mass accumulation in a storage term in the differential volume, the mass transport by in- and outgoing convection and dispersion as well as the mass transfer into the particles. Using a first-order Taylor series approximation for the outgoing streams,

$$\dot{m}^{x+dx} \approx \dot{m}^x + \frac{\partial \dot{m}^x}{\partial x} dx \qquad (6.2)$$

Eq. 6.1 can be written as:

$$\frac{\partial}{\partial}(m_{acc,i}) = -\frac{\partial(\dot{m}^x_{conv,i} + \dot{m}^x_{disp,i})}{\partial x} dx - \dot{m}_{mt,i} \qquad (6.3)$$

The superscript x is dropped in the following.

Mass transfer into the particle is equal to the overall accumulation of component i in the adsorbent.

$$\frac{\partial}{\partial t}(\overline{m}_{acc,ads,i}) = \dot{m}_{mt,i} \qquad (6.4)$$

Inside the adsorbent particles, mass transport takes place only due to pore and surface diffusion. The resulting mass balances including the adsorption kinetics in the reaction term are:

$$\frac{\partial}{\partial t}(m_{acc,pore,i}) = -\frac{\partial}{\partial r}(\dot{m}_{diff,pore,i}) dr - \dot{m}_{reac,i} \qquad (6.5)$$

$$\frac{\partial}{\partial t}(m_{acc,solid,i}) = -\frac{\partial}{\partial r}(\dot{m}_{diff,solid,i}) dr + \dot{m}_{reac,i} \qquad (6.6)$$

Alternatively, either Eq. 6.5 or 6.6 can be replaced by their sum, Eq. 6.7:

$$\frac{\partial}{\partial t}(m_{acc,pore,i} + m_{acc,solid,i}) = -\frac{\partial}{\partial r}(\dot{m}_{diff,pore,i} + \dot{m}_{diff,solid,i})dr \tag{6.7}$$

If adsorption equilibrium is assumed, the reaction kinetics are infinitively fast and the term $\dot{m}_{reac}$ is no longer well defined (Section 6.2.2.7). This makes independent mass balances for the pore and solid phase impractical as they are coupled through an isotherm equation. In this case Eq. 6.7 together with an isotherm equation (Section 6.2.2.7) must be used instead of adsorption kinetics (Section 6.2.2.6) and Eqs. 6.5 and 6.6.

To transfer these equations into mass balances based on concentrations, it is necessary to introduce different volumes. The overall differential volume dV_c is the sum of the mobile dV_{int} and the stationary phase dV_{ads}. By means of the void fraction (Eq. 2.6) and the cross section A_c of the column those are calculated as:

$$dV_{int} = \varepsilon A_c\, dx \tag{6.8}$$

$$dV_{ads} = (1 - \varepsilon)A_c\, dx \tag{6.9}$$

With

$$dV_c = A_c\, dx = dV_{int} + dV_{ads} \tag{6.10}$$

The same applies for the pore phase dV_{pore} and the solid phase dV_{solid} of the adsorbent, for which the phase distribution is given by the porosity (Eq. 2.7) of the adsorbent.

$$dV_{pore} = \varepsilon_p\, dV_{ads} = \varepsilon_p(1 - \varepsilon)A_c\, dx \tag{6.11}$$

$$dV_{solid} = (1 - \varepsilon_p)dV_{ads} = (1 - \varepsilon_p)(1 - \varepsilon)A_c\, dx \tag{6.12}$$

Introducing the concentration in the liquid (bulk) phase c_i, the mean overall adsorbent loading $\bar{q}_i^*$, the mean pore concentration $\bar{c}_{p,i}$ and the mean solid loading $\bar{q}_i$ the mass balances transform into:

$$m_{acc,i} = c_i\, dV_{int} = c_i \varepsilon A_c\, dx \tag{6.13}$$

$$\overline{m}_{acc,ads,i} = \bar{q}_i^*\, dV_{ads} = \bar{q}_i^*(1 - \varepsilon)A_c\, dx \tag{6.14}$$

$$\overline{m}_{acc,pore,i} = \bar{c}_{p,i}\, dV_{pore} = \bar{c}_{p,i}\varepsilon_p(1 - \varepsilon)A_c\, dx \tag{6.15}$$

$$\overline{m}_{acc,solid,i} = \bar{q}_i\, dV_{solid} = \bar{q}_i(1 - \varepsilon_p)(1 - \varepsilon)A_c\, dx \tag{6.16}$$

The overall adsorbent loading is equal to the sum of the pore concentration and the solid loading (Eq. 2.35):

$$\bar{q}_i^* = \varepsilon_p \bar{c}_{p,i} + (1 - \varepsilon_p)\bar{q}_i \tag{6.17}$$

The bar above the symbols in Eqs. 6.14–6.17 denotes average values. These average concentrations and the concentration in the liquid phase are only a function of the time t and the axial coordinate, x. The overall balance Eq. 6.17 has to be distinguished from the balance that takes into account a radial distribution each particle:

$$q_i^*(r) = \varepsilon_p c_{p,i}(r) + (1 - \varepsilon_p)q_i(r) \tag{6.18}$$

Note that these variables are also functions of time and the axial coordinate. Average concentrations and loadings within spherical particles are given by the following integrals:

$$\bar{c}_{p,i} = \frac{1}{\frac{4}{3}\pi r_p^3} \int_0^{r_p} c_{p,i}(r) \cdot 4\pi r^2 \, dr = \frac{3}{r_p^3} \int_0^{r_p} r^2 c_{p,i}(r) \, dr$$

$$\tag{6.19}$$

$$\bar{q}_i = \frac{3}{r_p^3} \int_0^{r_p} r^2 q_i(r) \, dr$$

In the general case, the mass balances have to take into account the number of particles per volume element:

$$N_p = \frac{dV_{ads}}{\text{volume particle}} = \frac{(1 - \varepsilon)A_c \, dx}{\frac{4}{3}\pi r_p^3} \tag{6.20}$$

As all particles at one axial position of the column are assumed to be identical, the differential balances in one particle can be multiplied by Eq. 6.20 to include all particles in the volume element. This yields the following accumulation terms, taking into account only the differential spherical shell (Fig. 6.3) of one particle:

$$m_{acc,pore,i}(r) = N_p c_{p,i}(r) \, \varepsilon_p 4\pi r^2 \, dr$$

$$= \frac{(1 - \varepsilon)A_c \, dx}{\frac{4}{3}\pi r_p^3} \cdot c_{p,i}(r) \, \varepsilon_p 4\pi r^2 \, dr \tag{6.21}$$

$$m_{acc,solid,i}(r) = N_p q_i(r) \, (1 - \varepsilon_p)4\pi r^2 \, dr \tag{6.22}$$

6.2.2.2 Convective Transport

After having defined the accumulation terms of the general mass balances (Eqs. 6.1–6.6), the transport and source terms are evaluated as follows. Mass transport in the mobile phase due to convection with the interstitial velocity u_{int} (Eq. 2.9) is given by

$$\dot{m}_{conv,i} = \varepsilon A_c u_{int} c_i \qquad (6.23)$$

6.2.2.3 Axial Dispersion

It is assumed that axial dispersion (Eq. 6.24) in the liquid phase can be defined in analogy to Fick's first law of diffusion.

$$\dot{m}_{disp,i} = -\varepsilon A_c D_{ax} \frac{\partial c_i}{\partial x} \qquad (6.24)$$

The axial dispersion coefficient D_{ax} depends only on the quality of the packing and represents any deviation of the fluid dynamics from plug flow. The contribution of molecular diffusion is generally negligible (Section 6.5.6.2).

6.2.2.4 Intraparticle Diffusion

Diffusion inside the adsorbent particles is also given by Fick's law. As for the accumulation terms (Eqs. 6.21 and 6.22), size and number of particles per unit volume (Eq. 6.20) are taken into account.

$$\dot{m}_{diff,pore,i}(r) = -N_p \varepsilon_p 4\pi r^2 D_{pore,i} \frac{\partial c_{p,i}(r)}{\partial r} \qquad (6.25)$$

$$\dot{m}_{diff,solid,i}(r) = -N_p (1 - \varepsilon_p) 4\pi r^2 D_{solid,i} \frac{\partial q_i(r)}{\partial r} \qquad (6.26)$$

In Eq. 6.25 the transport in the pore fluid is modeled as free diffusion in the macro- and mesopores, but the diffusion coefficient $D_{pore,i}$ is usually lower than in the liquid mobile phase because of the random orientation and variations in the diameter of the pores (tortuosity) (Section 6.5.8).

In Eq. 6.26 transport is assumed to be micropore or surface diffusion where the molecules are under the influence of a force field (e.g. electrostatic) of the inner adsorbent surface. Notably, surface diffusion especially is just a model to express the dependency of the molecular transport on surface loading. Discussions on the real transport mechanism still arouse controversy (Chapter 2.4.1).

6.2.2.5 Mass Transfer

According to the assumptions in Section 6.2.1, the liquid phase concentration changes only in axial direction and is constant in a cross section. Therefore, mass transfer between liquid and solid phase is not defined by a local concentration gradient around the particles. Instead, a general mass transfer resistance is postulated. A common method describes the (external) mass transfer $\dot{m}_{mt,i}$ as a linear function of the concentration difference between the concentration in the bulk phase and on the adsorbent surface, which are separated by a film of stagnant liquid (boundary layer). This so-called linear driving force model (LDF model) has proven to be sufficient in

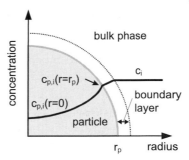

Fig. 6.4 Concentration profile for liquid film linear driving force models.

most cases (Guiochon et al., 1994b, Bellot and Condoret, 1991, Ruthven, 1984). Resultant concentration profiles are depicted in Fig. 6.4.

In the corresponding equation for the mass transfer rate of component i, calculated by Eq. 6.27, the film transfer coefficient k_{film} is based on the overall surface area dA_s of the adsorbent of all particles (Eq. 6.20) in the finite volume element (Eq. 6.28):

$$\dot{m}_{\text{mt},i} = k_{\text{film},i} \left[c_i - c_{\text{p},i}(r = r_{\text{p}}) \right] dA_s \tag{6.27}$$

$$dA_s = N_{\text{p}} 4\pi r_{\text{p}}^2 = \frac{3}{r_{\text{p}}} (1 - \varepsilon) A_{\text{c}} \, dx \tag{6.28}$$

A widely used characteristic value is the specific surface area a_s of the particles per unit volume, which can be derived from Eq. 6.28:

$$a_s = \frac{dA_s}{dV} = \frac{3}{r_{\text{p}}} (1 - \varepsilon) \tag{6.29}$$

Combining Eqs. 6.27 and 6.28 yields Eq. 6.30 for the mass transfer rate:

$$\dot{m}_{\text{mt},i} = k_{\text{film},i} \left[c_i - c_{\text{p},i}(r = r_{\text{p}}) \right] \frac{3}{r_{\text{p}}} (1 - \varepsilon) A_{\text{c}} \, dx \tag{6.30}$$

(some publications include the factor $3/r_{\text{p}}$ in the transfer coefficient).

According to Eq. 6.4 the mass flow $\dot{m}_{\text{mt}}$ is generally equal to the overall accumulation in the adsorbent. If concentrations and loadings change inside the particles, the overall accumulation (Eq. 6.14) must be calculated by integration (Eq. 6.19) after the radial concentration profiles are obtained. In this case it is reasonable to replace Eq. 6.4 with the continuity equation around the particles, as the mass flow through the outer liquid film is equal to the mass flow entering the adsorbent particle (Eqs. 6.25 and 6.26).

$$\dot{m}_{\text{mt},i} = - \left[\dot{m}_{\text{diff,pore},i}(r = r_{\text{p}}) + \dot{m}_{\text{diff,solid},i}(r = r_{\text{p}}) \right] \tag{6.31}$$

In this form, mass transfer by pore and surface diffusion instead of an overall mass transfer resistance is assumed inside the particle (internal mass transfer). Equation 6.31 represents a boundary condition of the general rate model, which is further discussed in Section 6.2.6.

6.2.2.6 **Adsorption Kinetics**

The final elements of the mass balances are the adsorption equilibria or the adsorption kinetics. According to Eqs. 6.5 and 6.6, adsorption and desorption steps are modeled as reactions with finite rate. The volumetric reaction rate based on the solid volume of all particles (Eq. 6.20) in the volume element is

$$\dot{m}_{reac,i} = (1 - \varepsilon_p) N_p \, \psi_{reac,i} \, 4\pi r^2 \, dr \tag{6.32}$$

Like isotherms, the net adsorption rate $\psi_{reac,i}$ is defined in quite different ways. One example is given in Eq. 6.33 where a first order equilibrium reaction with two rate constants for the adsorption (k_{ads}) and desorption (k_{des}) step (Ma et al., 1996) is specified.

$$\psi_{reac,i}(r) = k_{ads,i} \, q_{sat,i} \left(1 - \sum_{j=1}^{N_{comp}} \frac{q_j(r)}{q_{sat,j}} \right) c_{p,i}(r) - k_{des,i} \, q_i(r) \tag{6.33}$$

The terms $q_{sat,i}$ represent the maximum loadings for each component. Interaction between the different components is considered by the sum of all N_{comp} components. Equation 6.33 is the non-equilibrium form of the multi-component Langmuir isotherm (Eq. 2.44).

The non-equilibrium of the linear isotherm (Eq. 2.36) can be obtained by neglecting the sum in Eq. 6.33, leading to

$$\psi_{reac,i}(r) = k_{ads,i} \, q_{sat,i} \, c_{p,i}(r) - k_{des,i} \, q_i(r) \tag{6.34}$$

with

$$H = \frac{q_{sat} \, k_{ads}}{k_{des}} \tag{6.35}$$

6.2.2.7 **Adsorption Equilibrium**

Notably, adsorption and desorption steps are usually very fast. In the limit of the rate constants approaching infinity, $\psi_{reac,i}$ and $\dot{m}_{reac,i}$ are no longer well defined. In this case the pore concentration and the solid loading are in equilibrium, which is expressed through the isotherm equation:

$$q_i(r) = f \left[c_{p,1}(r), c_{p,2}(r), \ldots, c_{p,N_{comp}}(r) \right] \tag{6.36}$$

or simply

$$q_i = f(c_{p,1}, \ldots, c_{p,N_{comp}}) \tag{6.37}$$

Formally, Eq. 6.36 can be derived from Eq. 6.33 by setting $\psi_{reac,i}$ equal to zero, which corresponds to equally fast adsorption and desorption steps (a „dynamic" chemical equilibrium).

Table 6.1 Mass balance equations (general rate model).

Mobile phase

$$\frac{\partial}{\partial t}(m_{acc,i}) = -\frac{\partial(\dot{m}_{conv,i} + \dot{m}_{disp,i})}{\partial x}\,dx - \dot{m}_{mt,i}$$

Accumulation	$m_{acc,i} = c_i\,dV_{int} = c_i\,\varepsilon\,A_c\,dx$
Convection	$\dot{m}_{conv,i} = \varepsilon\,A_c\,u_{int}\,c_i$
Dispersion	$\dot{m}_{disp,i} = -\varepsilon A_c D_{ax}\dfrac{\partial c_i}{\partial x}$
Mass transfer	$\dot{m}_{mt,i} = k_{film,i}\left[c_i - c_{p,i}(r = r_p)\right]\dfrac{3}{r_p}(1-\varepsilon)A_c\,dx$

Stationary phase

$$\frac{\partial}{\partial t}(m_{acc,pore,i} + m_{acc,solid,i}) = -\frac{\partial}{\partial r}(\dot{m}_{diff,pore,i} + \dot{m}_{diff,solid,i})\,dr$$

Number of particles for volume element	$N_p = \dfrac{dV_{ads}}{\text{volume particle}} = \dfrac{(1-\varepsilon)A_c\,dx}{\frac{4}{3}\pi r_p^3}$
Accumulation	$m_{acc,pore,i}(r) = N_p c_{p,i}(r)\,\varepsilon_p\,4\pi r^2\,dr$
	$m_{acc,solid,i}(r) = N_p q_i(r)\,(1-\varepsilon_p)\,4\pi r^2\,dr$
Diffusion	$\dot{m}_{diff,pore,i}(r) = -N_p\,\varepsilon_p\,4\pi r^2\,D_{pore,i}\dfrac{\partial c_{p,i}(r)}{\partial r}$
	$\dot{m}_{diff,solid,i}(r) = -N_p(1-\varepsilon_p)\,4\pi r^2\,D_{solid,i}\dfrac{\partial q_i(r)}{\partial r}$
For adsorption equilibrium	$q_i(r) = f\left[c_{p,1}(r), c_{p,2}(r), \ldots, c_{p,N_{comp}}(r)\right]$
For adsorption kinetics	$\dfrac{\partial}{\partial t}(m_{acc,solid,i}) = -\dfrac{\partial}{\partial r}(\dot{m}_{diff,solid,i})\,dr + \dot{m}_{reac,i}$
	$\dot{m}_{reac} = (1-\varepsilon_p)N_p\psi_{reac,i}\,4\pi r^2\,dr$
	$\psi_{reac,i}(r) = k_{ads,i}\,q_{sat,i}\left(1 - \displaystyle\sum_{j=1}^{N_{comp}}\dfrac{q_j(r)}{q_{sat,j}}\right)c_{p,i}(r) - k_{des,i}\,q_i(r)$

Overall

$$\frac{\partial}{\partial t}(\overline{m}_{acc,ads,i}) = \dot{m}_{mt,i}$$

Accumulation	$\overline{m}_{acc,ads,i} = \overline{q}_i^*\,dV_{ads} = \overline{q}_i^*(1-\varepsilon)A_c\,dx$

Table 6.1 Mass balance equations (general rate model). (continued)

Overall loading	$q_i^*(r) = \varepsilon_p\, c_{p,i}(r) + (1 - \varepsilon_p)\, q_i(r)$
Average overall loading	$\bar{q}_i^* = \varepsilon_p\, \bar{c}_{p,i} + (1 - \varepsilon_p)\, \bar{q}_i$ $\bar{c}_{p,i} = \dfrac{3}{r_p^3} \displaystyle\int_0^{r_p} r^2 c_{p,i}(r)\, dr$ $\bar{q}_i = \dfrac{3}{r_p^3} \displaystyle\int_0^{r_p} r^2 q_i(r)\, dr$

Here the balance equation, Eq. 6.7, together with an isotherm equation must be used instead of adsorption kinetics (Section 6.2.2.6) and Eqs. 6.5 and 6.6.

All terms of the mass balance for liquid chromatography have now been specified in detail, which represents the most extended model discussed. Table 6.1 summarizes these elements.

In the following the most relevant models for liquid chromatography are derived in a bottom-up procedure related to Fig. 6.2. To illustrate the difference between these models their specific assumptions are discussed and the level of accuracy and their field of application are pointed out. The mass balances are completed by their boundary conditions (Section 6.2.7). For the favored transport dispersive model a dimensionless representation will also be presented.

6.2.3
Ideal Equilibrium Model

The simplest model takes into account convective transport and thermodynamics only. It assumes local equilibrium between mobile and stationary phase. This model, also called the ideal or basic model of chromatography, was described first by Wicke (1939) for the elution of a single component. Subsequently, De Vault (1943) derived the correct form of the mass balance.

The ideal equilibrium model neglects the influence of axial dispersion and all mass transfer and kinetic effects:

$$D_{ax} = 0$$
$$D_{pore,i} = D_{solid,i} = \infty \qquad\qquad (6.38)$$
$$k_{ads,i}, k_{des,i}, k_{film,i} = \infty$$

Consequently, the loading and concentration within the adsorbent are constant and no function of the particle radius. A further simplification arises from the absence of film transfer resistance. Hence the concentration in the liquid phase is identical to that in the particle pores.

$$\left. \begin{array}{c} c_{p,i} = \bar{c}_{p,i} \\ q_i = \bar{q}_i \end{array} \right\} \neq f(r)$$

$$c_i = c_{p,i}$$

(6.39)

Therefore, Eqs. 6.3 and 6.4 reduce to the material balance:

$$\frac{\partial}{\partial t}(m_{acc,i}) = -\frac{\partial(\dot{m}_{conv,i})}{\partial x}dx - \frac{\partial}{\partial t}(\overline{m}_{acc,ads,i})$$

(6.40)

Introducing the appropriate terms given by Eqs. 6.13–6.16 and 6.23 leads to the common form of this model:

$$\frac{\partial c_i}{\partial t} + u_{int}\frac{\partial c_i}{\partial x} + \frac{1-\varepsilon}{\varepsilon}\left[\varepsilon_p\frac{\partial c_i}{\partial t} + (1-\varepsilon_p)\frac{\partial q_i}{\partial t}\right] = 0$$

(6.41)

The only additional equation required is the thermodynamic equilibrium:

$$q_i = f(c_1, \ldots, c_{N_{comp}})$$

(6.42)

Another popular equivalent form of the equilibrium model (Eq. 6.41) is obtained by introducing the total porosity,

$$\varepsilon_t = \varepsilon + \varepsilon_p(1 - \varepsilon)$$

(2.8)

and the effective velocity u_m of a non-retained solute that enters the pore space:

$$u_m = \frac{L_c}{t_0} = \frac{\varepsilon}{\varepsilon_t}u_{int} = \frac{\varepsilon}{\varepsilon_t}\frac{L_c}{t_{0,int}}$$

(6.43)

Note that u_m is directly linked to the dead time t_0 (Eq. 2.10) while u_{int} is connected to $t_{0,int}$ (Eq. 2.11).

After some rearrangement, insertion of Eqs. 6.43 and 2.8 into Eq. 6.41 leads to

$$\frac{\partial c_i}{\partial t} + u_m\frac{\partial c_i}{\partial x} + \frac{1-\varepsilon_t}{\varepsilon_t}\frac{\partial q_i}{\partial t} = 0$$

(6.44)

The ideal model should be applied to get information about the thermodynamic behavior of a chromatographic column. Through work by Lapidus and Amundson (1952) and van Deemter et al. (1956) in the case of linear isotherms and by Glueckauf (1947, 1949) for nonlinear isotherms, considerable progress was made in understanding the influences of the isotherm shape on the elution profile. This work was later expanded to a comprehensive theory due to improved mathematics. Major contributions come from the application of nonlinear wave theory and the method of characteristics by Helfferich et al. (1970, 1996) and Rhee et al. (1970, 1986, 1989), who made analytical solutions available for Eqs. 6.41 and 6.42 for multi-component Langmuir isotherms.

The derivative of the loading can be eliminated by means of the partial differential for single component elution:

$$\frac{\partial q_i}{\partial t} = \frac{\partial q_i}{\partial c_i} \frac{\partial c_i}{\partial t}$$ (6.45)

Equation 6.44 can now be rearranged to

$$\frac{\partial c_i}{\partial t} + \frac{u_m}{\left(1 + \frac{1 - \varepsilon_t}{\varepsilon_t} \frac{\partial q_i}{\partial c_i}\right)} \frac{\partial c_i}{\partial x} = 0$$ (6.46)

Equation 6.46 predicts the propagation velocity w of an arbitrary concentration c^+ inside the column depending on the isotherm slope.

$$w(c_i^+) = \frac{u_m}{1 + \frac{1 - \varepsilon_t}{\varepsilon_t} \frac{\partial q_i}{\partial c_i}\Big|_{c_i^+}}$$ (6.47)

The velocity is connected to the observable retention time by

$$t_{R,i}(c_i^+) = \frac{L_c}{w(c_i^+)}$$ (6.48)

Combining Eqs. 6.43, 6.47 and 6.48 leads directly to the basic equation of chromatography, already presented in Chapter 2 (Eq. 2.16), for the injection of an ideal (Dirac) pulse.

$$t_{R,i}(c_i^+) = t_0 \left(1 + \frac{1 - \varepsilon_t}{\varepsilon_t} \frac{\partial q_i}{\partial c_i}\Big|_{c_i^+}\right)$$ (6.49)

In the special case of linear isotherms (Eq. 2.36) it reduces to

$$t_{R,\text{lin},i} = t_0 \left(1 + \frac{1 - \varepsilon_t}{\varepsilon_t} H_i\right)$$ (6.50)

Equation 6.49 is strictly valid only for the disperse part of the peak (Chapter 2.2.3). Depending on the shape of the isotherm, this is the rear part ("Langmuir") or the front part ("anti-Langmuir") of the peak (Fig 2.6). The sharp fronts ("Langmuir") or tails ("anti-Langmuir") of the peaks are called concentration discontinuities or shocks. To describe the movement of these shocks, the differential in Eq. 6.49 has to be replaced by discrete differences Δ, the secant of the isotherm, which describe the amplitudes of the concentration shocks in the mobile and stationary phases:

$$t_{R,i,\text{shock}} = t_0 \left(1 + \frac{1 - \varepsilon_t}{\varepsilon_t} \frac{\Delta q_i}{\Delta c_i}\Big|_{\text{shock}}\right)$$ (6.51)

For large injected pulses that cause a concentration breakthrough, Eq. 6.51 can be used to calculate the position of the shocks (Section 6.5.7.5).

Equation 6.49 is the basis for the practical interpretation of a given chromatogram and is also used to determine isotherm parameters. The relation between equilibrium theory and characteristic retention times in the chromatogram are as follows. First, Eq. 6.49 is of course only valid for retention times corrected for the dead time of the plant (Chapter 2.2.1 and Section 6.5.3.1). Second, the injection time t_{inj} of a real peak (cf. Fig. 6.11 and Eq. 6.94) also has to be considered when evaluating the characteristic retention times for components with a Langmuir-type (convex) isotherm, as the disperse part of the elution profile originates from the rear part of the rectangular injection pulse:

$$t_{R,i}(c_i^+) = t_{inj} + t_0 \left(1 + \frac{1 - \varepsilon_t}{\varepsilon_t} \frac{\partial q_i}{\partial c_i} \bigg|_{c_i^+} \right) \qquad (6.52)$$

For anti-Langmuir-type (concave) isotherms Eq. 6.49 without t_{inj} prevails because in this case the disperse front originates from the front of the injection pulse.

For multi-component mixtures and isotherms depending on more than one concentration, the partial differentials are replaced by total differentials. For two components they are:

$$\frac{dq_1}{dc_1} = \frac{\partial q_1}{\partial c_1} + \frac{\partial q_1}{\partial c_2} \frac{dc_2}{dc_1}$$

$$\frac{dq_2}{dc_2} = \frac{\partial q_2}{\partial c_2} + \frac{\partial q_2}{\partial c_1} \frac{dc_1}{dc_2} \qquad (6.53)$$

Here the elution behavior of both components is coupled through the concentration dependence of both isotherm equations. The impact of the concentration of one component on the propagation velocity of the other is included in the so-called coherence condition introduced by Helfferich and Klein (1970):

$$w_1(c_1^+, c_2^+) = w_2(c_1^+, c_2^+) \qquad (6.54)$$

One of the main advantages of equilibrium theory is the capability to predict some fundamental phenomena that occur in multi-component chromatography such as the displacement effect and the tag-along effect (Chapter 2.6.2). Another application is the use as short-cut methods for preliminary process design. As no effects causing band spreading are included, it is not possible to predict the system behavior exactly.

6.2.4
Models with One Band-broadening Effect

Basically, models using only one effect to describe band spreading lump all effects in one model parameter, which is straightforward for linear isotherms (Section 6.5.3.1) but is also commonly applied in the nonlinear range. Of these models, listed in the second level from the bottom in Fig. 6.2, the equilibrium dispersive model plays a prominent role.

$$\sigma_t^2(x = L_c) = t_{R,\text{lin},i}^2 \cdot \frac{2D_{\text{app},i}}{u_{\text{int}} L_c} \tag{6.64}$$

As this approximate profile is symmetrical, the first moment is identical to the position of the maximum.

Although Eq. 6.61 is different from the exact analytical solution of the equilibrium dispersive model (Pallaske, 1984, Levenspiel and Bischoff, 1963, Guiochon et al., 1994b, Guiochon and Lin, 2003), the resulting moments derived from this analytical solution are equal to Eqs. 6.63 and 6.64.

When comparing the moments of Eqs. 6.63 and 6.64 with the definition of the number of stages in Eq. 2.29, a correlation between dispersion coefficient and the number of stages N or the HETP can be derived:

$$N_i = \frac{\mu_t^2}{\sigma_t^2} = \frac{L_c}{\text{HETP}_i} = \frac{L_c u_{\text{int}}}{2D_{\text{app},i}} \tag{6.65}$$

If HETP has been determined (Section 6.5.3.1), $D_{\text{app},i}$ is easily estimated by Eq. 6.65. Although Eq. 6.65 is, strictly, valid only for linear or nearly linear isotherms it is often used for nonlinear isotherms, too.

6.2.4.2 Transport Model

The transport model has been used by some authors to simulate SMB processes, when mass transfer resistance is assumed to be the limiting factor so that axial dispersion can be neglected without altering the resulting concentration profiles significantly (Hashimoto et al., 1983a and 1993a). This model considers adsorption equilibrium, convection and mass transfer as the only rate limiting and band broadening step. In analogy to the equilibrium dispersive model, all mass transfer effects are lumped into an effective mass transfer coefficient, which is modeled in analogy to film transfer:

$$\tilde{k}_{\text{eff},i} = f(D_{\text{ax}}, D_{\text{pore}}, D_{\text{solid}}, k_{\text{film}}, c_i, \dots, c_{N_{\text{comp}}}, u_{\text{int}}) \tag{6.66}$$

The mass balance for the mobile phase is similar to Section 6.2.3, but the concentrations in the pores c_p is no longer equal to c:

$$\frac{\partial c_i}{\partial t} + u_{\text{int}} \frac{\partial c_i}{\partial x} + \frac{1 - \varepsilon}{\varepsilon} \left(\varepsilon_p \frac{\partial c_{p,i}}{\partial t} + (1 - \varepsilon_p) \frac{\partial q_i}{\partial t} \right) = 0 \tag{6.67}$$

An additional equation for the material balance for the particle phase is necessary (Eqs. 6.4, 6.14 and 6.30), where the accumulation in the stationary phase equals the mass transfer stream.

$$\varepsilon_p \frac{\partial c_{p,i}}{\partial t} + (1 - \varepsilon_p) \frac{\partial q_i}{\partial t} = \tilde{k}_{\text{eff},i} \frac{3}{r_p} (c_i - c_{p,i}) \tag{6.68}$$

The isotherm equation is given by Eq. 6.37.

6.2.4.3 Reaction Model

This model was first used by Thomas (1944) to simulate ion-exchange processes. It postulates a rate-limiting adsorption kinetic as the only effect causing band broadening. Originally, homogeneous particles without pores ($\varepsilon_p = 0$) were assumed, but this can be neglected to fit in the framework presented here. It need only be stated that the concentration in the bulk phase is the same as in the particle pores. Thus, the mobile phase balance is still the same as in the ideal model (Eq. 6.41). Compared with the transport model, the right-hand side of the stationary phase balance equation Eq. 6.68 is replaced by an adsorption kinetic such as Eq. 6.33 rewritten for constant concentrations inside the particles (Eq. 6.39):

$$\varepsilon_p \frac{\partial c_i}{\partial t} + (1 - \varepsilon_p)\frac{\partial q_i}{\partial t} = \tilde{k}_{ads,i}\, q_{sat,i}\left(1 - \sum_{j=1}^{N_{comp}} \frac{q_j}{q_{sat,j}}\right) c_i - \tilde{k}_{des,i}\, q_i \tag{6.69}$$

In Eq. 6.69 $\tilde{k}_{ads}$ and $\tilde{k}_{des}$ are the lumped rate constants of the adsorption and desorption steps, respectively. Equation 6.33 is just one example how to express the kinetic equation – others have been published by Bellot and Condoret (1991) and Ma et al. (1996).

Notably, for „standard" adsorption chromatography, the assumption of only a rate-determining kinetic step is unrealistic because this step is generally much faster than other influencing effects (Guiochon et al., 1994b and Ruthven, 1984). However, when considering ion exchange or protein adsorption this simplification may be justified.

6.2.5
Lumped Rate Models

The next level of detail in the model hierarchy of Fig. 6.2 is the so-called „lumped rate models" (third from the bottom). They are characterized by a second parameter describing rate limitations apart from axial dispersion. This second parameter subdivides the models into those where either mass transport or kinetic terms are rate limiting. No concentration distribution inside the particles is considered and, formally, the diffusion coefficients inside the adsorbent are assumed to be infinite.

$$\left. \begin{aligned} D_{solid,i} &= D_{pore,i} = \infty \\ c_{p,i} &= \bar{c}_{p,i} \\ q_i &= \bar{q}_i \end{aligned} \right\} \neq f(r) \tag{6.70}$$

The basic material balance of the mobile phase for all lumped rate models is based on Eqs. 6.3, 6.4 and 6.13–6.17 and can be derived in the same manner as the equilibrium dispersive model (Eq. 6.58):

$$\frac{\partial c_i}{\partial t} + u_{int}\frac{\partial c_i}{\partial x} + \frac{1 - \varepsilon}{\varepsilon}\frac{\partial q_i^*}{\partial t} = D_{ax}\frac{\partial^2 c_i}{\partial x^2} \tag{6.71}$$

The lumped rate models are distinguished on the basis of different equations for the particle phase, considering either adsorption kinetics or mass transfer. In the former case the concentration inside the particle pores c_p is identical to the mobile phase concentration c, while they are different for the latter.

The dispersion coefficient D_{ax} (Section 6.5.6.2) is assumed to depend only on the packing properties and flow conditions (Eq. 6.24) and, therefore, generally differs from the apparent dispersion coefficient D_{app} defined in Section 6.2.4.1.

6.2.5.1 Transport Dispersive Model

The transport dispersive model (TDM) is an extension of the transport model and summarizes the internal and external mass transfer resistance in one lumped film (= effective) transfer coefficient, k_{eff} (compare Eq. 6.30):

$$k_{eff,i} = f\left[D_{solid}, D_{pore}, k_{film}, c_i, \ldots, c_{N_{comp}}, (u_{int})\right] \tag{6.72}$$

In contrast to the transport model $k_{eff,i}$ is theoretically assumed to be independent of axial dispersion and thereby of the packing quality (Sections 6.5.3.1 and 6.5.8). Compared, for example, with D_{ax}, k_{eff} is much less dependent on fluid velocity (Section 6.5.8).

As mentioned in Section 6.2.2, the mass transfer term in Eq. 6.3 is defined by the linear driving force approach. Therefore, the transport dispersive model consists of the balance equations in the mobile phase (Eq. 6.71) written with the pore concentration

$$\frac{\partial c_i}{\partial t} + u_{int}\frac{\partial c_i}{\partial x} + \frac{1-\varepsilon}{\varepsilon}\left[\varepsilon_p\frac{\partial c_{p,i}}{\partial t} + (1-\varepsilon_p)\frac{\partial q_i}{\partial t}\right] = D_{ax}\frac{\partial^2 c_i}{\partial x^2} \tag{6.73}$$

as well as in the stationary phase (Eq. 6.74),

$$\varepsilon_p\frac{\partial c_{p,i}}{\partial t} + (1-\varepsilon_p)\frac{\partial q_i}{\partial t} = k_{eff,i}\frac{3}{r_p}(c_i - c_{p,i}) \tag{6.74}$$

which is derived from Eqs. 6.4 and 6.14–6.17 as well as the assumptions of Eq. 6.70. As discussed in Section 6.2.2, the driving force in Eq. 6.74 is the difference between the concentration c_i in the bulk phase and the concentration $c_{p,i}$ on the particle surface, which in case of the lumped rate model is identical to the constant concentration inside the particle pores. The local adsorption equilibrium is given by

$$q_i = f(c_{p,1}, \ldots, c_{p,N_{comp}}) \tag{6.37}$$

In Eq. 6.74 the main resistance is modeled to lie within the liquid boundary layer surrounding every particle.

Notably, concerning the overall peak shape, as with the equilibrium dispersive model (Section 6.2.4.1), the analytical solution of the transport dispersive model is always an asymmetric peak, and the asymmetry is enhanced by increasing D_{ax} as well as decreasing k_{eff} (Lapidus and Amundson 1952).

Several modifications of this model can be found in the literature. One that is frequently used considers the mass transfer resistance in the solid phase to be dominant. As proposed by Glueckauf and Coates (1947), an analog linear driving force approach for the mass transfer in the solid can then be applied and Eq. 6.74 as well as Eq. 6.37 are replaced by Eqs. 6.75 and 6.76. Mathematically, this linear driving force is modeled as the difference between the overall solid loading of Eq. 6.17 and an additional hypothetical loading q_{eq}, which is in equilibrium with the liquid phase concentration.

$$\frac{\partial q_i^*}{\partial t} = k_{eff,s,i} \frac{3}{R_p} (q_{eq,i}^* - q_i^*)$$ (6.75)

$$q_{eq,i}^* = f(c_1, \ldots, c_{N_{comp}})$$ (6.76)

For linear isotherms ($q_i = H_i\, c_{p,i}$), Eqs. 6.74 and 6.75 are equivalent, using the following relationships between the transport coefficients:

$$k_{eff,i} = \left[\varepsilon_p + (1 - \varepsilon_p)H_i\right] k_{eff,s,i}$$ (6.77)

and

$$q_{eq,i}^* = \left[\varepsilon_p + (1 - \varepsilon_p)H_i\right] c_i$$ (6.78)

In this book, „transport dispersive model" always refers to the liquid film linear driving force model (Eqs. 6.71, 6.74 and 6.37).

6.2.5.2 Reaction Dispersive Model

The other subgroup of the lumped rate approach consists of the reaction dispersive model where the adsorption kinetic is the rate-limiting step. It is an extension of the reaction model (Section 6.2.4.3). Like the mass transfer coefficient in the transport dispersive model, the adsorption and desorption rate constants are considered as effective lumped parameters, $k_{ads,eff}$ and $k_{des,eff}$. Since no film transfer resistance exists ($c_{p,i} = c_i$), the model can be described by Eq. 6.79:

$$\varepsilon_p \frac{\partial c_i}{\partial t} + (1 - \varepsilon_p)\frac{\partial q_i}{\partial t} = k_{ads,eff,i}\, q_{sat,i} \left(1 - \sum_{j=1}^{N_{comp}} \frac{q_j}{q_{sat,j}}\right) c_i - k_{des,eff,i}\, q_i$$ (6.79)

and the material balance of the mobile phase (Eq. 6.71) with $c_{p,i} = c_i$

$$\frac{\partial c_i}{\partial t} + u_{int}\frac{\partial c_i}{\partial x} + \frac{1-\varepsilon}{\varepsilon}\left(\varepsilon_p \frac{\partial c_i}{\partial t} + (1 - \varepsilon_p)\frac{\partial q_i}{\partial t}\right) = D_{ax}\frac{\partial^2 c_i}{\partial x^2}$$ (6.80)

6.2.6

General Rate Models

General rate models (GRM) are the most detailed models. In addition to axial dispersion they are characterized by a minimum of two other parameters describing mass transport effects. These two parameters may combine mass transfer in the liquid film and inside the pores as well as surface diffusion and adsorption kinetics in various kinds. Only a small representative selection of the abundance of different reported models is given here in order to characterize the main fields of application.

The underlying equations of a comprehensive approach using film transport, pore diffusion, surface diffusion and adsorption kinetics are discussed in Section 6.2.2 (Berninger et al., 1991, Whitley et al., 1993 and Ma et al., 1996).

As a consequence, the model equations can be derived without further simplifications. Compared with models presented in the previous sections, radial mass transport inside the particle pores is taken into account, which results in concentration and loading distributions along the particle radius and hence the average concentrations in Eqs. 6.14–6.16 have to be calculated by Eqs. 6.18 and 6.19.

The mass balance in the liquid phase (Eq. 6.3) includes accumulation within the liquid (Eq. 6.13), convection (Eq. 6.23), axial dispersion (Eq. 6.24) and (external) mass transfer through the liquid film outside the particles (Eq. 6.30):

$$\frac{\partial c_i}{\partial t} + u_{\text{int}} \cdot \frac{\partial c_i}{\partial x} + \frac{1-\varepsilon}{\varepsilon}\frac{3}{r_{\text{p}}}k_{\text{film},i}\left[c_i - c_{\text{p},i}(r=r_{\text{p}})\right] = D_{\text{ax}}\frac{\partial^2 c_i}{\partial x^2} \tag{6.81}$$

The differential mass balance for the pores (Eq. 6.5) and the solid (Eq. 6.6) are obtained together with Eqs. 6.21 and 6.22, 6.25 and 6.26 and 6.32 and 6.33 as:

$$\frac{\partial c_{\text{p},i}(r)}{\partial t} = \frac{1}{r^2}\frac{\partial}{\partial r}\left(r^2 D_{\text{pore},i}\frac{\partial c_{\text{p},i}(r)}{\partial r}\right) - \frac{1-\varepsilon_{\text{p}}}{\varepsilon_{\text{p}}}\psi_{\text{reac},i}(r) \tag{6.82}$$

$$\frac{\partial q_i(r)}{\partial t} = \frac{1}{r^2}\frac{\partial}{\partial r}\left(r^2 D_{\text{solid},i}\frac{\partial q_i(r)}{\partial r}\right) + \psi_{\text{reac},i}(r) \tag{6.83}$$

$$\psi_{\text{reac},i}(r) = k_{\text{ads},i}\, q_{\text{sat},i}\left(1 - \sum_{j=1}^{N_{\text{comp}}}\frac{q_j(r)}{q_{\text{sat},j}}\right)c_{\text{p},i}(r) - k_{\text{des},i}\, q_i(r) \tag{6.33}$$

The internal mass transfer is modeled with Fick's diffusion inside the (macro) pores (Eq. 6.82) as well as surface or micropore diffusion in the solid phase (Eq. 6.83). Note that Eqs. 6.82 and 6.83 no longer include the number of particles (Eq. 6.20) and therefore represent the balance in one particle.

Equation 6.33 is the net adsorption rate based on the solid volume, exemplified here as a first-order kinetic with saturation capacities $q_{\text{sat},i}$.

If no solutes with complex adsorption behavior (e.g. in bio-separations) are considered, the assumption of adsorption equilibrium (Section 6.2.2.7) is normally valid. In this case Eq. 6.33 formally reduces to a relationship between an isotherm for pore concentration and solid loading (Eq. 6.36), which for Eq. 6.33 is the multi-component Langmuir isotherm (Eq. 2.44).

As mentioned in Section 6.2.2.7, the two balances in the stationary phase (Eq. 6.5 and Eq. 6.6) and the adsorption kinetics (Eq. 6.33) must then be replaced by an isotherm equation Eq. 6.36 and the overall balance in one particle (Eq. 6.7). The latter can be derived analogous to Eqs. 6.82–6.83, leading to:

$$\varepsilon_p \frac{\partial c_{p,i}}{\partial t} + (1 - \varepsilon_p) \frac{\partial q_i}{\partial t} = \frac{1}{r^2} \frac{\partial}{\partial r} \left[r^2 \left(\varepsilon_p D_{pore,i} \frac{\partial c_{p,i}}{\partial r} + (1 - \varepsilon_p) D_{solid,i} \frac{\partial q_i}{\partial r} \right) \right] \tag{6.84}$$

Gu et al. (1990a and 1990b) proposed an even more reduced general rate model, which only considers pore diffusion inside the particles (pore diffusion model). Thus, Eq. 6.84 is replaced by Eq. 6.85:

$$\varepsilon_p \frac{\partial c_{p,i}}{\partial t} + (1 - \varepsilon_p) \frac{\partial q_i}{\partial t} = \varepsilon_p \frac{1}{r^2} \frac{\partial}{\partial r} \left(r^2 D_{pore,i} \frac{\partial c_{p,i}}{\partial r} \right) \tag{6.85}$$

The original model considered adsorption equilibrium between $c_{p,i}$ and q_i, but adsorption kinetics were included later (Gu et al. 1991, 1993 and 1995). The approach by Gu et al. to neglect surface diffusion is justified for so-called „low-affinity" adsorbents in adsorption chromatography, where transport by free pore diffusion dominates surface diffusion (Furuya et al., 1989). Only for adsorbents with a pro-nounced micropore system („high-affinity" adsorbents) can surface diffusion domi-nate pore diffusion, and a model consisting of Eqs. 6.81–6.83 must be used. In these high affinity adsorbents, the loadings are several orders of magnitudes higher than in „normal" adsorption chromatography. The resulting high loading gradients lead to a dominance of surface diffusion (Ma et al., 1996), although D_{solid} is lower, by two orders of magnitude, than D_{pore} (Suzuki, 1990).

Note that Eqs. 6.85 and 6.84 are identical, if the pore diffusion coefficient is taken as a lumped concentration dependent parameter that includes pore and surface dif-fusion:

$$D_{app,pore,i} = D_{pore,i} + \frac{1 - \varepsilon_p}{\varepsilon_p} D_{solid,i} \frac{\partial q_i}{\partial c_{p,i}} \tag{6.86}$$

As mentioned in Section 6.2.2, with general rate models, boundary conditions for the adsorbent phase are necessary in addition to the conditions at the column inlet and outlet (Section 6.2.7). The choice of appropriate boundary conditions is mathemati-cally subtle and often a cause for discussion in the literature. The following is restricted to the form of the boundary condition derived by Ma et al. (1996) for a „complete" general rate model.

Owing to symmetry, the concentration and loading gradients vanish at the particle centre:

$$\left. \frac{\partial c_{p,i}}{\partial r} \right|_{r=0} = \left. \frac{\partial q_i}{\partial r} \right|_{r=0} = 0 \tag{6.87}$$

The links between liquid, pore and solid phase are given by mass balances at the par-ticle boundary. Equation 6.31 connects the external mass transfer rate and the diffu-sion inside all particles, which after insertion of Eqs. 6.25, 6.26 and 6.30 results in

$$k_{film,i} \left[c_i - c_{p,i}(r - r_p) \right] = \varepsilon_p D_{pore,i} \left. \frac{\partial c_{p,i}}{\partial r} \right|_{r=r_p} + (1 - \varepsilon_p) D_{solid,i} \left. \frac{\partial q_i}{\partial r} \right|_{r=r_p} \tag{6.88}$$

The set of boundary conditions is completed by recognizing that no surface diffusion exists outside the particles and, therefore, the gradient of the flux is zero:

$$\frac{\partial}{\partial r}\left(-\varepsilon_i D_{\text{solid},i}\frac{\partial q_i}{\partial r}\right)\bigg|_{r=r_p} = 0 \Leftrightarrow \frac{\partial}{\partial r}\left(\frac{\partial q_i}{\partial r}\right)\bigg|_{r=r_p} = \frac{\partial^2 q_i}{\partial r^2}\bigg|_{r=r_p} = 0 \tag{6.89}$$

A substitute for Eq. 6.89 is the combination of Eq. 6.83 at the particle boundary with Eq. 6.89:

$$\frac{\partial q_i}{\partial t}\bigg|_{r=r_p} = D_{\text{solid},i}\frac{2}{r_p}\frac{\partial q_i}{\partial r}\bigg|_{r=r_p} + \psi_i\bigg|_{r=r_p} \tag{6.90}$$

6.2.7
Initial and Boundary Conditions of the Column

Mathematically, all models form a system of (partial) differential and algebraic equations. For the solution of those systems initial and boundary conditions for the chromatographic column are necessary. The initial conditions for the concentration and the loading specify their values at time $t = 0$. Generally, zero values are assumed:

$$c_i = c_i(t = 0, x) = 0$$
$$c_{p,i} = c_{p,i}(t = 0, x, r) = 0 \tag{6.91}$$
$$(q_i = q_i(t = 0, x, r) = 0)$$

(NB the initial profiles have to fulfill the isotherm equation.)

Since no adsorbent enters or leaves the column, the task is to find suitable inlet and outlet boundary conditions for the mass balance of the mobile phase (Eq. 6.3).

One form frequently encountered at the column inlet is the classic „closed boundary" condition for dispersive systems derived by Danckwerts (1953):

$$c_i(t, x = 0) = c_{\text{in},i}(t) - \frac{D_{\text{ax}}}{u_{\text{int}}}\cdot\frac{\partial c_i(t, x = 0)}{\partial x} \tag{6.92}$$

In general, the overall balance for the mass transport streams (Eqs. 6.23 and 6.24) at the column inlet and outlet has to be fulfilled. In Eq. 6.92 the closed boundary condition is obtained by setting the dispersion coefficient outside the column equal to zero. In open systems, the column stretches to infinity and in these limits concentration changes are zero.

For real chromatographic systems, the dispersion coefficient is usually small as the number of stages is very high ($N > 100$) and convection dominates. Therefore, Eq. 6.92 may be simplified to

$$c_i(t, x = 0) = c_{\text{in},i}(t) \tag{6.93}$$

which is the same condition as for open systems.

A common condition for of the inlet function is a rectangular pulse (Section 6.3.2.1), i.e. an injection of a constant feed concentration c_{feed} for a given time period, t_{inj}:

$$c_{in,i}(t) = \begin{cases} c_{feed,i} & t \leq t_{inj} \\ 0 & t > t_{inj} \end{cases} \tag{6.94}$$

The outlet boundary condition is generally assumed to be a zero gradient of the fluid concentration (Danckwerts, 1953):

$$\frac{\partial c_i(t, x = L_c)}{\partial x} = 0 \tag{6.95}$$

Notably, Eq. 6.93 is in any case the „correct" inlet condition for all models without axial dispersion and therefore often convenient to use. In practice the difference between the solutions for different boundary conditions is often irrelevant (Guiochon et al., 1994b). With numerical simulations, the validity of the boundary conditions should be determined by computing the mass balance at the inlet and outlet.

6.2.8
Stage Models

An entirely different approach to describe a chromatographic column leads to the class of equilibrium stage or plate models. Instead of a dynamic microscopic balance, the column is modeled as a sequence of a finite number N of similar stages. Each stage is filled with liquid and solid that are completely mixed. Two general groups of models have been reported. The so-called Craig model is based on the assumption of a constant residence time in each stage (Craig 1944), but is of little importance today. Another stage model was introduced by Martin and Synge (1941) and is equal to the concept of stirred tank cascades common in reaction technology. This model is used here to illustrate the application of „number of stages" in chromatography.

A constant flow of mobile phase through a cascade of N ideally stirred tanks (C.S.T.) is assumed, each tank having a total volume equal to V_c/N. Inside each tank, a fraction $(1 - \varepsilon_t)$ is occupied by the solid phase and the concentration inside the liquid is the same in the bulk and in the pore phase. This leads to the following mass balance for the kth tank, where accumulation is equal to difference between the inlet and the outlet stream:

$$\frac{V_c}{N}\left(\varepsilon_t \frac{\partial c^k}{\partial t} + (1 - \varepsilon_t)\frac{\partial q^k}{\partial t}\right) = \dot{V}(c^{k-1} - c^k) \tag{6.96}$$

Additionally, equilibrium between c and q is assumed.

Band broadening effects such as dispersion and mass transfer resistance are represented by the number of tanks (or stages) N. This can be explained by evaluating the moments of the analytical solution of Eq. 6.96. For linear isotherms and the injection of an ideal Dirac pulse of one component, this yields a Gamma density function as the concentration profile. With the retention time $t_{R,lini}$ of Eq. 6.50 one obtains the elution profile as the concentration in the last tank ($k = N$):

$$c^N(t) = \frac{m_{inj}}{\dot{V}} \frac{N}{t_{R,lin}} \left(\frac{Nt}{t_{R,lin}} \right)^{N-1} \frac{1}{(N-1)!} \exp\left(-\frac{Nt}{t_{R,lin}} \right) \tag{6.97}$$

The injected amount m_{inj} is given by Eq.6.62.
Calculation of the first moment μ_t (Eq. 2.27) and the second moment σ_t (Eq. 2.28) results in the following:

$$\mu_t = t_{R,lin} \tag{6.98}$$

$$\sigma_t = \frac{t_{R,lin}}{\sqrt{N}} \tag{6.99}$$

$$N = \left(\frac{t_{R,lin}}{\sigma_t} \right)^2 = \left(\frac{\mu_t}{\sigma_t} \right)^2 \tag{6.100}$$

Equation 6.100 is identical to Eq. 2.29 and is used as the definition of the number of stages N (Van Deemter et al., 1956), which is equal to the number of tanks in this model. As the second moment (variance) is directly related to the „width" of the peak (Section 2.4.2), the stage number is an appropriate value to describe band broadening. This model's behavior is consistent with the requirement that a high number of stages lead to a highly efficient separation (low band broadening). For high N the Gamma density function can be approximated by a Gaussian distribution (Eq. 6.61).

Different numbers of stages for each component must be used to account for individual band broadening (the main disadvantage of stage models), making it impossible to get a proper description of the elution behavior of every component in a multicomponent separation.

6.2.9
Assessment of Different Model Approaches

All models presented in the previous sections fulfill the general requirement of compatibility with different modes of operation and adsorption behaviors. Also, appropriate analytic or numeric methods are available to solve all model equations. Because of the many combinations of solutes, mobile and stationary phases, it might be difficult to determine *a priori* which model is suitable for a problem. A guideline for model selection is the number of stages N (Golshan-Shirazi and Guiochon, 1992, Guiochon et al., 1994b and Seidel-Morgenstern 1995), which can be obtained either from the column manufacturer or by a few initial experiments (Chapter 2.4.2 and in Section 6.5.3.1).

The ideal model provides good accuracy only if the column efficiency is very high ($N \gg 1000$). If band broadening cannot be neglected the ideal model as well as the stage models, which can only reproduce the number of stages for a single component, are not suited to describe multi-component separations.

The equilibrium dispersive model offers an acceptable accuracy only for $N \gg 100$, which may be sufficient for many practical cases. However, sometimes the number of stages is considerably lower and more sophisticated models have to be applied.

Ma et al. (1996) and Whitley et al. (1993) have provided methods to decide if effects such as pore and surface diffusion or adsorption kinetics have to be considered in a model. Their approach is based on the qualitative assessment of breakthrough curves, which are the result of a step input for different feed concentrations and flow rates. When the physical parameters are known or can be estimated, the value of dimensionless parameters defined in these publications may be used to select a model.

Since the main focus of this book is on adsorption chromatography (low affinity adsorbents), adsorption kinetics as well as surface diffusion can be neglected. Thus, the reaction models and models including surface diffusion are not considered in the following.

Consequently, the two remaining groups to consider for columns with a low number of stages are the lumped and the general rate models (GRM). Under the simplification mentioned above, general rate models take into account the individual mass transfer resistance in the liquid film/boundary layer (Eq. 6.81) and inside the pore system of the particles (Eq. 6.85). Based on theoretical studies, Ludemann-Hombourger et al. (2000a) concluded that, even for very small particle diameters ($d_p = 2\,\mu m$) and extremely low fluid velocities, pore diffusion is the limiting mass transfer step in adsorption chromatography (Section 6.5.8). For most practical applications it is therefore unnecessary in terms of accuracy to use two parameters to describe mass transfer. Thus, a lumped rate models using only one parameter (e.g. Section 6.2.5.1) will be sufficient.

The decision for a certain model has also to include considerations of methods to measure or estimate the model parameters. For the general rate model (Eq. 6.81 and Eq. 6.85) it is not possible to derive independently different transport parameters such as D_{pore} and k_{film} for a given column from a chromatogram. Therefore, one parameter has to be calculated (e.g. by correlations in Section 6.5.8) and used to determine the other.

The transport dispersive model (TDM, Section 6.2.5.1) is thus appropriate to simulate systems with considerable band broadening (Section 6.6), using only two different parameters to characterize packing properties (D_{ax}) and mass transfer (k_{eff}). From theoretical viewpoint, Kaczmarski and Antos et al. (1996) provide rules, in which case both TDM and GRM give identical results.

As increasing computational power and sophisticated numerical solvers are now available it is no longer necessary to use even more simplified models to reduce computing time. For these reasons the rest of this chapter is restricted to the transport dispersive model.

Table 6.2 gives recommendations for the application of the models discussed above.

Table 6.2 Fields of application of different models.

Model	Recommended application
Ideal model	• Analysis of thermodynamic effects only ($N \gg 1000$) • Initial guess
Equilibrium dispersive model	• Adsorption chromatography for products with low molecular weights • High accuracy only if $N \gg 100$
Transport dispersive model	• Adsorption chromatography for products with low molecular weights • Generally high accuracy • Chiral separation
General rate model	Chromatographic separation of solutes with complex mass transfer and adsorption behavior (e.g. bio separations or ion exchange chromatography)
Equilibrium stage model	• Adsorption chromatography for products with low molecular weights • Accuracy only for single components or small differences in N_i for all components

6.2.10
Dimensionless Model Equations

To reduce the number of parameters and to analyze their interdependence it is recommended that the model equations as well as the boundary conditions be converted into a dimensionless form. The following definition of dimensionless variables is used: Feed concentrations $c_{\text{feed},i}$ should be selected as a reference for concentrations and loadings:

• Dimensionless fluid concentration:

$$C_{\text{DL},i} = \frac{c_i}{c_{\text{feed},i}} \quad \Rightarrow \quad \partial c_{\text{DL},i} = \frac{1}{c_{\text{feed},i}} \cdot \partial c_i \tag{6.101}$$

• Dimensionless concentration in the particle:

$$C_{\text{p,DL},i} = \frac{c_{\text{p},i}}{c_{\text{feed},i}} \quad \Rightarrow \quad \partial C_{\text{p,DL},i} = \frac{1}{c_{\text{feed},i}} \cdot \partial c_{\text{p},i} \tag{6.102}$$

• Dimensionless loading

$$Q_{\text{DL},i} = \frac{q_i}{c_{\text{feed},i}} \quad \Rightarrow \quad \partial Q_{\text{DL},i} = \frac{1}{c_{\text{feed},i}} \partial q_i \tag{6.103}$$

The dimensionless axial coordinate Z is obtained by division through the column length

$$Z = \frac{x}{L_c} \quad \Rightarrow \quad \partial Z = \frac{1}{L_c} \partial x \tag{6.104}$$

and the dimensionless time τ is defined with the residence time $t_{0,int}$ (Eq. 6.43) based on the interstitial velocity:

$$\tau = \frac{t}{t_{0,int}} = \frac{u_{int}}{L_c} t \quad \Rightarrow \quad \partial \tau = \frac{u_{int}}{L_c} \partial t \tag{6.105}$$

Dimensionless parameters represent ratios of different mass transport and reaction phenomena. One example is the Bodenstein number (Bo, sometimes called the axial Peclet number), which is the ratio of convection rate to axial dispersion:

$$\mathrm{Bo} = \frac{u_{int} L_c}{D_{ax}} \tag{6.106}$$

Additional parameters arise according to the model selected and hence the physical effects that are taken into account. Parameters appropriate for the general rate model are given by Berninger et al. (1991) and Ma et al. (1996).

Introducing Eqs. 6.101–6.105 into the equations of the transport dispersive model leads to the following dimensionless mass balances:

$$\frac{\partial C_{DL,i}}{\partial \tau} + \frac{\partial C_{DL,i}}{\partial Z} + \frac{1-\varepsilon}{\varepsilon} \left[\varepsilon_p \frac{\partial C_{p,DL,i}}{\partial \tau} + (1-\varepsilon_p) \frac{\partial Q_{DL,i}}{\partial \tau} \right] = \frac{D_{ax}}{u_{int} L_c} \frac{\partial^2 C_{DL,i}}{\partial Z^2} \tag{6.107}$$

$$\varepsilon_p \frac{\partial C_{p,DL,i}}{\partial \tau} + (1-\varepsilon_p) \frac{\partial Q_{DL,i}}{\partial \tau} = \frac{L_c}{u_{int}} \frac{6}{d_p} k_{eff,i} (C_{DL,i} - C_{p,DL,i}) \tag{6.108}$$

The dimensionless parameter in Eq. 6.107 is the Bodenstein number (Eq. 6.106). The remaining parameter in Eq. 6.108 determines the ratio of effective mass transport to convection, which is defined by the modified (effective) Stanton number (St_{eff}):

$$St_{eff,i} = k_{eff,i} \frac{6}{d_p} \frac{L_c}{u_{int}} \tag{6.109}$$

This finally leads to a system of differential equations that depend only on these two dimensionless parameters.

$$\frac{\partial C_{DL,i}}{\partial \tau} + \frac{\partial C_{DL,i}}{\partial Z} + \frac{1-\varepsilon}{\varepsilon} \left(\varepsilon_p \frac{\partial C_{p,DL,i}}{\partial \tau} + (1-\varepsilon_p) \frac{\partial Q_{DL,i}}{\partial \tau} \right) = \frac{1}{Bo} \frac{\partial^2 C_{DL,i}}{\partial Z^2} \tag{6.110}$$

$$\varepsilon_p \frac{\partial C_{p,DL,i}}{\partial \tau} + (1-\varepsilon_p) \frac{\partial Q_{DL,i}}{\partial \tau} = St_{eff,i} (C_{DL,i} - C_{p,DL,i}) \tag{6.111}$$

In addition to the mass balances, the isotherm equation and the boundary condition have to be transformed into a dimensionless form. Equation 6.112, for example, presents the dimensionless form of the multi-component Langmuir equation (Eq. 2.44) for a binary mixture:

$$Q_{\text{DL},i} = \frac{H_i C_{\text{p,DL},i}}{1 + b_1 c_{\text{feed},1} C_{\text{p,DL},1} + b_2 c_{\text{feed},2} C_{\text{p,DL},2}} \quad \text{with } H_i = q_{\text{sat},i} \cdot b_i \qquad (6.112)$$

The dimensionless parameters of this equation are the Henry coefficients H_i and the dimensionless Langmuir parameters $b_1 c_{\text{feed},1}$ and $b_2 c_{\text{feed},2}$. This is the mathematical indication that the feed concentrations are parameters for a certain separation problem, as the H_i and b_i are constant for the chromatographic system.

Conversion of the boundary condition at the column inlet (Eq. 6.94) leads to:

$$C_{\text{DL,in},i}(t) = \begin{cases} 1 & \tau \leq \dfrac{t_{\text{inj}}}{t_{0,\text{int}}} = \dfrac{t_{\text{inj}} \dot{V}}{\varepsilon \cdot V_c} \\[3mm] 0 & \tau > \dfrac{t_{\text{inj}}}{t_{0,\text{int}}} \end{cases} \qquad (6.113)$$

where $t_{\text{inj}}/t_{0,\text{int}}$ is the dimensionless injection time.

In summary, chromatographic batch separation depends on the following dimensionless parameters: Peclet and Stanton numbers, dimensionless injection time, Henry coefficients and dimensionless Langmuir parameters.

6.3
Modeling HPLC Plants

6.3.1
Experimental Set-up and Simulation Flowsheet

Section 6.2 presents various models for chromatographic columns. But it has to be kept in mind that these models only account for effects occurring within the packed bed. A HPLC-plant, however, consists of several additional equipment and fittings besides the column. Therefore, the effect of this extra column equipment has to be accounted for to obtain reasonable agreement between experimental results and process simulation. Peripheral equipment (for example pipes, injection system, pumps and detectors) causes dead times and mixing. Thus, it can contribute considerably to the band broadening measured by the detector.

So-called „plant dispersion" or „extra column effects" have to be taken into account by additional mathematical models rather than including them indirectly in the model parameters of the column, e.g. by altering the dispersion coefficient. The combination of peripheral and column models is easily implemented in a modular simulation approach. In a flowsheeting approach the boundary conditions of different models are connected by streams (node balances) and all material balances are solved simultaneously.

The upper part of Fig. 6.5 shows the standard set-up of an HPLC-plant. The injection of a rectangular pulse is performed via a three-way valve and a subsequent pump

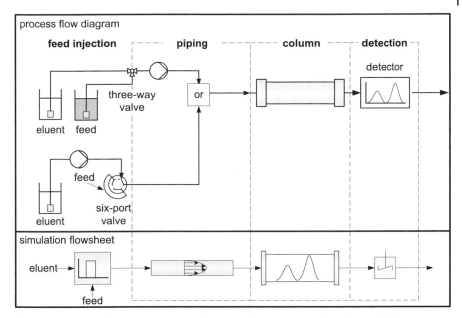

Fig. 6.5 Comparison between process flow diagram and simulation flowsheet for an HPLC plant.

or a six-port valve with a sample loop. The feed passes the connecting pipes and a flow distributor before entering the packed bed. At the exit there are another flow distributor and connecting pipes before a detector records the chromatogram. Because of these additional elements, an exact rectangular pulse will not enter the chromatographic column and the detected chromatogram is not identical to the concentration profile at the column exit.

As mentioned in Chapters 2 and 3, peak distortion is caused not only by non-ideal equipment outside but also inside the column. Although sophisticated measurements such as NMR (Tallarek et al., 1998) allow the investigation of the packed bed only, from a practical viewpoint the observable performance of a column always includes the effects of walls, internal distributors and filters. Using the method described in this chapter, these are always contained in an appropriate column parameter (e.g. D_{ax}). The column manufacturer has to ensure a proper bed packing and flow distribution (Chapter 3) and, thus, their negative influence on column performance, and hence on D_{ax}, can assumed to be small. However, these facts should be kept in mind when doing scale-up.

Transformation of the process diagram into a corresponding simulation flowsheet is illustrated in the lower part of Fig. 6.5. In principle all plant elements may be represented by a separate model. For practical applications, though, it is sufficient to take into account only a time delay as well as the dispersion of the peak until it enters the column. This can be achieved by a pipe-flow model that includes axial dispersion. The detector (including some connecting pipes) can be represented by a stirred-tank model.

6.3.2
Modeling Extra Column Equipment

6.3.2.1 Injection System

The injection system is sufficiently described by setting appropriate boundary conditions at the entry of the pipe:

$$c_{pipe,i}(x = 0, \ 0 \leq t \leq t_{inj}) = c_{feed,i}$$
$$c_{pipe,i}(x = 0, \ t > t_{inj}) = 0 \tag{6.114}$$

Injection volume and the injection time are related by:

$$V_{inj} = \dot{V} \, t_{inj} \tag{6.115}$$

6.3.2.2 Piping

If the piping only contributes to the dead time of the plant the delay can be described by a pipe model assuming an ideal plug-flow.

$$\frac{\partial c_{pipe,i}}{\partial t} = -u_{0,pipe} \frac{\partial c_{pipe,i}}{\partial x} \tag{6.116}$$

When the plant behavior without the column shows non-negligible backmixing, a dispersed plug-flow model might be used (Eq. 6.117).

$$\frac{\partial c_{pipe,i}}{\partial t} = D_{ax,pipe} \frac{\partial^2 c_{pipe,i}}{\partial x^2} - u_{0,pipe} \frac{\partial c_{pipe,i}}{\partial x} \tag{6.117}$$

The fluid velocity u_0 inside the pipe is given by the continuity equation:

$$u_{0,pipe} = \frac{\dot{V}}{A_{pipe}} \tag{6.118}$$

It is not necessary to model individual pipes, if the cross sections are represented by a typical diameter. In any case the primary parameter of interest is the volume (Eq. 6.119) or the dead time (Eq. 6.120) of the unit.

$$V_{pipe} = A_{pipe} L_{pipe} \tag{6.119}$$

$$t_{0,pipe} = \frac{V_{pipe}}{\dot{V}} \tag{6.120}$$

If $t_{0,pipe}$ is determined from experiments and A_{pipe} is set, all other parameters are known from Eqs. 6.118–6.120.

6.3.2.3 Detector

Detectors contain measuring cells that exhibit a backmixing behavior that dominates the influence of the pipe system behind the chromatographic column. Therefore, the whole system behind the column is modeled as an ideal continuously stirred tank (C.S.T.).

$$\frac{\partial c_{\text{tank},i}}{\partial t} = \frac{\dot{V}}{V_{\text{tank}}} (c_{\text{in,tank},i} - c_{\text{tank},i}) \qquad (6.121)$$

The dead time of the tank (Eq. 6.122) is

$$t_{0,\text{tank}} = \frac{V_{\text{tank}}}{\dot{V}} \qquad (6.122)$$

Note that the dead time of the plant (Eq. 6.123) is defined as the sum of the dead time of both elements:

$$t_{\text{plant}} = t_{0,\text{pipe}} + t_{0,\text{tank}} \qquad (6.123)$$

6.4
Numerical methods

6.4.1
General Solution Procedure

The balance equations described in the previous sections include both space and time derivatives. Apart from a few simple cases, the resulting set of coupled partial differential equations (PDE) cannot be solved analytically. The solution (the concentration profiles) must be obtained numerically, either using self-developed programs or commercially available dynamic process simulation tools. The latter can be distinguished in general equation solvers, where the model has to be implemented by the user, or special software dedicated to chromatography. Some providers are given in Tab. 6.3.

The generalized numerical solution procedure involves the following steps:

1. Transformation of the PDE system into ordinary differential equations (ODE) with respect to time by discretization of the spatial derivatives
2. Solving the ODEs using numerical integration routines, which generally involve another discretization into a nonlinear algebraic equation system and subsequent iterative solution

The stability, accuracy and speed of the solution process depend on the choice and parameter adjustment of the individual mathematical methods in each step as well as their proper combination. Notably, in simulating a batch column, computational time is hardly an issue using today's PC systems.

Detailed discussion of discretization techniques and numerical solution methods is beyond the scope of this book and, therefore, only some general procedures are presented. Numerical methods are discussed in detail by, for example, Finlayson (1980), Davis (1984) and Du Chatcau and Zachmann (1989). Summaries of different discretization methods applied in the simulation of chromatography are given

Table 6.3 Examples of dynamic process simulation tools.

Examples for dynamic process simulation tools	
Aspen Engineering Suite™ (e.g. Aspen Custom Modeler®) (Aspen Technology, Inc., USA)	http://www.aspentech.com
gPROMS® (Process Systems Enterprise Limited (PSE), UK)	http://www.psenterprise.com

Examples for application software for liquid chromatography	
Aspen Chromatography® (Aspen Technology, Inc., USA)	http://www.aspentech.com
ChromSim® for Windows™ (Wissenschaftliche Gerätebau Dr. Ing. Herbert Knauer GmbH)	http://www.knauer.net
BatchChrom Lehrstuhl für Anlagentechnik, Universität Dortmund	http://atwww.bci.uni-dortmund.de
SMBOpt Lehrstuhl für Anlagensteuerungstechnik, Universität Dortmund	http://astwww.bci.uni-dortmund.de
Chromulator (Tingyue Gu's Chromatography Simulation)	http://www.ent.ohiou.edu/~guting/
VERSE (The Bioseparations Group, School of Chemical Engineering, Purdue University, West Lafayette)	http://atom.ecn.purdue.edu/~biosep/research.html

by Guiochon et al. (1994b) and Guiochon and Lin (2003). For an introduction into numerical programming procedures see, for example, Press et al. (2002, http://www.nr.com) or Ferziger (1998).

6.4.2
Discretization

Discretization „replaces" the continuous space time domain by a rectangular mesh or grid of discrete elements and points (Fig. 6.6). Note that initial and boundary conditions of the system must also be considered.

As a result of the discretization, numerical solutions are only approximations of the „true" continuous solutions at discrete points of the space time domain. The quality of the numerical solution depends on the structure of the discretization method and the number of discretes.

An example of the „errors" introduced by the approximation is the effect of „numerical dispersion", which leads to an additional artificial band broadening.

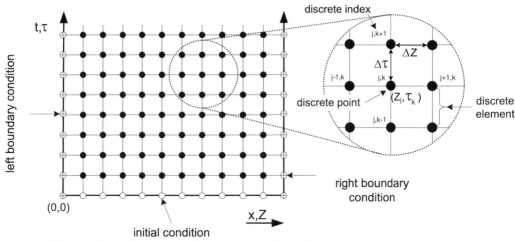

Fig. 6.6 Scheme of discretized space–time domain using a uniform grid.

Another example is the occurrence of nonphysical oscillations in the profile. A general assumption is that the numerical approximation becomes more accurate and, finally, approaches the true solution the finer the grid is. Naturally, smaller grid spacings mean more equations and longer calculation times. Practically, the grid is „fine enough" if the simulated elution profiles do not change noticeably on using a smaller grid size. Other simple tests for accuracy involve comparison with analytical solutions in special cases or control of the mass balance. Numerical integration (Eq. 6.62) of the simulated peak area should not deviate by more than 1% from the injected amount (Guiochon and Lin, 2003).

For brevity, further discussion is restricted to the spatial discretization used to obtain ordinary differential equations. Often the choice and parameters selection for this methods is left to the user of commercial process simulators, while the numerical (time) integrators for ODEs have default settings or sophisticated automatic parameter adjustment routines. For example, using finite difference methods for the time domain, an adaptive selection of the time step is performed that is coupled to the iteration needed to solve the resulting nonlinear algebraic equation system. For additional information concerning numerical procedures and algorithms the reader is referred to the literature.

Finite difference methods (FDM) are directly derived from the space time grid. Focusing on the space domain (horizontal lines in Fig. 6.6), the spatial differentials are replaced by discrete difference quotients based on interpolation polynomials. Using the dimensionless formulation of the balance equations (Eq. 6.107), the convection term at a grid point j (Fig. 6.6) can be approximated by assuming, for example, the linear polynomial.

$$\frac{\partial C_{DL}(Z = Z_j)}{\partial Z} \approx \frac{C_{DL,j+1} - C_{DL,j}}{\Delta Z} \quad \text{with } Z_j = j\Delta Z \tag{6.124}$$

Figure 6.7 illustrates this approach.

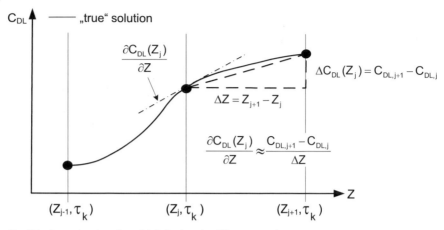

Fig. 6.7 Approximation of spatial derivatives by difference quotients.

Equation 6.124 is a forward FDM, as grid points „after" j (here $j+1$) are used for evaluation. It is a first-order scheme, as only one point is used. Higher order schemes involve more neighboring points for the approximation and are generally more accurate and have greater numerical stability. Other FD schemes involve grid points with lower indexes (backward FDM) or points on both sides of j (central FDM). Equation 6.124 only illustrates the principle of discretization. The continuous coordinates are replaced by the corresponding numbers of grid points (n_p) and the continuous profile $C_{DL}(Z)$ by a vector, with n_p elements $C_{DL,j}$.

The major drawback of (simple) FD schemes is that a rather fine grid, resulting in a large number of equations and hence high computational time, is necessary to obtain a stable and accurate solution, as chromatography is a convection dominated system. Fewer grid points and shorter computational times can be realized by using UPWIND schemes for the convection term (Strube, 1997, Du Chateau and Zachmann, 1989 and Shih, 1984). These are either higher order backward FDM or more complex schemes (Leonard, 1979).

FD methods are point approximations, because they focus on discrete points. In contrast, finite element methods focus on the concentration profile inside one grid element. As an example of these segment methods, orthogonal collocation on finite elements (OCFE) is briefly discussed below.

The collocation method is based on the assumption that the solution of the PDE system can be approximated by polynomials of order n. In this method the whole space domain (the column), including the boundary conditions, is approximated by one polynomial using $n + 2$ collocation points. Polynomial coefficients are determined by the condition that the differential equation must be satisfied at the collocation points. This approach allows the spatial derivatives to be described by the known derivatives of the polynomials and transforms the PDE into an ODE system.

Mathematical methods exist that guarantee an „optimal" placement of the collocation points. In orthogonal collocation (OC), the collocation points are equal to the zero points of the orthogonal polynomials.

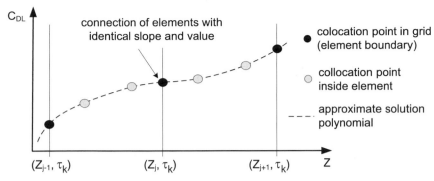

Fig. 6.8 Representation of the numerical solution by the OCFE method for a second-order polynomial.

The OCFE method (Villadsen and Stewart, 1967) is derived straightforwardly by dividing the column into n_e elements (Fig. 6.8). For each element an own orthogonal solution polynomial is used, leading to an ODE system for the column with $(n+2)n_e$ equations. The element boundaries are connected by setting equal values and slopes for the adjacent polynomials at the boundary points, which guarantees the continuity of the concentration profile.

Element methods proved to be a good choice for various column models (Guiochon et al. 1994b, Seidel-Morgenstern, 1995, Kaczmarski et al., 1997, Lu and Ching 1997, and Berninger et al., 1991). Comparisons show that, for the type of models discussed here, the finite element methods are to be preferred to difference methods, especially if high accuracy is needed (Dünnebier, 2000, Finlayson, 1980 and Guiochon and Lin, 2003).

6.5
Parameter Determination

After a suitable model for process simulation has been chosen the model parameters have to be determined.

6.5.1
Parameter Classes for Chromatographic Separations

6.5.1.1 Design Parameters
The models presented in Section 6.2 contain a set of independent and dependent parameters. Designing a chromatographic process starts with the selection and specification of the chromatographic system (Chapter 4) and, subsequently, the kind of the chromatographic process as well as its hardware (Chapter 5). These are characterized by design parameters.

- Stationary phase (particle diameter d_p)
- Mobile phase (composition, gradient elution)
- Length (L_c) and diameter (d_c) of the chromatographic bed
- Maximum allowable pressure drop (Δp)
- Temperature
- Process specific design parameters (e.g. number of columns per section in SMB processes)

Design parameters specify the chromatographic plant. Geometrical design parameters are objects of optimization if no already existing plant is used for the separation.

6.5.1.2 Operating Parameters

The second class consists of the operating parameters for a given plant, which can be changed during plant operation:

- Flow rate ($\dot{V}_{feed}$)
- Feed concentration (c_{feed})
- Amount of feed (V_{inj}, m_{inj})
- Additional degrees of freedom are, for example, switch times (fraction collection) in batch chromatography or flow rates in each section of the SMB process

Like geometrical design parameters, operating parameters are part of the degrees of freedom for optimizing chromatographic separations. Their impact is discussed in Chapter 7.

6.5.1.3 Model Parameters

System inherent physical and chemical parameters that specify the chromatographic system within the column as well as the plant operation make up the third set of parameters. These model parameters are not known *a priori*.

- Plant parameters (describing the residence time distribution of the plant peripheral)
- Packing parameters (void fraction, porosity and pressure drop)
- Axial dispersion coefficient
- Equilibrium isotherm
- Mass transfer coefficient

The above list refers to the transport dispersive model and, if other models are selected, the model equations and hence the number of parameters are expanded or reduced.

Physical properties and especially the isotherms depend on temperature as well as eluent composition. Feed and eluent composition influence the viscosity and therefore the fluid dynamics. However, these effects have already to be taken into account when selecting the chromatographic system (Chapter 4). The operating temperature for preparative processes is commonly selected to be to close room temperature for cost and stability reasons. Consequently, temperature and eluent composition are fixed parameters that are not explicitly considered during process design.

The time and money invested in model parameter determination has to be in balance with the aim of a certain chromatographic separation. Very often, only sample products are needed for further studies with the pure substances and, to evaluate a separation process, only a few milligrams of the feed mixture are available. Additionally, time is an important factor and „quick and dirty"/short-cut methods will be applied to find conservative operating parameter values for a safe separation. A quite different situation is the design of commercial large-scale production plants. Here process economics come to the fore. Therefore, precise parameters are necessary for process optimization. As some parameters may change during operation, repeated determination may be necessary to reestablish optimal production.

Research has a third focus. Here results have to be as accurate and reliable as possible. Consequently, more time and money are invested to acquire the data. Research aims, for instance, to better understand the process and to improve methods for process design, optimization and control. This cannot be done experimentally only. Therefore, verified models are needed to perform the investigations, with the aid of process simulation as virtual plant experiments.

Against this background, both precise and straightforward short-cut methods for model parameter determination are presented below. Based on the individual task one has to decide which procedures are most efficient. In this context it must be mentioned that the adsorption equilibrium has the most influence on the profile and, therefore, the isotherm should be determined with the greatest care.

6.5.2
Determination of Model Parameters

The method of consistent parameter determination is depicted in Fig. 6.9. It was first published by Altenhöner et al. (1997) for the transport dispersive model. The basic idea is to start with the simplest parameters and to use them subsequently to determine more complex ones. The procedure is structured into the following steps:

1. Determine extra column effects such as dead volume and the backmixing of the plant
2. Detector calibration
3. Determine void fraction, porosity and axial dispersion of the packed bed as well as pressure drop parameters.
4. Determine the adsorption equilibrium for the pure component and mixtures
5. Determine transport coefficients for the (adsorbable) solutes

The experimental techniques are mainly based on the injection of small (peaks) or large pulses (breakthrough curves) and analysis of the obtained chromatogram. Although the presented approach tries to limit the impact of measurement errors, experimental conditions and, especially, the material (adsorbent, eluent) used should

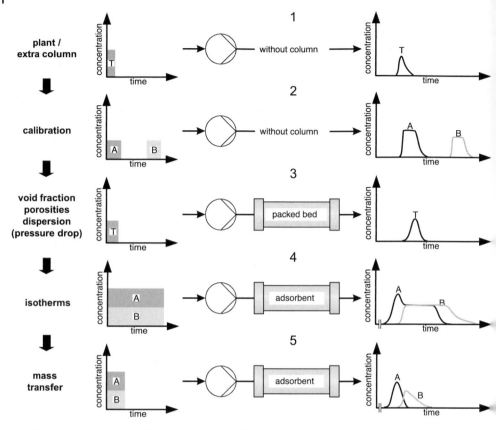

Fig. 6.9 Concept to determine the model parameters (T = tracer; A, B = solutes).

be the same or at least comparable to the envisaged preparative application to ensure reliable results. The effects of plant peripherals should be kept low. Another common source of errors is the calibration of the detector.

Probably the simplest theoretical methods for determining the parameters from the experimental data involve the use of analytical solutions of simple column models and moment analysis, e.g. to determine the dead time or the Henry coefficient. In some cases where less accuracy is acceptable, parameters such as axial dispersion might be estimated by means of empirical correlations from literature.

To determine isotherms, numerical integration and differentiation in combinations with overall mass balances may be used. More advanced methods involve curve fitting of the measured peak. Especially, the accuracy of moment analysis can be increased by „fitting" an analytical equation to the measured values. The most sophisticated and versatile method is to use a parameter estimation tool for curve fitting, which is the method of choice to obtain consistent and accurate data of (almost) all model parameters. Parameter estimation routines are included in some commercially available simulation programs (Section 6.4) or can be linked to one's own simulation software. By solving an optimization problem these tools minimize the difference between the measured data and the simulation results by varying the model parameters. The result is an optimal set of parameter values. Since it is the preferred method, „parameter estimation" is used, if not otherwise noted, in the following to denote curve fitting by means of a simulation program.

Table 6.4 gives an overview of different methods for parameter determination and illustrates the contradictory influences of accuracy and speed. Figure 6.10 describes the work flow for parameter determination. Different methods for the individual parameters are discussed in detail below. First, some general methods are presented, which are helpful as a tool box in evaluating a given chromatogram.

6.5.3
Evaluation of Chromatograms

The section above gives an overview of the experiments necessary to determine the model parameters. This section describes general procedures for evaluating parameter values from experimental data. Model parameters are defined in Chapter 2, while

Table 6.4 Parameter determination methods.

General remarks
Use preparative materials for the adsorbent and, if possible, „semi-preparative" columns.
Always check consistency of the results with those of the initial tests (pressure drop, k').
Use one plant for all measurements and try to keep extra column effects low.
Take care in calibrating the detector.
Check influence of volume flow to determine magnitude of kinetic effects.
With fluid mixtures as eluent check sensitivity to eluent composition.

Method:	High accuracy and experimental effort	Medium accuracy and experimental effort	Initial guess („quick and dirty") and low experimental effort
Plant/ extra column	Experiment + moment analysis + parameter estimation: V_{pipe}, V_{tank} (t_{plant}) if necessary $D_{ax,pipe}$	Experiment + moment analysis: t_{plant}	Experiment + moment analysis: t_{plant}
Packing: void fraction and porosity	Experiment + moment analysis + parameter estimation: ε and ε_t	Experiment + moment analysis: ε_t, set ε	ε_t (from manufacturer), set ε
Packing: D_{ax}	Experiment + parameter estimation	Experiment + moment analysis	Use correlation
Isotherm	Measure by frontal analysis, perturbation	Measure by peak-maximum method	Measure with ECP
Mass transfer: k_{eff}	Experiment (overload injection) + parameter estimation	Experiment + moment analysis (or HETP)	Use correlation

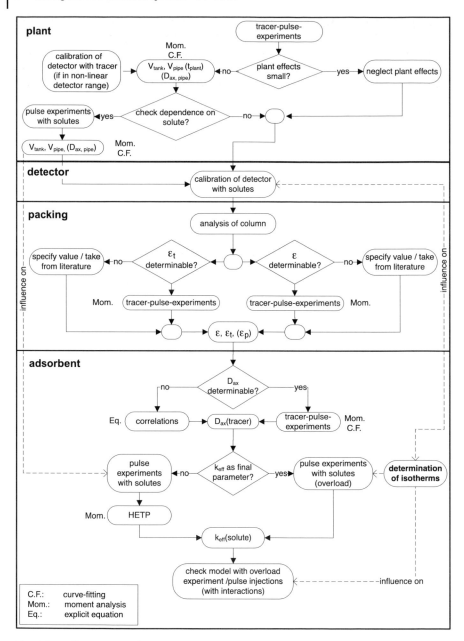

Fig. 6.10 Different ways to determine model parameters.

Section 6.2 shows how they are placed within the model equations. Based on the assumptions for these models it follows that all plant effects as well as axial dispersion, void fraction and mass transfer resistance are independent of the adsorption/desorption within the column. Modeling always results in a virtual image of the real world, i.e. in reality the parameters might be influenced by adsorption, but with a reliable model this is of minor importance.

The following general methods are appropriate for the evaluation of model parameters: Moment analysis and HETP plots, peak fitting and parameter estimation. They extract the information given by measured chromatograms to differing extents, which correspond to the reliability of the calculated values.

6.5.3.1 Moment Analysis and HETP Plot

Moment analysis and HETP plots are described in Chapter 2. Here it will be shown how they can be used in estimating the model parameters. If the detector signal is a linear function of concentration, analysis can be carried out directly with the detector signal without calibration. Since the concentration profile is available as n_p discrete concentration vs. time values, the momentums have to be calculated by numerical integration.

$$\mu_t = \frac{\int_0^\infty t\, c(t)\, dt}{\int_0^\infty c(t)\, dt} \approx \frac{\sum_{j=1}^{n_p} t_j\, c_j \Delta t}{\sum_{j=1}^{n_p} c_j \Delta t} \tag{6.125}$$

$$\sigma_t^2 = \frac{\int_0^\infty (t - \mu_t)^2\, c(t)\, dt}{\int_0^\infty c(t)\, dt} \approx \frac{\sum_{j=1}^{n_p} (t_j - \mu_t)^2\, c_j \Delta t}{\sum_{j=1}^{n_p} c_j \Delta t} \tag{6.126}$$

The signal quality (= low detector noise) and sample rate should be sufficiently high to obtain correct results, otherwise post processing of the data is necessary, such as smoothing, baseline correction, etc. Even then the result of the integration in Eq. 6.126 depends very much on the extension of the baseline, and the obtained value of the second moment can be very inaccurate (Chapter 2.7, Section 6.5.3.3).

To determine the model parameters, a minimum of three injection experiments with tracers plus one for each solute have to be carried out. The evaluation of these experiments is sketched in Fig. 6.11, together with the symbols of the measured first moments. The injected signal for all experiments is represented by a rectangular pulse. The first tracer experiment detects the dead time of the plant while the column is replaced by a zero volume connector. The other experiments are carried out with the column in place, using a tracer that cannot get into the pores (Tracer1) and

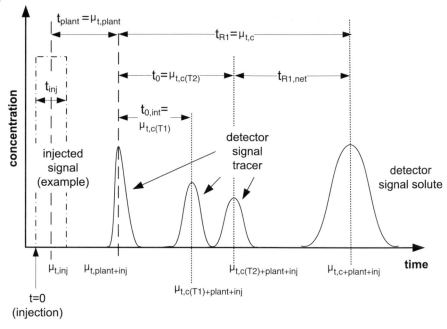

Fig. 6.11 Representation of measured peaks from four different experiments and the characteristic times.

another one that penetrates the pores (Tracer2). These experiments are necessary to determine the dead times $t_{0,\text{int}}$ and t_0. Finally, a solute peak is analyzed, but it has to be noted that meaningful results are only obtained if the concentration is within the linear range of the isotherm. Figure 6.11 gives an overview of the different first moments and the respective times that are evaluated. Second moments are determined analogously.

Contributions to the moments from all parts of the chromatographic plant (Section 6.3.1) are additive in linear chromatography (Ashley and Reilley, 1965). Assuming the hypothetical injection of a Dirac pulse „prior" to the injector (the injector „transforms" this into the rectangular pulse of Fig. 6.11) the model parameters can be extracted from the measurements. For plant peripherals, equations for the resulting moments can be found in standard chemical engineering textbooks (e.g. Levenspiel, 1999 and Baerns et al., 1999).

Injector (Eq. 6.114): $\quad \mu_{t,\text{inj}} = \dfrac{t_{\text{inj}}}{2}; \quad \sigma^2_{t,\text{inj}} = \dfrac{t_{\text{inj}}^2}{12}; \quad t_{\text{inj}} = \dfrac{V_{\text{inj}}}{\dot{V}}$ $\qquad$ (6.127)

Pipe (Eq. 6.117): $\quad \mu_{t,\text{pipe}} = \dfrac{V_{\text{pipe}}}{\dot{V}}; \quad \sigma^2_{t,\text{pipe}} = 2\left(\dfrac{V_{\text{pipe}}}{\dot{V}}\right)^2 \dfrac{D_{ax,\text{pipe}}}{u_{0,\text{pipe}} L_{\text{pipe}}}$ $\qquad$ (6.128)

Detector (Eq. 6.121): $\quad \mu_{t,\text{tank}} = \dfrac{V_{\text{tank}}}{\dot{V}}; \quad \sigma^2_{t,\text{tank}} = \left(\dfrac{V_{\text{tank}}}{\dot{V}}\right)^2$ $\qquad$ (6.129)

Plant:
$$\mu_{t,\text{plant}} = \mu_{t,\text{pipe}} + \mu_{t,\text{tank}}$$
$$\sigma^2_{t,\text{plant}} = \sigma^2_{t,\text{pipe}} + \sigma^2_{t,\text{tank}} \qquad (6.130)$$

As the parameters for the injector in Eq. 6.127 are known, the plant characteristics can be calculated from the experimentally determined values of $\mu_{t,\text{plant+inj}}$ and $\sigma_{t,\text{plant+inj}}$.

$$t_{\text{plant}} = \mu_{t,\text{plant}} = \mu_{t,\text{plant+inj}} - \mu_{t,\text{inj}} = \mu_{t,\text{plant+inj}} - \frac{t_{\text{inj}}}{2}$$
$$\sigma^2_{t,\text{plant}} = \sigma^2_{t,\text{plant+inj}} - \sigma^2_{t,\text{inj}} = \sigma^2_{t,\text{plant+inj}} - \frac{t^2_{\text{inj}}}{12} \qquad (6.131)$$

If axial dispersion in the piping can be neglected, Eqs. 6.127–6.131 allow all plant parameters to be determined from one experiment only. Plant characterization is discussed in more detail later (Section 6.5.5).

Experimental determination of the first and second moment for the column is straightforward, using a pulse injection with and without the column and the parameters previously determined.

$$\mu_{t,c} = \mu_{t,c+\text{plant+inj}} - \mu_{t,\text{plant+inj}} = \mu_{t,c+\text{plant+inj}} - \mu_{t,\text{plant}} - \frac{t_{\text{inj}}}{2}$$
$$\sigma^2_{t,c} = \sigma^2_{t,c \mid \text{plant} \mid \text{inj}} - \sigma^2_{t,\text{plant} \mid \text{inj}} = \sigma^2_{t,c \mid \text{plant} \mid \text{inj}} - \sigma^2_{t,\text{plant}} - \frac{t^2_{\text{inj}}}{12} \qquad (6.132)$$

Note that Eq. 6.132 is valid for tracer and solute injections.

In the following the calculated values for the moments are related to the model parameters (e.g. ε, ε_t, D_{ax}, H and k_{eff}) of the column. For this purpose the resulting theoretical solutions for the first and second moment derived for different column models are presented. Besides the sole purpose of parameter determination, this approach allows us to demonstrate one justification of „lumped parameter" models.

Only systems in the linear range of the isotherm are discussed, more complex equations have been published by Kucera (1965) and Kubin (1965) and Ma et al. (1996).

For a general rate model (GRM) including axial dispersion (Eq. 6.81), mass transfer (Eq. 6.81), (apparent) pore diffusion (Eqs. 6.83, 6.85 and 6.86) and linear adsorption kinetics (Eqs. 6.34 and 6.35), Kucera (1965) derived the moments by Laplace transformation, assuming the injection of an ideal Dirac pulse. If axial dispersion is not too strong (Bo $\gg$ 4), the equations for the first and second moment can be simplified to (Ma et al., 1996):

$$\mu_{t,c,\text{GRM}} = t_{R,\text{lin}} = \frac{L_c}{u_{\text{int}}} \left[1 + \frac{1-\varepsilon}{\varepsilon}(\varepsilon_p + (1-\varepsilon_p)H) \right] = \frac{L_c}{u_{\text{int}}}(1 + \tilde{k}') \qquad (6.133)$$

$$\sigma_{t,c,\text{GRM}}^2 = 2 \left(\frac{L_c}{u_{\text{int}}}\right)^2 \left(\frac{D_{\text{ax}}}{u_{\text{int}} L_c}(1 + \tilde{k}')^2 + \tilde{k}' \frac{\varepsilon}{(1-\varepsilon)} \left(\frac{u_{\text{int}} r_p^2}{15 \, \varepsilon_p D_{\text{app,pore}} L_c} + \frac{r_p u_{\text{int}}}{3 k_{\text{film}} L_c}\right)\right)$$

$$+ 2 \left(\frac{L_c}{u_{\text{int}}}\right)^2 \tilde{k}'^2 \frac{\varepsilon}{(1-\varepsilon)} \cdot \left(\frac{\dfrac{1-\varepsilon_p}{\varepsilon_p} H_i}{1 + \left(\dfrac{1-\varepsilon_p}{\varepsilon_p}\right) H_i}\right)^2 \cdot \frac{u_{\text{int}}}{(1-\varepsilon_p) q_{\text{sat}} k_{\text{ads}} L_c} \qquad (6.134)$$

Note that the first moment in this case is independent of any mass transfer and dispersion coefficients and thus equivalent to the retention time obtained for the ideal model (Eq. 6.50). The factor $\widetilde{k}'$ is a modified retention factor that is zero for non-pore-penetrating tracers. It is connected to k' (Eq. 2.1) by:

$$\tilde{k}' = \frac{1-\varepsilon}{\varepsilon} \left[\varepsilon_p + (1-\varepsilon_p)H\right]$$

$$= \frac{\varepsilon_t}{\varepsilon}(1+k') - 1 \qquad (6.135)$$

Whereas, in principle, two simple experiments with tracers and one for each solute (explained below) allow the determination of ε, ε_t and H from the experimentally determined $\mu_{t,c}$, the other model parameters are not so simply extracted from the second moment. Dispersion (D_{ax}), liquid film mass transfer (k_{film}), diffusion inside the particles ($D_{\text{app,pore}}$) as well as adsorption kinetics (k_{ads}) contribute to the overall band broadening described by $\sigma_{t,c}$.

Therefore, an independent determination of these four parameters is not possible from Eq. 6.134 only. In principle, additional equations could be obtained from higher moments (Kucera, 1965 and Kubin, 1965). However, as the effect of detector noise etc. (Chapter 2.7) on the accuracy of the moment value strongly increases the higher the order of the moment, a meaningful measurement of the third, fourth and fifth moment is practically impossible. Equation 6.134 is thus not directly suited for parameter determination, but establishes the important connection between the model parameters of transport and their influence on band broadening.

The corresponding equations for the first and second moments for the transport dispersive model (TDM) (Lapidus and Amundson, 1952 and van Deemter et al., 1956) are:

$$\mu_{t,c,\text{TDM}} = t_{R,\text{lin}} = \frac{L_c}{u_{\text{int}}}(1 + \tilde{k}') \qquad (6.136)$$

$$\sigma_{t,c,\text{TDM}}^2 = 2 \left(\frac{L_c}{u_{\text{int}}}\right)^2 \left[\tilde{k}'^2 \frac{\varepsilon}{(1-\varepsilon)} \frac{r_p u_{\text{int}}}{3 k_{\text{eff}} L_c} + \frac{D_{\text{ax}}}{u_{\text{int}} L_c}(1 + \tilde{k}')^2\right] \qquad (6.137)$$

The results for the first moment (Eq. 6.136) are identical to Eq. 6.133 for the GRM. Comparison of Eqs. 6.137 and 6.134 shows that both models describe similar band broadening if k_{eff} is given by:

$$\frac{1}{k_{\text{eff}}} = \frac{r_p}{5 \varepsilon_p D_{\text{app,pore}}} + \frac{1}{k_{\text{film}}} + \left[\frac{\dfrac{1-\varepsilon_p}{\varepsilon_p} H}{1 + \left(\dfrac{1-\varepsilon_p}{\varepsilon_p}\right) H}\right]^2 \frac{3}{r_p(1-\varepsilon_p) q_{\text{sat}} k_{\text{ads}}} \qquad (6.138)$$

Equation 6.138 defines a formal connection between the effective mass transport and the film transport, the pore diffusion and the adsorption rate coefficient. It illustrates that k_{eff} is a „lumped parameter", composed of several transport effects connected in series. This also gives reasons for the use lumped rate models as it proves that the impact of the lumped parameters on the most important peak characteristics, retention time and peak width, is identical to the effect described by general rate model parameters in linearized chromatography.

The second moment (Eq. 6.137) still consists of two transport parameters, k_{eff} and D_{ax}, but with additional assumptions about their dependence on the flow rate, Eq. 6.137 can be used to determine both (see below).

For completeness, a similar deduction can be drawn for the equilibrium dispersive model (EDM) (Section 6.2.4.1). The first moment (Eq. 6.63) is again identical to Eq. 6.133 and the second moment of the EDM is given by Eq. 6.64. By comparing Eqs. 6.64 and 6.137, the following relationship between the apparent dispersion coefficient (D_{app}), k_{eff} and D_{ax} can be derived:

$$D_{app} = D_{ax} + \left(\frac{\tilde{k}'}{1 + \tilde{k}'} \right)^2 \frac{\varepsilon}{(1 - \varepsilon)} \frac{r_p \, u_{int}^2}{3 \, k_{eff}} \tag{6.139}$$

Equation 6.139 illustrates the meaning of the apparent dispersion coefficient as a lumped parameter.

In addition to the interpretation of lumped parameters, Eqs. 6.133–6.139 may be used to calculate the model parameters from experimentally determined moments $\mu_{t,c}$ and $\sigma_{t,c}$. This method is described here for the TD model but can also be applied to the ED model.

Based on tracer and solute experiments (Fig. 6.11) the model parameters are determined step-by-step, beginning with the void fraction, total porosity and the axial dispersion coefficient (Section 6.5.6). All experimental data must be corrected for plant effects (Eq. 6.132).

This section deals with moment analysis only and refers to the specific sections where the determination of each parameter is extensively discussed. Table 6.5 summarizes the different steps from the point of view of moment analysis. Owing to the limitations mentioned above, the derived parameters often posses approximate character. The experimental procedure is listed in chronological order, where each step uses data from the previous ones.

Porosities are obtained from two experiments with the different tracers, using Eqs. 6.135 and 6.136 (Section 6.5.6.1). If D_{ax} is assumed to be a linear function of the fluid velocity (Section 6.5.6.2), its value is derived from experiments with different volume flows. Subsequently, the Henry (Section 6.5.7.2) and mass transfer coefficients (linear isotherm range, Section 6.5.8) are calculated based on experiments with the solutes. In the framework of Fig. 6.9, k_{eff} is used as the final fitting parameter and may differ from the value determined for the linear case (Antos et al., 2003, Section 6.5.8).

Table 6.5 also includes the application of the HETP–u_{int} plot as an alternative to direct analysis of the second moment. Transport parameters are derived from this plot by the following method.

Table 6.5 Determination of parameters of the transport dispersive model based on moment analysis only.

Experiment	Measurement	Equation	Unknown
Defined by experimentalist	t_{inj} (V_{inj})	6.127 (6.115)	$\mu_{t,inj}$, $\sigma_{t,inj}$
Plant effects	$\mu_{t,plant+inj}$, $\sigma_{t,plant+inj}$	6.131	$\mu_{t,plant}$ (t_{plant}), $\sigma_{t,plant}$
Tracer 1 (does not enter pores)	$\mu_{t,c(T1)+plant+inj}$	6.132 (6.136, 6.135, 2.9)	ε
Tracer 2 (does enter pores)	$\mu_{t,c(T2)+plant+inj}$	6.132 (6.136, 6.135, 2.9)	ε_t
Tracer 1 at different flow rates	$\sigma_{t,c(T1)+plant+inj} = f(u_{int})$; or HETP–$u_{int}$ plot	6.132 (6.137, 6.135, 2.9)	$D_{ax} = f(u_{int})$ (with $k_{eff} = \infty$)
Solutes (linear isotherm)	$\mu_{t,c+plant+inj}$	6.132, 6.137 (6.135, 2.9)	H
Solutes (linear isotherm); at different flow rates	$\sigma_{t,c+plant+inj}$; or HETP–u_{int} plot	6.132, 6.137 (6.135, 2.9)	k_{eff}

Using the definition of the number of stages N (Eq. 2.29) and the equations for the first and second moment (Eqs. 6.136 and 6.137) one obtains for the transport dispersive model:

$$\frac{1}{N_{TDM}} = \left(\frac{\sigma_{t,c,TDM}}{\mu_{t,c,TDM}}\right)^2 = 2\left(\frac{\tilde{k}'}{1+\tilde{k}'}\right)^2 \frac{\varepsilon}{(1-\varepsilon)} \frac{r_p\, u_{int}}{3\, k_{eff}\, L_c} + 2\frac{D_{ax}}{u_{int}\, L_c} \qquad (6.140)$$

The resulting equation for the equilibrium dispersive model is given by Eq. 6.65.

Assuming D_{ax} is proportional to u_{int}, and k_{eff} is independent of the velocity, as explained in Sections 6.5.6.2 and 6.5.8 respectively, HETP can be expressed as a function of the model parameters using Eqs. 2.19 and 6.140:

$$
\begin{aligned}
HETP_{TDM} = \frac{L_c}{N_{TDM}} &= 2\frac{D_{ax}}{u_{int}} + 2\left(\frac{\tilde{k}'}{1+\tilde{k}'}\right)^2 \frac{\varepsilon}{(1-\varepsilon)} \frac{r_p\, u_{int}}{3\, k_{eff}} \\
&= A + \frac{C}{u_{int}} + \qquad\quad B\, u_{int} \\
&\approx \quad A \quad + \qquad\quad B\, u_{int}
\end{aligned}
\qquad (6.141)
$$

The second line in Eq. 6.141 represents the van Deemter equation (Eq. 2.31) and shows how the model parameters A, B and C are related to D_{ax} and k_{eff}. For preparative chromatography the C term can be neglected (Section 6.5.6.2), which results in a direct relation between A and D_{ax} as well as B and k_{eff}.

Using these simplifications, Eq. 6.141 can be used as an alternative approximation method for D_{ax} and k_{eff}, if the van Deemter parameters A and B are determined as shown in Fig. 6.12.

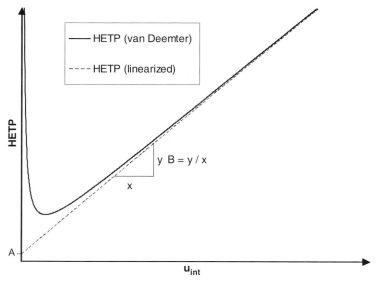

Fig. 6.12 HETP curve according to the van Deemter approximation and in simple linearized form.

As pointed out in Tab. 6.5, the HETP plot is needed for tracers and solutes. With tracers the theoretical HETP plot would be a straight horizontal line ($B = 0$, $k_{eff} = \infty$) that determines A and, thereby, D_{ax}/u_{int}. To fit within the framework presented here, for every solute only the slope B is determined and k_{eff} is subsequently evaluated. Basically, A can also be derived from these plots, but in practice different axis intercepts A might be observed for solutes and tracers. Because of the limited precision of the second moment and the linearized HETP plot, the accuracy of the derived parameters is limited and the values for A and B (D_{ax}/u_{int} and k_{eff}) should be taken as „initial guesses" only (Chapter 2.7, Section 6.5.3.3). Thus, the differences in the observed axis intercepts can be safely ignored and are often negligible. For further discussions of HETP plots see, for example, Van Deemter et al. (1956), Grushka et al. (1975), Weber and Carr (1989) or recent publications on NMR measurements by, for example, Tallarek et al. (1998).

In summary, it is not advised to determine all parameters from moment analysis or HETP alone, especially if one is interested in a consistent set of reliable model parameters. In this case simulation-based parameter estimation should be used.

6.5.3.2 Parameter Estimation

Moment analysis offers only two global parameters to characterize the peak while the exact peak shape is not taken into account, which makes this method more sensitive to signal distortions. By „fitting" of either analytical equations (Section 6.5.3.3) or simulation results to the peak shape this drawback may be overcome. It also allows an easy comparison between the calculated peak and the measured concentration profile.

As suitable analytical solutions are not available for most of the column models, simulation-based parameter estimation using simulation software is recommended. This method is very versatile in terms of the number and complexity of models. One example for an estimation task is the gEST tool included in the gPROMS program package (PSEnterprise, UK). An additional advantage of the simulation based approach is the consistency of the obtained data, if the same models and simulation tools are used for subsequent process analysis and optimization.

As „parameter estimation" is the recommended method the results presented in this book always refer to simulation-based curve fitting.

The fitting procedure results in a set of model parameters, minimizing the difference between theoretical and measured concentration profile. A discussion of statistical based objective functions and optimization procedures is beyond the scope of this book. For further information see, for example, Lapidus (1962), Barns (1994), Korns (2000) and Press et al. (2002, http://www.nr.com/).

Objective functions O often contain some kind of least-squares method. In this case O is a function of the absolute or relative squared error of n_p measured concentration values, c_{exp}, and the theoretical values, c_{theo}.

$$O = f \left[\sum_{j=1}^{n_p} (c_{exp,j} - c_{theo,j})^2 \right] \text{ or } f \left[\sum_{j=1}^{n_p} \left(\frac{c_{exp,j} - c_{theo,j}}{c_{exp,j}} \right)^2 \right] \qquad (6.142)$$

In general, objective functions can consider more than one set of experimental data.

To solve the arising optimization problem, the solution (in this case the shape of the simulated chromatogram) must be sensitive to the variable in question, otherwise no meaningful parameter set can be obtained. It is not recommended to determine more than two or maximal three parameters simultaneously, because this increases the chances of finding more than one set of parameters that fulfill the optimization criteria. The concept in Fig. 6.9 follows this consideration as not all column parameters are obtained from one measured chromatogram. Instead, estimation is performed step by step, using different experimental data and increasing the complexity of the model with each step.

Another consideration is the quality of the measurement, as all errors (for example detector noise) affect the accuracy of the determined parameters. Notably, in contrast to these statistical errors, systematic measurement errors cannot be minimized by repeated measurements.

In addition, it has to be remembered that initial guesses of the parameters must be provided by the users, either coming from initial tests, experience or short-cut calculations. If the equations are nonlinear, the initial guesses influence the result of the optimization and it is recommended to test the sensitivity of the obtained parameters on the initial guesses.

In any case, the simulated profiles must be compared with the measured data to check the validity of the determined data as well as the assumed model. Statistical methods to quantify the goodness of the fit are given, for example, in Lapidus (1962), Barns (1994), Korns (2000) and Press et al. (2002).

6.5.3.3 Peak Fitting Functions

A simplified approach to peak fitting is to use a suitable analytical function to approximate the measured peak. As in the previous section, the function parameters are adjusted to obtain an optimal fit to the experimental data. In a next step the adjusted function (or its parameters) is used to calculate the first and second moment. This procedure may, for example, help to overcome the inaccuracy of moment analysis in case of asymmetric peaks (Chapter 2.7). It also offers the benefit that standard software such as spread sheets can be used instead of special parameter estimation systems.

The simplest method is the representation of an ideal chromatographic peak by a Gaussian function according to Eq. 6.143:

$$c_g(x = L_c, t) = \frac{m_{inj}}{\dot{V} t_g} \frac{t_g}{\sigma_g \sqrt{2\pi}} \exp\left\{-\frac{t_g^2 \left(\frac{t}{t_g} - 1\right)^2}{2\sigma_g^2}\right\} \qquad (6.143)$$

The retention time $t_{R,lin}$ and the second moment for the Gaussian profile (Eq. 6.61) have been replaced by variables indexed with „g". These parameters t_g and σ_g must be optimized by curve fitting. Equation 6.143 is only suitable for symmetric peaks. Analytical solutions of, for example, the transport dispersive model (which describes asymmetric band broadening only for a very low number of stages) are not suited to describing the asymmetry often encountered in practical chromatograms. Thus, many different, mostly empirical functions have been developed for peak modeling. A recent extensive review by Marco and Bombi (2001) lists over 90 of them.

One most frequently used equation is called the Exponential-Modified-Gauss (EMG) function (Jeansonne and Foley, 1991 and 1992; Foley and Dorsey 1983 and 1984). It is defined as a Gaussian peak (Eq. 6.143) superimposed by an exponential decay function h_{exp}

$$h_{exp}(t) = \begin{cases} \dfrac{1}{\tau_{EMG}} \exp\left\{-\dfrac{t}{\tau_{EMG}}\right\} & (t \geq 0) \\ \\ 0 & (t < 0) \end{cases} \qquad (6.144)$$

where τ_{EMG} is used to specify the peak skew in the EMG function. The exponential decay function Eq. 6.144 is equivalent to the residence time distribution of a stirred tank. Notably, the use of any empirical function to describe peak tailing does not provide any additional physical information about the nature of this phenomenon. Thus, it is recommended to apply these equations to, for example, increase the accuracy in moment analysis in the case of linearized chromatography. In fact Eq. 2.30 is an approximate correlation derived from the EMG model (Foley and Dorsey, 1983).

Another common application of fitting functions is in deconvoluting partially resolved peaks (Marco and Bombi, 2001).

The superposition (convolution) of Eqs. 6.143 and 6.144 gives the EMG function:

$$c_{EMG}(t) = \int_0^t h_{exp}(t - t') c_g(t') \, dt' \qquad (6.145)$$

Solution of Eq. 6.145 employs the error function (erf)

$$\text{erf}(t) = \frac{2}{\sqrt{\pi}} \int\limits_0^t \exp\{-\gamma^2\}\, d\gamma \qquad (6.146)$$

and is given by

$$
c_{\text{EMG}}(t) = \frac{m_{\text{inj}}}{\dot{V}\, t_g} \frac{t_g}{\tau_{\text{EMG}}} \exp\left\{ \frac{1}{2}\left(\frac{\sigma_g}{\tau_{\text{EMG}}}\right)^2 - \frac{t - t_g}{\tau_{\text{EMG}}} \right\}
$$
$$
\cdot \left[1 + \text{erf}\left\{ \frac{1}{\sqrt{2}}\left(\frac{t - t_g}{\sigma_g} - \frac{\sigma_g}{\tau_{\text{EMG}}}\right) \right\} \right] \qquad (6.147)
$$

This function is easily implemented in standard software tools. When erf is not available, polynomial approximations exist (Foley and Dorsey, 1984).

The moments of this function are listed below:

$$\text{Area} = \frac{m_{\text{inj}}}{\dot{V}} \qquad (6.148)$$

$$\mu_t(x = L_c) = t_g + \tau_{\text{EMG}} \qquad (6.149)$$

$$\sigma_t^2(x = L_c) = \sigma_g^2 + \tau_{\text{EMG}}^2 \qquad (6.150)$$

Due to the mass balance the area Eq. 6.148 is identical to that of the Gaussian peak. Parameters t_g, σ_g and τ_{EMG} are determined by curve fitting and thus the values of the moments are directly available.

To illustrate the peak shape resulting from the EMG function, two peaks are shown in Fig. 6.13. One plot represents a Gaussian peak with $N_g = 500$ stages while the other stands for the EMG function with $\tau_{\text{EMG}}/t_g = 0.05$, resulting in $N_{\text{EMG}} = 245$. This EMG-peak is calculated for the same t_g and σ_g of the Gaussian peak and both have the same area. The concentrations are normalized by the maximum concentration of the Gaussian peak (Eq. 6.143) and the time axis is scaled with t_g. For the EMG function considerable peak tailing occurs and the position of the maximum differs from that of the Gaussian peak.

Finally, Fig. 6.14 compares different methods of calculating the number of stages N. The squares represent the data calculated in Chapter 2.7 (Tab. 2.2). Due to the numerical integration (Eq. 6.126) the second moment decreases for increasing integration boundaries t_2 (Fig. 2.22, Tab. 2.2). The value calculated for $t_2 = 12.2$ min is taken as a reference for comparison with the other procedures.

Additionally, Fig. 6.14 presents the number of stages calculated by the peak width at a certain peak height (Eq. 2.25, Eq. 2.30) as well as peak fitting by the Gaussian function (Eq. 6.143) and the EMG function (Eq. 6.147). All functions are fitted to the peak given in Fig. 2.22 for the different integration intervals. In contrast to the stage number determined from moment analysis, all other methods provide stage numbers nearly independent of the integration boundary but of quite different values.

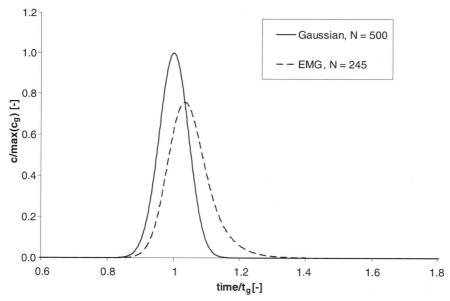

Fig. 6.13 Comparison of peak profiles resulting from the Gaussian equation ($N_g = 500$) and the derived EMG function ($N_g = 500$, $\tau_{EMG}/t_g = 0.05$, $N_{EMG} = 245$) (concentrations normalized by the maximum concentration of the Gaussian peak).

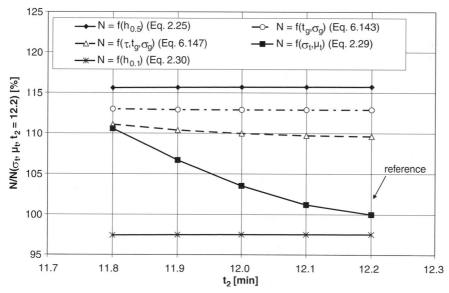

Fig. 6.14 Comparison of different methods to determine the stage number (for further data see Fig. 2.22 and Tab. 2.2).

This confirms that the same calculation method should always be used for comparing different chromatographic systems. In view of the limited accuracy of any of these methods mentioned in Fig. 6.14, a more or less robust, but always the same, procedure should be used. In most cases this will be a method implemented into a peak detection software or one that can be evaluated easily, e.g. taking the peak width at a certain peak height (Eq.2.25, Eq. 2.30).

6.5.4
Detector Calibration

Detector calibration is of basic importance, as it directly influences the accuracy of the measured data. Another major practical problem is the availability of pure components.

When adsorption isotherms are determined by breakthrough experiments these data are also taken for the detector calibration. If this is not possible, breakthrough curves should be injected into the plant without the column, which consumes considerably less sample. Since in this case the plateau concentration is the known injected feed concentration, the plot of c_{feed} vs. the detector signal u is the sought-after relationship and a calibration function is fitted to these values.

If the form of the calibration function

$$c = f_{cal}(u) \tag{6.151}$$

is known *a priori*, a pulse experiment may also be used to obtain the calibration curve. For a known amount of injected sample

$$m_{inj} = V_{inj}\, c_{feed} = \dot{V}\, t_{inj}\, c_{feed} \tag{6.152}$$

The mass balance (Eq. 6.62) for the detected peak is:

$$m_{inj} = \dot{V} \int_0^\infty c(L_c, t)\,dt = \dot{V} \int_0^\infty f\,[u(t)]\,dt \tag{6.153}$$

In the special case of a linear relationship,

$$c = F_{cal}\, u \tag{6.154}$$

Equations 6.152 and 6.153 together with the numerical integration of the detector signal allow the determination of the calibration factor F_{cal}:

$$F_{cal} = \frac{V_{inj}\, c_{feed}}{\dot{V} \displaystyle\int_0^\infty u(t)\,dt} = \frac{t_{inj}\, c_{feed}}{\displaystyle\int_0^\infty u(t)\,dt} \tag{6.155}$$

For nonlinear detectors the calibration curve may be determined by mathematical curve fitting. For complex calibration functions with more than one unknown parameter, additional information (e.g. about the physical theory of the measuring method) or pulse experiments with different concentrations are necessary. In any case, it is advised to check the low and high concentration range by separate experiments with different feed concentrations. There should be no detector overflow at maximum concentration.

6.5.5
Plant Parameters

Following the concept of Fig. 6.9, the parameters describing the fluid dynamic and residence time distribution of the plant peripherals are determined by means of pulse experiments. For this purpose, a small amount of tracer is injected into the plant without the column and the output concentration is measured.

The overall retention time of the plant $\mu_{t,\text{plant+inj}}$ can be obtained by moment analysis of a chromatogram. Using the known volume of the injected sample V_{inj}, the dead volume V_{plant} as well as the dead time t_{plant} are evaluated using Eq. 6.131:

$$V_{\text{plant}} = \dot{V}(\mu_{t,\text{plant+inj}} - u_{t,\text{inj}}) = \dot{V}\,\mu_{t,\text{plant+inj}} - \frac{V_{\text{inj}}}{2}$$
$$= \dot{V}\,\mu_{t,\text{plant}} \tag{6.156}$$

$$t_{\text{plant}} = \mu_{t,\text{plant}} = \frac{V_{\text{plant}}}{\dot{V}} \tag{6.157}$$

Using a parameter estimation tool is another possible way to determine the parameters for the plant. Volumes of the pipe V_{pipe}, detector system V_{pipe} and if necessary the axial dispersion coefficient $D_{\text{ax,pipe}}$ are estimated based on the model equations in Sections 6.3.2.2 and 6.3.2.3 (Eqs. 6.117 and 6.121).

An initial estimate for the tank volume can be obtained from moment analysis by Eqs. 6.129 and 6.131, if axial dispersion in the pipe is neglected and $\sigma_{t,\text{plant+inj}}$ is measured:

$$\sigma_{t,\text{tank}}^2 \approx \sigma_{t,\text{plant}}^2 = \sigma_{t,\text{plant+inj}}^2 - \sigma_{t,\text{inj}}^2$$

$$\Leftrightarrow \left(\frac{V_{\text{tank}}}{\dot{V}}\right)^2 \approx \sigma_{t,\text{plant+inj}}^2 - \frac{t_{\text{inj}}^2}{12} \tag{6.158}$$

$$\Leftrightarrow V_{\text{tank}} \approx \dot{V}\sqrt{\sigma_{t,\text{plant+inj}}^2 - \frac{t_{\text{inj}}^2}{12}}$$

As Eq. 6.159 is always valid, due to Eq. 6.130 (Eq. 6.123), an initial value for the pipe volume is obtained.

$$V_{\text{plant}} = V_{\text{pipe}} + V_{\text{tank}} \tag{6.159}$$

The dispersion coefficient and the parameter „pipe length" in Eq. 6.119 are mainly adjusted parameters, especially since the actual cross section of the pipe is not known exactly and therefore a representative value has to be chosen. For the axial dispersion coefficient, initial guesses using a very low value (e.g. $10^{-5}\,\mathrm{cm^2\,s^{-1}}$) should be used and the pipe length is set according to Eq. 6.119 and the specified cross section.

In Fig. 6.15 two different models for parameter estimation are used and the resulting simulated concentration profiles are compared with the measurements. In one case ideal plug-flow (Eq. 6.116) and in the other axial dispersive flow (Eq. 6.117) is assumed for the pipe system, while both models use the C.S.T. model (Eq. 6.121) to describe the detector system. Figure 6.15 shows that the second model using axial dispersion provides an excellent fit for this set-up, while the other cannot predict the peak deformation. Because of the asymmetric shape a model without a tank would also be inappropriate.

When comparing the simulated profile with the measured profile, the desired level of accuracy has to be specified by the user and has to take into account the relevance of the plant effects in comparison to the column effects. As a rule the dead time of the plant always has to be determined. The necessity of the other parameters depends on the design of the plant. Especially for analytical-scale plants, careful parameter estimation is recommended. Finally, the influence of volume flow on the plant parameters should be checked. With deviations it has to be remembered that, upon changing the fluid patterns in the piping and detector cells, the finite response time of the detector or slight changes in the temperature might alter the measured signal and, as a result, different parameters are obtained. In practice, deviations in volume flow are often unimportant, especially if the plant parameters are determined at typical flow rates envisaged for later production.

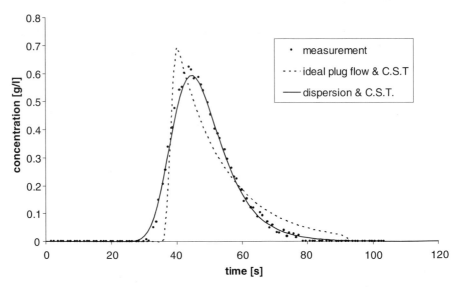

Fig. 6.15 Comparison of experimental and simulated profiles (Feed: *R*-enantiomer EMD53986 at 4 mg ml^{-1}, $\dot{V}$ = 20 ml min^{-1}, preparative scale; for additional data see Appendix B.1).

6.5.6
Determination of Packing Parameters

When the model parameters of the plant are known, the packing parameters i.e. void fraction, pressure drop and dispersion are determined (step 3 in Fig. 6.9).

6.5.6.1 Void Fraction and Porosity of the Packing

To calculate the void fraction (Eq. 2.6) and the (total) porosity (Eq. 2.7) a pulse of non-penetrating tracer (T1) and a non adsorbable tracer (T2) which penetrates the pore system (t_0 marker), respectively, are injected. When the measured profiles are evaluated by means of moment analysis (Section 6.5.3.1, Fig. 6.11) the retention time μ_t has to be corrected with the dead time of the plant (Eq. 6.132). Using the definition of the dead times (Eqs. 6.43 and 6.136, and Eqs. 2.10 and 2.11) one obtains:

$$\varepsilon = t_{0,\text{int}} \frac{\dot{V}}{V_c} \quad \text{with} \quad t_{0,\text{int}} = \mu_{t,c(T1)+\text{plant}+\text{inj}} - t_{\text{plant}} - \frac{t_{\text{inj}}}{2} \tag{6.160}$$

$$\varepsilon_t = t_0 \frac{\dot{V}}{V_c} \quad \text{with} \quad t_0 = \mu_{t,c(T2)+\text{plant}+\text{inj}} - t_{\text{plant}} - \frac{t_{\text{inj}}}{2} \tag{6.161}$$

If no suitable tracers are available, other methods (Chapter 2.2.2) may be used. In practice it may be difficult to determine the void fraction. In this case a value of about $\varepsilon = 0.4$ can be estimated, since the theoretical value for ideal packings of spheres lies between 0.32 and 0.42.

6.5.6.2 Axial Dispersion

The axial dispersion coefficient is determined from the concentration profile of a non-penetrating tracer (T1). A reasonable approximation for its velocity dependence goes back to van Deemter et al. (1956). The axial dispersion coefficient is the sum of the contributions of eddy diffusion and molecular diffusion (Chapter 2.3.4):

$$D_{\text{ax}} = \gamma D_m + \lambda d_p u_{\text{int}} \tag{6.162}$$

where D_m is the molecular diffusion coefficient while γ and λ are the (external) tortuosity and the characterization factor of the packing respectively. For typical values,

$$\gamma \approx 0.7; \qquad \lambda \approx 1;$$
$$D_m \approx 10^{-6} \text{--} 10^{-5} \text{ cm}^2 \text{ s}^{-1}; \quad d_p u_{\text{int}} \approx (10^{-3} \times 0.1) = 10^{-4} \text{ cm}^2 \text{ s}^{-1} \tag{6.163}$$

Clearly, the contribution of molecular diffusion is negligibly small and D_{ax} approximately becomes a linear function of the velocity:

$$D_{\text{ax}} \approx \lambda d_p u_{\text{int}} \tag{6.164}$$

The constant λ should be determined for different velocities and a mean value should be evaluated.

Several methods for determining D_{ax} have been described in the literature. If a suitable tracer is available, chromatograms at different flow rates should be measured. If parameter estimation is performed to obtain the axial dispersion coefficient or the factor λ, the complete model (including the plant peripherals and the column) as well as all parameters determined so far have to be taken into account. Another possibility is moment analysis (Eq. 6.132), which is based on the connection between the axial dispersion coefficient and the second moment (Section 6.5.3.1).

$$\lambda = \text{const.} \approx \frac{D_{ax}}{u_{int}d_p} = \frac{\sigma_{t,c(T1)}^2}{\mu_{t,c(T1)}^2}\frac{L_c}{2d_p} = \frac{\sigma_{t,c(T1)}^2}{t_{0,int}^2}\frac{L_c}{2d_p} \tag{6.165}$$

$$D_{ax} = \frac{\sigma_{t,c(T1)}^2}{t_{0,int}^2}\frac{L_c u_{int}}{2} \tag{6.166}$$

All momentums are corrected by the dead time of the plant and the injection volume (Section 6.5.3.1). Alternatively, λ is taken from Fig. 6.12.

$$\lambda = \frac{A}{2d_p} = \frac{\text{HETP}_{T1}(u_{int} = 0)}{2d_p} \tag{6.167}$$

Note that the graph for a tracer that does not enter the particle pores is a horizontal line.

Chung and Wen (1968) and Wen and Fan (1975) have proposed a dimensionless equation using the dependency of the dispersion coefficient on the (particle) Reynolds number Re (Eq. 6.169) for fixed and expanded beds. It is an empirical correlation based on published experimental data and correlations from other authors that covers a wide range of Re. Owing to two different definitions of the Reynolds number, the actual appearance varies in the literature. Since the particle diameter d_p, is the characteristic value of the packing, Eq. 6.168 based on the (particle) Peclet number Pe (Eq. 6.170) is used here:

$$\text{Pe} = \frac{0.2}{\varepsilon} + \frac{0.011}{\varepsilon}(\varepsilon\text{Re})^{0.48} \Rightarrow (10^{-3} \leq \text{Re} \leq 10^3) \tag{6.168}$$

$$\text{Re} = \frac{u_{int}d_p\rho_l}{\eta_l} = \frac{u_{int}d_p}{\nu_l} \tag{6.169}$$

$$\text{Pe} = \frac{u_{int}d_p}{D_{ax}} \tag{6.170}$$

Notably, the Peclet number differs from the Bodenstein number (or axial Peclet number) defined in Eq. 6.106 by the ratio of particle diameter to column length:

$$\text{Bo} = \frac{u_{int}L_c}{D_{ax}} = \text{Pe}\frac{L_c}{d_p} \tag{6.171}$$

Transforming Eq. 6.168 leads to Eq. 6.172

$$D_{ax} = \frac{u_{int}\,d_p\,\varepsilon}{0.2 + 0.011(\varepsilon\text{Re})^{0.48}} \tag{6.172}$$

which allows a good estimation of the axial dispersion, especially for high quality packings. For these applications, Re has little influence as the value is considerably smaller than one. As an example, a particle Re of 0.01 is calculated from the data of Eq. 6.163 and a typical viscosity of about $0.01\,\mathrm{cm^2\,s^{-1}}$.

6.5.6.3 Pressure Drop

Pressure drop and flow rate are usually linearly related – expressed by Darcy's law (Chapter 3.1.4). If unknown from initial tests, the pressure drop is measured for different volume flows, once with column in the plant and once without the column in the plant, using a zero-volume connector. The difference between the two values yields the pressure drop characteristic Δp_c of the column alone. By plotting Δp_c vs. u_{int} the unknown coefficient k_0 is readily determined from the slope of the curve by rearranging Eq. 3.7.

$$\Delta p_c = \frac{\eta\, L_c}{k_0\, d_p^2}\, \Delta u_{int}$$

$$\Leftrightarrow k_0 = \frac{\Delta u_{int}}{\Delta p_c}\, \frac{\eta\, L_c}{d_p^2}$$

(6.173)

6.5.7
Isotherms

6.5.7.1 Determination of Adsorption Isotherms

The adsorption isotherm has the most influence on the chromatogram. Consequently, single- and multi-component isotherms have to be determined with high accuracy to achieve good agreement between simulation and experiment, using all model parameters measured so far (Fig. 6.9).

The goal is to obtain the unknown parameters for a selected isotherm equation. Special parameters of nearly all types of isotherms are the Henry coefficient as well as the saturation capacities for large concentrations. It is advisable to check the validity of the single-component isotherm equation before determining the component interaction parameters. In general the decision on a certain isotherm equation should be made on the basis of the ability to predict the experimental overloaded concentration profiles rather than fitting the experimental isotherm data. In any case, consistency with the Henry coefficient determined from initial pulse experiments with very low sample amounts must be fulfilled.

This chapter discusses some of the most used methods. For further information, see the reviews by Seidel-Morgenstern (2004) and Seidel-Morgenstern and Nicoud (1996) as well as the monograph by Guiochon et al. (1994b). Table 6.6 gives an overview of the different methods.

These techniques not only differ in terms of accuracy and experimental effort, but are also limited by the chromatographic system or the available equipment. For instance, not all methods are suitable for columns with low efficiency, because any peak deformation will be attributed to the isotherm parameters. The actual set-up and experimental conditions should be as close as possible to the operating conditions for production. However, to save time and solutes, the column dimensions are

Table 6.6 Methods for isotherm determination.

General aspects
Use preparative adsorbents and if possible „semi-preparative" columns.
Use one plant for all measurements and try to keep extra column effects low.
Determine all dead volumes of plant and column.
Control column temperature within 1 °C.
Calibrate the detector carefully.
For multi-component systems: Component specific detectors available?
Favor measurement methods with packed columns.
Check concentration range of the process (for SMB > batch).
Check applicability of measurement method.
Determine single component parameters first.
Check consistency with initial tests, especially Henry coefficient!
Select isotherm equation (careful data analysis).
Try to simulate mixture experiments with single component isotherms.
If necessary determine component interaction.
Check agreement between theoretical and experimental chromatogram.

Static methods

Name	Methods	Isotherm type	Comments
Batch method (closed vessel)	• Immersion of adsorbent in solution • Mass balance	Single and multi-component	• No packed column • No detector calibration necessary (only HPLC analytics) • Tedious • Limited accuracy
Adsorption/desorption method	• Loading and unloading of column • Mass balance	Single and multi-component	• No detector calibration necessary (only HPLC analytics) • Tedious but accurate
Circulation method	• Closed system • Stepwise injection and circulation of components till equilibrium • Mass balance	Single and multi-component	• Lower amount of sample • Accumulation of errors for the injection • Possibility for automation

---

Dynamic methods

Name	Methods	Isotherm type	Comments
Frontal analysis	• Integration of step response (breakthrough-curves) • Numerical integration and mass balance	Single and multi-component	• High accuracy • Easy automation • Detector calibration necessary (directly obtainable from breakthrough experiments) • Component specific detectors or fractionated analysis necessary for multi-component isotherms only • High amounts of sample • Suitable for low efficient columns

Table 6.6 Methods for isotherm determination. (continued)

Name	Methods	Isotherm type	Comments
			• No kinetic errors • Impurities must not cause significant detector signals
Analysis of disperse fronts: ECP (elution by characteristic point) FACP (frontal analysis by characteristic point)	• Pulse or step injection (high concentration) • Slope of dispersive front	Single component	• Small sample amounts • Highly efficient columns and small plant effects necessary • Phase equilibrium is required (sensitive to kinetics) • Precise detector calibration necessary
Peak-maximum	• Peak injection with systematic overload • Peak maximum equal to retention time	Single component	• Small sample amounts • Less sensitive to low efficiency columns than ECP/FACP • For special cases no detector calibration necessary
Perturbation method (minor disturbance method)	• Determination of the retention times of small (minor) disturbances • Column with different loading states	Single and multi-component	• No detector calibration necessary • Can deal with small impurities • Equilibrium is required • Can deal with low efficient columns • Isotherm models necessary to calculate multi-component isotherms • Potential for automation
Curve-fitting of the chromatogram	• Parameters estimation	• Single and binary (difficult!) • Multi-component not feasible	• Column and plant model necessary • Isotherm model necessary • Isotherm parameters must be sensitive

Determination of component interactions from single component isotherms

Name	Methods	Isotherm type	Comments
IAS (etc.)	• Thermo-dynamics	• Multi-component	• Measurements for single component isotherms only

usually only of analytical scale, or „semi-preparative" at maximum. The similarity between experiment and production plant also implies the use of the preparative adsorbents – different particle sizes or packed and unpacked materials have different effects. As another consequence, temperature control is crucial, since adsorption may show strong temperature dependence. Sometimes the cost of solutes also forbids the use of methods that need a high amount of samples.

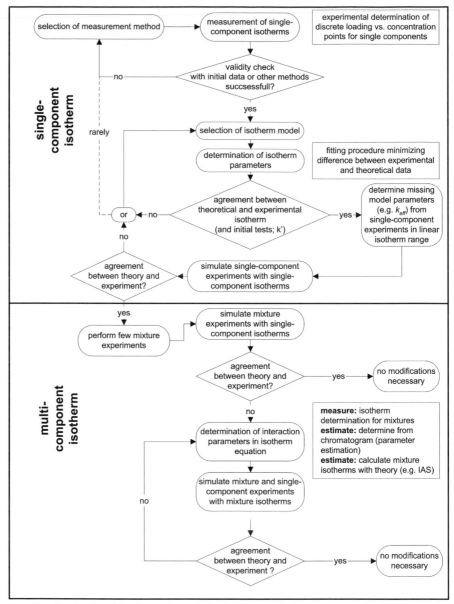

Fig. 6.16 Graphical guideline for isotherm determination.

The graphical guideline in Fig. 6.16 illustrates the general steps of isotherm determination.

6.5.7.2 Determination of the Henry Coefficient

Henry coefficients are generally determined independently from other isotherm parameters, by pulse injections with very low amounts of solutes to ensure linear isotherm behavior (Section 6.5.3.1). This can be tested by two or three pulses with different concentrations. If the results for the determined Henry coefficients are identical, the system is linear. H is calculated by moment analysis using the measured $\mu_{t,c+inj+plant}$ and Eqs. 6.135, 6.136 and 6.132:

$$H = \frac{1}{(1 - \varepsilon_t)}\left(\mu_{t,c}\frac{\dot{V}}{V_c} - \varepsilon_t\right) \text{ with } \mu_{t,c} = t_{R,lin} = \mu_{t,c+plant+inj} - t_{plant} - \frac{t_{inj}}{2} \quad (6.174)$$

6.5.7.3 Static Methods

Static methods (Fig. 6.17) determine phase equilibrium data based on overall mass balances. They are often more time consuming and less accurate than dynamic procedures.

Batch method: A known amount of adsorbent V_{ads} is added to a solution of the volume V_l containing the solute with the concentration $c_{0,i}$. The mixture is then agitated in a closed vessel until equilibrium is reached. The final concentration in the solution ($c_{eq,i}$) is determined by standard analytical methods. From the following mass balance the appropriate equilibrium loading, $q(c_{eq})$, is calculated:

$$V_l c_{0,i} = V_l c_{eq,i} + V_{ads}\left[(1 - \varepsilon_p)c_{eq,i} + \varepsilon_p q_i(c_{eq,i})\right] \quad (6.175)$$

Using different initial concentrations or adsorbent amounts, the relevant concentration range is covered. The method is easily expanded to multi-component mixtures, where the loading is a function of all components present. Drawbacks are the time consuming preparations of the different mixtures and the transferability of the results to packed columns (e.g. uncertainty in phase ratio/porosity). Because of the numerous steps of manual work and the uncertainty when equilibrium is reached, the accuracy is not too high.

Adsorption–desorption method: An initially unloaded ($q = 0$) column is equilibrated by a feed concentration, c_{feed}, which may be a single- or multi-component mixture. Equilibrium is achieved by pumping a sufficient quantity of feed through this column. The plant is then flushed without the column to remove the solute solution. Afterwards, all solute is eluted from the column, collected and analyzed to obtain the

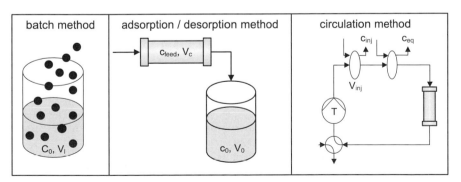

Fig. 6.17 Principle of different static methods for isotherm determination.

desorbed amount $m_{des,i}$. The equilibrium loading $q(c_{feed})$ for each component i can be calculated according to:

$$m_{des,i} = \varepsilon V_c c_{feed,i} + (1 - \varepsilon) V_c \left[\varepsilon_p c_{feed,i} + (1 - \varepsilon_p) q(c_{feed,i})\right] \tag{6.176}$$

The experimental effort to include different concentrations is considerable but the obtained equilibrium values are reliable.

Circulation method: Another static method is based on a closed fluid circuit that includes the chromatographic column. A known amount m_{inj} of solute or a mixture of several solutes is injected into this circuit and pumped around until equilibrium is established. Samples are taken and analyzed to determine the resulting equilibrium concentration c_{eq}. The mass balance for the equilibrium loading accounts for the hold-up of the complete plant:

$$m_{inj,i} = V_{plant} c_{eq,i} + \varepsilon V_c c_{eq,i} + (1 - \varepsilon) V_c \left[\varepsilon_p c_{eq,i} + (1 - \varepsilon_p) q(c_{eq,i})\right] \tag{6.177}$$

with

$$m_{inj,i} = V_{inj} c_{inj,i} \tag{6.178}$$

Subsequently, new injections are made to change the concentration step-by-step. The successive nature of this method saves solute but as a drawback the inaccuracies in each injection and the determined plant volumes accumulate.

6.5.7.4 Dynamic Methods

Dynamic methods extract information about the isotherm from the measured concentration profile. The basic principle is to inject disturbances in an equilibrated column and to analyze the column response. Based on the type of disturbance three major groups can be distinguished. Injection of a large sample, with a concentration different to the existing equilibrated state, results in a breakthrough curve (frontal analysis, method of step response). The complementary case induces small disturbances to an equilibrated chromatographic system. Another possibility is the injection of overloaded pulses with concentrations in the nonlinear range of the isotherm but injection volumes small enough not to reach breakthrough. The last two methods are also referred to as pulse-response techniques.

6.5.7.5 Frontal Analysis

Frontal analysis is one of the most popular methods to determine isotherms. At $t = 0$ a step signal with concentration c^{II} is injected into the column until $t_{inj} = t_{des}$, where the feed concentration is once again lowered to the initial feed concentration. The outgoing concentration is detected. The injected volume has to be large enough to reach a plateau concentration, resulting in a concentration profile as depicted in Fig. 6.18 for a pure component and the column equilibrated initially with the concentration c^I. The component index i is omitted for brevity.

It takes a certain time for the outlet profile to reach a plateau. During this adsorption period a new equilibrium is established, with a liquid concentration being the feed concentration c^{II}. Likewise, during the desorption step the initial equilibrium is

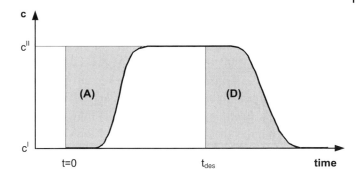

Fig. 6.18 Typical breakthrough curve for adsorption and desorption of a pure component. (Reproduced from Seidel-Morgenstern, 1995.)

restored with a delay of t_{des}. This experimental procedure is easy to implement and to automate if a gradient delivery system is available.

A very effective way to evaluate the equilibrium is to calculate the overall mass balances by numerical integration. Area „A" in Fig. 6.18 is equivalent to the solute accumulated inside the plant and the column, which is split into the liquid and the adsorbent phase. The integral mass balance allows the calculation of the loading, $q^{II} = q(c^{II})$, if the status I and the total porosity are known:

$$V_{plant}(c^{II} - c^{I}) + V_c \left\{ \varepsilon_t(c^{II} - c^{I}) + (1 - \varepsilon_t) \left[q(c^{II}) - q(c^{I}) \right] \right\}$$
$$= \dot{V} \int_0^{t_{des}} \left[c^{II} - c(t) \right] dt \tag{6.179}$$

If the solute is injected in the plant without the column, the dead time of the plant can be estimated at the inflection point of the breakthrough curve. The resulting plateau also allows one to verify that the signal resulting from the feed concentration does not exceed the detector range.

Frontal analysis is straightforward when starting from an unloaded column ($c^{I} = 0$). A modification to reduce the amount of solute is the stepwise increase of the feed concentration, starting from the unloaded column. This results in successive plateaus. Desorption steps are obtained after the highest concentration plateau has passed through the column, if the concentrations are reduced inversely to the adsorption steps. To consume even less feed mixture, this procedure can be performed in closed-loop or circulation operation (Fig. 6.17).

Independent of the method applied, the grey area „D" in Fig. 6.18 has to be equal to the area „A" and results from the desorption step back to c^{I}. Comparison of both values can be used as a consistency check. Another possibility to verify the result is to stop the flow after the plateau is reached and analyze the desorption front according to the adsorption–desorption method (Section 6.5.7.3).

An extension to multi-component systems is straightforward, by injecting more than one component. A prerequisite is the measurement of the concentration profile of each solute during the elution. This can be achieved either by using solute-specific detectors or by collection of multiple fractions and subsequent chemical analysis

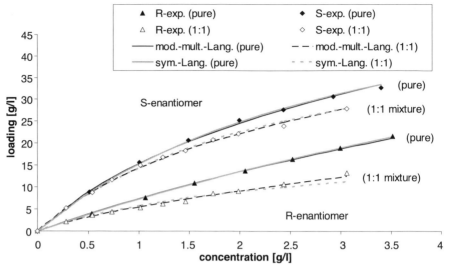

Fig. 6.21 Comparison of experimental data (rhombuses and triangles) and fitted isotherm equations (lines) for R and S-enantiomers of EMD53986 (pure components and the 1:1 mixture on Chiralpak AD in ethanol at 25 °C, mod-multi-Langmuir = Eq. 6.181, sym-Langmuir = Eq. 2.44; for additional data see Appendix B.1).

$$q_{glu} = 0.27\, c_{glu} + 0.000122\, c_{glu}^2 + 0.103\, c_{glu}\, c_{fru}$$
$$q_{fru} = 0.47\, c_{fru} + 0.000119\, c_{fru}^2 + 0.248\, c_{glu}\, c_{fru} \tag{6.182}$$

Another aspect to keep in mind when performing frontal analysis is that drastic changes in the general shape of the breakthrough curve for different concentrations may hint at mass transfer or adsorption kinetic effects. As similar changes occur when changing the flow rate, the variation of these two parameters can be used to further discriminate kinetic effects (Ma et al., 1996). Although analogue changes are observed for peak injection too, breakthrough curves are easier to analyze because of the defined conditions for the equilibrium plateau.

Finally, a few characteristics of the frontal analysis method are summarized:

- Only equilibrium states are measured, thereby errors due to kinetic effects are eliminated
- Amount of solute and the experimental effort is high, as only one equilibrium point is determined per injection. This is increasingly import if kinetic effects are strong and thus the time needed to reach the phase equilibrium is rather long. Closed-loop operation may be used to reduce the consumption of solutes
- It generally provides the best accuracy compared with the other methods mentioned
- Detector calibration is directly obtained from the signal values at the plateau
- Multi-component isotherms can be determined if suitable off-line analytic methods or solute-specific detectors are available

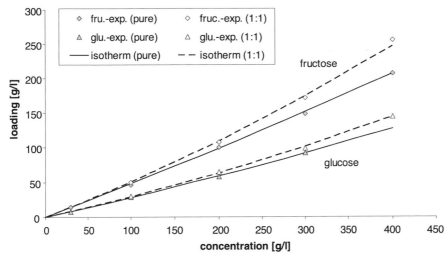

Fig. 6.22 Measured and calculated isotherms for pure components and mixtures of glucose and fructose (adsorbent: ion exchange resin Amberlite CR 1320 Ca, d_p = 325 µm, eluent: water; for additional data see Appendix B.3).

6.5.7.6 Analysis of Disperse Fronts (ECP/FACP)

Analysis of disperse fronts is based on the equilibrium theory of chromatography (Section 6.2.3). It generally reduces the experimental effort as well as the sample amount needed compared with frontal analysis. Because of the complex mathematical solutions in the case of mixtures, it is only suitable for single-component isotherms. Two different forms are described in literature. After injection of a pulse that is not wide enough to cause a concentration breakthrough, the disperse part of the overloaded concentration profile is analyzed. For Langmuir-type isotherms, this is the rear part of the peak. This method is called „Elution by Characteristic Point" (ECP method). If the injected volume is high enough to get a breakthrough, it is termed „Frontal Analysis by Characteristic Points (FACP).

The principle of both methods is explained in Fig. 6.23.

According to equilibrium theory, the retention time t_R at a characteristic concentration c^+ correlates to the slope of the isotherm at this specific concentration. Rearranging Eq. 6.49, the slope can be calculated by

$$\left.\frac{dq}{dc}\right|_{c^+} = \frac{t_R(c^+) - t_0}{t_0} \frac{\varepsilon_t}{1 - \varepsilon_t} \tag{6.183}$$

The measured retention times must be corrected by the dead time t_{plant} of the plant and for Langmuir-shaped isotherms additionally by the injection time t_{inj} (Eq. 6.52).

If this analysis is repeated at different positions of the profile, the whole isotherm can be estimated based on a single chromatogram. The concentration range is limited by the value at the peak maximum. Isotherm parameters can be determined from the slope vs. concentration values in two ways.

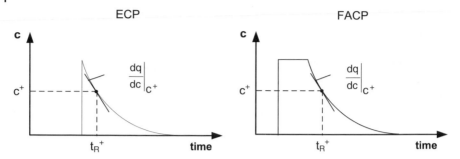

Fig. 6.23 Principle of the ECP and FACP methods.

First, numerical integration, which leads to the desired relationship between loading and concentration according to

$$q(c^+) = \int_0^{c^+} \frac{dq}{dc} dc = \int_0^{c^+} \frac{t_R(c) - t_0}{t_0} \frac{\varepsilon_t}{1 - \varepsilon_t} dc \qquad (6.184)$$

The isotherm equation is then fitted as described in Section 6.5.7.11.

Second, the derivative of the isotherm equation can be fitted directly to the slope values. For this optimization problem, the error of the loading values is replaced by the errors of the slopes.

Regardless of which method is used, the initial slope of the isotherm must be determined with great care. Either the lower bound for integration must be approximated with the lowest recorded concentration or the slope for zero concentration must be determined separately. As the latter task is identical to the determination of the Henry constant, this value can be obtained from pulse experiments with very low injection concentrations using momentum analysis (Section 6.5.7.2).

ECP and FACP are limited to single-component systems. They save time and sample as the whole isotherm is derived from one experiment only. However, experimental inaccuracies limit the application. This results from the assignment of discrete retention times to discrete concentrations, which may include several experimental and systematic errors:

- Imprecise detector calibration directly influences the concentration profile and thus the isotherm
- Peak deformation through extra column effects of the plant cannot be accounted for
- Since these methods depend on the assumption of immediate adsorption equilibrium, it requires high efficiency columns with several thousand theoretical stages to avoid kinetic effects
- Determination of the slopes for very low concentrations includes a high experimental error and may deviate from the Henry values obtained from pulse experiments

For the FACP method, frontal analysis might be used to check the determined isotherm at the breakthrough concentration.

6.5.7.7 Peak-maximum Method

A possibility to reduce the influence of column efficiency on the results obtained by the ECP method is to detect the position of the peak maximum only, which is called the peak-maximum or retention-time method. Graphs like Fig. 6.23 are then achieved by a series of pulse injections with different sample concentrations. The concentration and position of the maximum is strongly influenced by the adsorption equilibrium due to the compressive nature of either the front or the rear of the peak (Chapter 2.2.3). Thus, the obtained values are less sensitive to kinetic effects than in the case of the ECP method. The isotherm parameters can be evaluated in the same way as described in Section 6.5.7.6, but the same limitations have to be kept in mind. For some isotherm equations, analytical solutions of the ideal model can be used to replace the concentration at the maximum (Golshan-Shirazi and Guiochon, 1989 and Guiochon et al., 1994b). Thus, only retention times must be considered and detector calibration can be omitted in these cases.

6.5.7.8 Minor Disturbance/Perturbation Method

The minor disturbance or perturbation method relies on equilibrium theory too and was first suggested by Reilley et al. (1962). As known from linear chromatography, the retention time of a small pulse injected into a column filled with pure eluent can be used to obtain the initial slope of the isotherm. This approach is expanded to cover the whole isotherm range. For the example of a single-component system (Fig. 6.24) the procedure is as follows: The column is equilibrated with a concentration c_a and, once the plateau is established, a small pulse is injected at a time $t_{start,a}$ and a pulse of a different concentration is detected at the corresponding retention time $t_{R,a}$.

The injected concentration can be either higher or lower than the plateau concentration, c_a. To maintain equilibrium conditions inside the column, the concentration should not deviate too much from c_a and the injected volume should be small, but care has to be taken that the resulting peak is large enough to be distinguished from the signal noise. If for that reason the injected amount must be comparatively high, it is recommended to average the results obtained with concentrated and diluted injection. In practice, pure eluent with very small injected volumes often provides sufficient accuracy.

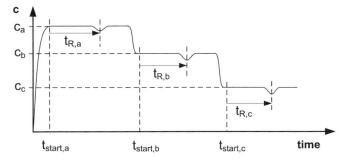

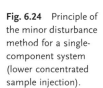

Fig. 6.24 Principle of the minor disturbance method for a single-component system (lower concentrated sample injection).

According to equilibrium theory, elution of the small disturbance depends on the isotherm slope at the plateau concentration. Because the perturbation peak is almost Gaussian, the time at the peak maximum respectively minimum can be taken to calculate simply the isotherm slope by Eq. 6.49:

$$\left.\frac{dq}{dc}\right|_{c_a} = \frac{t_{R,a} - t_0}{t_0}\frac{\varepsilon_t}{1 - \varepsilon_t} \tag{6.185}$$

The characteristic retention times have to be corrected by the dead time of the plant and the injected volume (Section 6.5.3.1). Using values at different plateau concentrations, the isotherm parameters are derived similar to the manner in Section 6.5.7.6.

To determine a multi-component isotherm, an equation for the isotherm has to be chosen *a priori*. For example, an injection of pure mobile phase on a column equilibrated with a two-component mixture results in two peaks, one for each component. Using Eq. 6.47 and the definition of the retention times (Eq. 6.49), for two-component systems together with the coherence condition Eq. 6.54, the two retention times can be used to calculate the four partial differentials of Eq. 6.53 for a given isotherm equation (see below). The complexity of these calculations increases rapidly with increasing number of components. Additionally, detector noise can make it difficult to distinguish the peak for the second, later-eluting component of binary mixtures, because the longer residence time in the column causes a broader and lower peak than for the early-eluting component.

An advantage of this method compared with ECP is that the assumption of equilibrium between mobile and stationary phase is more realistic for the injection of small disturbances than for a large overloaded peak. Thus, the perturbation method is not as sensitive to the number of stages of a column. Additionally, detector calibration is not necessary and the analysis does not necessarily require solute-specific detectors.

The method is advantageously combined with the frontal analysis method, which also requires a concentration plateau and thus shares the disadvantage of high sample consumption if operated in open mode. As indicated in Fig. 6.24, the measurement procedure starts at maximum concentration. This concentration plateau is reduced step-by-step by diluting the solution. To reduce the amount of samples needed for the isotherm determination the experiments can be done in a closed loop arrangement (Fig. 6.17). It is also possible to automate this procedure.

As an example, Fig. 6.25a gives the results of the isotherm determination for Tröger's base enantiomer on Chiralpak AD ($d_p = 20\,\mu m$) from perturbation measurements (Mihlbachler et al., 2001). Theoretical retention times for the pure components and racemic mixtures (lines) were fitted to the measured data (symbols) by means of Eq. 6.185 to determine the unknown parameter in Eq. 6.186. Total differentials for the mixture (Eq. 6.53) were evaluated using the coherence condition Eq. 6.54, resulting in the isotherm equation Eq. 6.186. Note that the Henry coefficients were independently determined by pulse experiments and were fixed during the fitting procedure.

$$R\text{-enantiomer: } q_R = \frac{0.0311\, c_R(54 + 0.732\, c_S)}{1 + 0.0311\, c_R} + \frac{0.732 \times 0.0365\, c_R\, c_S}{1 - 0.0365\, c_R}$$

(6.186)

$$S\text{-enantiomer: } q_S = \frac{27\, c_S(0.1269 + 2 \times 0.0153\, c_S)}{1 + 0.1269\, c_S + 0.0153\, c_S^2}$$

The resulting isotherms are illustrated as lines in Fig. 6.25b. The stronger adsorbing S-enantiomer isotherm exhibits an inflection point and is assumed to be independent of the R-enantiomer concentration. The R-enantiomer isotherm shows typical Langmuir behavior and minor interaction with the S-enantiomers. The rather uncommon case of higher loadings of the R-enantiomer in mixtures can be explained with multi-layer adsorption processes (Mihlbachler et al., 2001).

Figure 6.25b also shows the isotherm data obtained from frontal analysis experiments (symbols), which agree sufficiently with those from the perturbation analysis. Because of impurities contained in the original substance, which could not be removed, the obtained data may be more inaccurate than those from perturbation.

For later use in process simulation, these data were fitted to the model of concave isotherms with saturation (Hill, 1960, Lin et al., 1989, Mihlbachler et al. 2001, and Jupke, 2004):

$$R\text{-enantiomer: } q_R = \frac{54\, c_R(0.035 + 0.0046\, c_S)}{1 + 0.035\, c_R + 0.062\, c_S + 0.0046\, c_R\, c_S + 0.0052\, c_S^2}$$

(6.187)

$$S\text{-enantiomer: } q_S = \frac{54\, c_S(0.062 + 0.0046\, c_R + 2 \times 0.0052\, c_S)}{1 + 0.035\, c_R + 0.062\, c_S + 0.0046\, c_R\, c_S + 0.0052\, c_S^2}$$

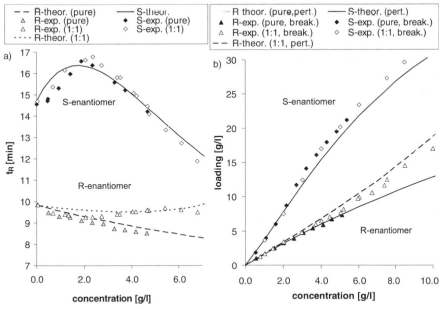

Fig. 6.25 (a) Results for isotherm determination for Tröger's base on Chiralpak AD for pure components a mixtures (symbols = experimental data, lines = retention times calculated by Eqs. 6.185 and 6.186, (t_0 = 5.144 min, $\dot{V}$ = 1 ml min^{-1}; for additional data see Appendix B.2). (b) Comparison of Eq. 6.186 with experimental data from frontal analysis. (Data taken from Mihlbachler et al., 2001.)

6.5.7.9 **Curve Fitting of the Chromatogram**

The approach of parameter estimation based on process models may also be applied to isotherm parameters. In the framework of Fig. 6.9 this requires the selection of plant (Section 6.3.2) and column models (Section 6.2) together with an isotherm equation. This is important in eliminating all sources of band broadening that do not result from thermodynamic equilibrium. If the plant parameters, packing parameters and calibration curves are determined, the transport coefficient (Section 6.5.8) can be estimated by small pulse injections in the linear range of the isotherm for the pure components. The Henry coefficient should be obtained from these experiments using Eq. 6.174 and momentum analysis (Section 6.5.3.1). The isotherm parameters are then estimated using parameter estimation or curve fitting (Section 6.5.3.2) from a series of overloaded pulse or breakthrough injections first for single-components systems and then for mixtures. As for all parameter estimation tools, initial guesses have to be provided. To cover a broad concentration range for the solute, the experimental peaks should be as high as possible.

The drawback of this approach compared with that described in Section 6.5.2 is that all errors are lumped into the isotherm parameters rather than the effective mass transfer coefficient, because either the „wrong" column or isotherm model is chosen. This approach is thus recommended to get a quick first idea of system behavior using only little amounts of sample, and not for a complete analysis, especially if binary mixtures with component interactions are investigated. The significance of the results decreases even further if some plant and packing parameters are only guessed or even neglected.

6.5.7.10 **Calculation of Mixture Behavior from Single Component Data**

As mentioned in Chapter 2.5.2.3, methods exist to calculate data for multi-component isotherms if the single-component equations are known. When the resulting accuracy is acceptable, this reduces the experimental effort for parameter determination considerably. As an example, the experimental data shown in Fig. 6.21 are used to derive the multi-component isotherm based on the single-component equation and IAS theory. Experimental data of the EMD system (rhombuses for *S*-enantiomer and triangles for *R*-enantiomer) as well as the data calculated by the IAS theory (straight lines) are plotted in Fig. 6.26. Data for the modified multi-Langmuir isotherm (Fig. 6.21) are added for reference as dashed lines. Figure 6.26 shows satisfying agreement between IAS theory and experimental data, although the deviation increases for higher concentrations due to the extrapolation of pure-component data in the IAS calculations. Here the mixture spreading pressure is equal to a hypothetical pure component spreading pressure of the lower affinity component, which cannot be reached in reality.

As with the selection of an isotherm model, a final decision about the accuracy and validity of this approach has to be based upon a comparison between simulated and measured concentration profiles.

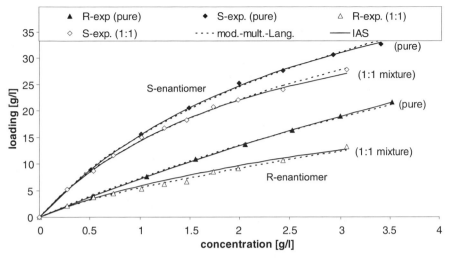

Fig. 6.26 Multi-component isotherm derived from single-component data using the IAS-theory as well as reference data for the EMD-system (for other data see Fig. 6.21).

6.5.7.11 Data Analysis and Accuracy

When dynamic methods are applied, the measured profiles should include enough points and sufficiently low detector noise to perform numerical calculations (e.g. integration). This is discussed in another context in Section 6.5.3.2 and Chapter 2.7. The number of data points in the measured profile can be increased by changing the flow rate of the pump or the sample rate.

Through repetition of the same experiment and subsequent evaluation of the equilibrium data, the accuracy may be increased by averaging the values, if no systematic errors occur. Too high a deviation between equivalent measurements indicates problems in either the data evaluation or the experiment itself. In the latter case, it should be checked if the pumps deliver a constant flow rate and that the temperature is constant in the range of a few tenths of one centigrade. If the eluent consists of a fluid mixture, the influence of slight changes in eluent composition must be taken into account (Section 6.5.7.1).

As already mentioned, all plant and packing parameters, as well as the calibration curve of the detector, must be determined with care to give proper results. As Seidel-Morgenstern (2004) points out, simulation and experiment might still agree sufficiently if the „wrong" void fractions and porosities are determined and, consequently, the isotherm parameters are inaccurate. This is because not the isotherm but rather the isotherm multiplied by the porosities appears in the model equations and thus both inaccuracies cancel each other out to a certain degree.

Most determination methods finally lead to discrete loading versus concentration data that have to be fitted to a continuous isotherm equation. For this purpose it is advised to use a least-squares method to obtain the parameters of the isotherm. Nonlinear optimization algorithms for such problems are implemented in standard spreadsheet programs. To select an isotherm equation and obtain a meaningful fit, the number of data points should not be too low and their distribution should be such that changes in the slope or curvature are properly represented. A suitable objective function to minimize the overall error δ includes the sum of the weighed relative error

$$\delta = \sqrt{\sum_{j=1}^{n_p} f_j \cdot \left(\frac{q_{\exp,j} - q_{\text{theo},j}}{q_{\exp,j}}\right)^2} \qquad (6.188)$$

where $q_{\text{theo},j}$ is the value obtained from the isotherm equation, $q_{\exp,j}$ the experimental value and f_j describes a weighting factor. Often f_j is a constant and depends only on the number of experimentally determined points (n_p) and the number of isotherm parameters (n_{para}):

$$f_j = \frac{1}{n_p - n_{\text{para}}} \qquad (6.189)$$

This allows a comparison of the fitting quality for certain isotherm equations with different numbers of adjustable parameters. Further equations for statistical analysis can be found, for example, in Barns (1994) and Press et al. (2002). The use of relative instead of absolute errors is necessary to increase the fit in the low concentration region of the isotherm. Otherwise, the isotherm is inaccurate in the low concentration region and the calculated band profile is inaccurate at the rear boundary.

A problem often encountered in nonlinear optimization is the necessity to provide suitable initial guesses. Therefore, it should be tested if different initial guesses lead to drastically different sets of isotherm parameters. This is connected to the sensitivity problem, which is pronounced in the case of multi-component systems where several parameters need to be fitted at once. Substantial initial guesses are often difficult to find and the sensitivity is often low, which demands much experimental data or leads to the selection of other isotherm equations.

For the linear part of the isotherm, the Henry coefficient may be determined separately by pulse experiments (Section 6.5.7.2). If a significant deviation from the value obtained with the isotherm equation is encountered, additional experiments in the low concentration range should be carried out. These can be used to clarify whether the isotherm equation (e.g. adding an extra linear term as in Eq. 2.47) or the method of isotherm determination must be changed.

As the overall aim of parameter determination is the (simulation-based) prediction of the process behavior the final decision about the suitability of the isotherm equation can only be made by comparing experimental and theoretical elution profiles (Section 6.6). Depending on the desired accuracy, this may involve iteration loops for the selection of isotherm equations or in some cases even the methods (Fig. 6.16). In this context it should be remembered that, according to the model equations (e.g. Eq. 6.47), the migration velocity and thus the position of the profiles is a function of the isotherm slope. Therefore, it is most important for the reliability of process simulation that the slopes of the measured and calculated elution profile fit to each other. If the deviations are unacceptable, another isotherm equation should be tested.

6.5.8
Mass Transfer

In the framework of the TDM model, the transport coefficient is the last parameter to be determined according to Fig. 6.9. All prior experimental errors and model inaccuracies are lumped into this parameter. In addition it cannot be excluded that the mass transfer depends on concentration because of surface diffusion or adsorption kinetics. However, in many cases, e.g. for the target solutes discussed in this book, the transfer coefficient can be assumed to be independent of operating conditions (especially flow rate) with reasonable accuracy.

As k_{eff} acts also as a final tuning parameter, it should be determined by a simulation-based parameter estimation that is performed with the full set of model equations, including the parameters for the plant as well as the column (Fig. 6.5). The estimated value of k_{eff} might be verified by measurements at different volume flows and injection amounts.

An initial guess of k_{eff} can be obtained from moment analysis (Section 6.5.3.1) or empirical correlations. For peak injections in the linear range of the isotherm k_{eff} is calculated by Eq. 6.137 if the axial dispersion coefficient is already known (Section 6.5.6.2). Another rough estimation is based on the slope B of the simplified HETP curve (Eq. 6.141):

$$k_{eff} = 2 \left(\frac{\tilde{k}'}{1 + \tilde{k}'} \right)^2 \frac{\varepsilon}{(1 - \varepsilon)} \frac{r_p}{3} \frac{1}{B} \tag{6.190}$$

If mass transfer in the film and diffusion inside the pores are taken into account the effective mass transfer coefficient is given as a series connection of the internal ($1/k_{pore}$) and external ($1/k_{film}$) mass transfer resistance (Eq. 6.138):

$$\frac{1}{k_{eff}} = \frac{d_p}{10 \varepsilon_p D_{pore}} + \frac{1}{k_{film}} = \frac{1}{k_{pore}} + \frac{1}{k_{film}} \tag{6.191}$$

For the film transfer coefficient k_{film}, Wilson and Geankoplis (1966) developed the following correlations:

$$Sh = \frac{1.09}{\varepsilon} (\varepsilon Re)^{0.33} Sc^{0.33} \quad (0.0015 < \varepsilon Re < 55)$$

$$Sh = \frac{0.25}{\varepsilon} (\varepsilon Re)^{0.69} Sc^{0.33} \quad (55 < \varepsilon Re < 1050) \tag{6.192}$$

where the Sherwood number Sh and the Schmidt number Sc are defined as:

$$Sh = \frac{k_{film} d_p}{D_m} \tag{6.193}$$

$$Sc = \frac{\eta_l}{\rho_l D_m} = \frac{\nu_l}{D_m} \tag{6.194}$$

The (particle) Reynolds number is given by Eq. 6.169

$$Re = \frac{u_{int} d_p \rho_1}{\eta_l} = \frac{u_{int} d_p}{\nu_l} \tag{6.169}$$

Using the same assumptions as in Section 6.5.6.2 and typical viscosity values ν_l of about 0.01 cm^2 s^{-1}, the Schmidt number for small molecules is of the order of 1000. The Reynolds number is about 0.01 so that only the first equation in Eq. 6.192 is applicable. For a typical void fraction ε of 0.37–0.4, the Sherwood number is about 4.5 and the film transfer coefficient is then calculated by

$$k_{film} = \frac{1.09}{\varepsilon} \frac{D_m}{d_p} \left(\varepsilon \frac{u_{int} d_p}{D_m} \right)^{0.33} \tag{6.195}$$

According to the approximations above the film coefficient k_{film} is about 4.5×10^{-2} cm s^{-1} for a 10 µm particle.

The intraparticle (pore) diffusion coefficient defined in Section 6.2.2.4 may estimated by the Mackie-Meares correlation (Mackie and Meares, 1955):

$$D_{pore} = \frac{\varepsilon_p}{(2 - \varepsilon_p)^2} D_m \tag{6.196}$$

The factor in front of D_m is an approximation of the internal tortuosity factor. For a porosity of 0.9 to 0.5, D_{pore} is 5 times smaller than D_m or even lower, so the contribution to the effective mass transfer coefficient Eq. 6.191 is

$$k_{pore} = \frac{10 \, \varepsilon_p D_{pore}}{d_p} = \frac{10}{d_p} \frac{\varepsilon_p^2}{(2 - \varepsilon_p)^2} D_m \tag{6.197}$$

which gives a k_{pore} of about 1×10^{-3} cm s^{-1} for a small particle (Eq. 6.163). Intraparticle transport is an order of magnitude slower than film transfer and is, therefore, the rate-limiting step (see also Ludemann-Hombourger et al., 2000a).

Thus, k_{eff} is mainly defined by Eq. 6.197. This simplified analysis is also a justification to take the effective transfer coefficient as independent of volume flow and inversely proportional to particle diameter.

6.6
Validation of Column Models

After all parameters for the plant (Section 6.5.5) and column (Sections 6.5.6–8) models are determined, either from experimental data or from empirical correlations, the validity of the model should be checked using different experiments than those for parameter determination. Keeping in mind the separation task at hand, the position of the adsorption and desorption fronts may be taken as an indicator for the purity of the products. Therefore, the position of these fronts is often more important for process optimization than the exact height of the simulated profile.

The following examples illustrate a few effects encountered in model validation based on our research (Epping, 2005 and Jupke, 2004). All process simulations are based on the transport dispersive model. Model equations were solved by the gPROMS® Software (PSenterprise, UK) using the OCFE method (Section 6.4).

Experimental data were obtained with on-line detector set-ups. Figure 6.27 shows the good consistency of the on-line and off-line analysis (= multiple fractions and HPLC analysis) for the injection of the EMD53986 racemic mixture. In this case a two-detector set-up of polarimeter and UV detector was used, which permits the solute-specific detection of both components (Epping, 2005, Jupke, 2004 and Mannschreck, 1992). As the collection and analysis of multiple fractions with a high sample rate is tedious, the on-line measurement method should be used whenever possible.

Figure 6.28 compares measured and simulated profiles for the batch separation of EMD53986. Very good agreement between theory (solid lines) and experiment (symbols) is achieved using the multi-component modified-Langmuir isotherm (Fig. 6.21). Also shown are the simulation results neglecting component interaction by using only the single-component isotherms (dashed line), which deviate strongly from the observed mixture behavior. Typical for competitive adsorption is the displacement of the weaker retained *R*-enantiomer and the peak expansion of the stronger adsorbed *S*-enantiomer.

Figure 6.29 shows a comparison between two different isotherm models (Fig. 6.21) used for validation: agreement between theory and experiment is good using the modified multi-component Langmuir model (Fig. 6.21), while the symmetrical Langmuir isothcrm leads to profiles shifted to earlier retention times. The structural

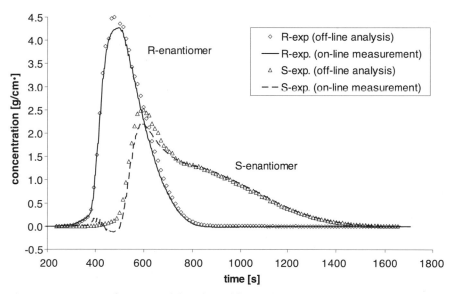

Fig. 6.27 Comparison of experimental data obtained from on-line measurement (lines) and off-line analysis for the EMD53986 racemic mixture.

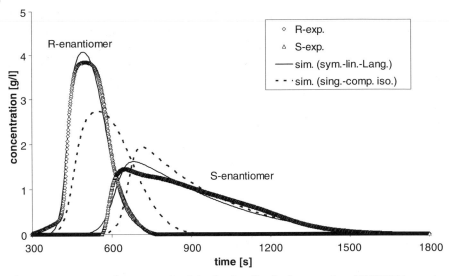

Fig. 6.28 Comparison of experimental and simulated profiles for the separation of EMD53986 racemic mixture using single- and multi-component isotherms (exp: = experimental; sim. = simulated; $c_{feed} = 6\,g\,l^{-1}$ $\dot{V} = 20\,ml\,min^{-1}$, $V_{inj} = 80\,ml$, $V_c = 54\,ml$; for additional data see Appendix B.1).

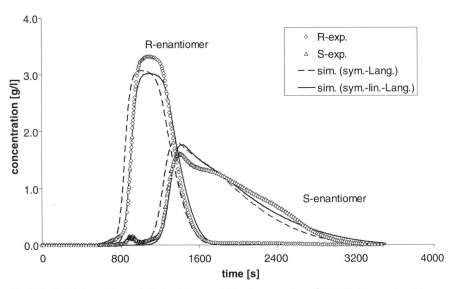

Fig. 6.29 Validation of the overall simulation model for the separation of EMD53986 racemic mixture using two different isotherm equations (exp: = experimental; sim. = simulated; $c_{feed} = 4.4\,g\,l^{-1}$ $\dot{V} = 10\,ml\,min^{-1}$, $V_{inj} = 120\,ml$, $V_c = 54\,ml^{-1}$; for additional data see Appendix B.1).

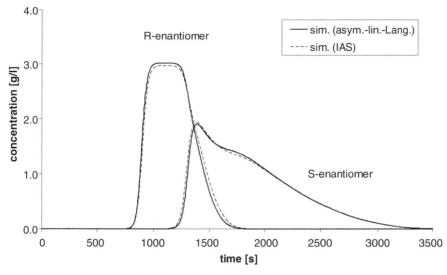

Fig. 6.30 Comparison of the simulated profiles for the modified multi-component Langmuir isotherm and the IAS equation (Fig. 6.26) ($c_{feed} = 4.4\,\mathrm{g\,l^{-1}}$ $\dot{V} = 10\,\mathrm{ml\,min^{-1}}$, $V_{inj} = 120\,\mathrm{ml}$, $V_c = 54\,\mathrm{ml}$; for additional data see Appendix B.1).

disadvantages of the symmetric Langmuir model for EMD 53986 are less pronounced for lower loadings (not shown here for brevity).

Finally, simulated profiles using an experimentally determined multi-component isotherm and the respective data calculated according to the IAS theory (Fig. 6.26) are compared.

Figure 6.30 shows the very close agreement between these methods for this enantiomer system. Especially when considering the effort necessary to measure the multi-component isotherms, IAS theory or its extensions may provide a good estimate for the component interaction in the case of competitive adsorption. Therefore, it is advisable to simulate elution profiles using the IAS theory after single-component isotherms have been measured. These calculations should then be compared with a few separation experiments to decide if measurements of the multi-component isotherm are still necessary.

The validity of the transport dispersive model was further confirmed by experiments with other chromatographic systems such as Tröger's base (Mihlbachler et al., 2001 and Jupke, 2004), the WMK-Keton (Epping, 2005) and fructose–glucose (Jupke, 2004).

As one example, Fig. 6.31 shows the results for Tröger's base using isotherm data determined from breakthrough curves (Eq. 6.187).

Even for the fructose–glucose isomer system, with liquid phase concentrations that are an order of magnitude higher than those of the enantiomer system, the transport dispersive model is valid. Figure 6.32 shows that experimental and theoretical concentration profiles for a pulse experiment match very well using the isotherm data of Eq. 6.182. Eight semi-preparative columns were connected in series to

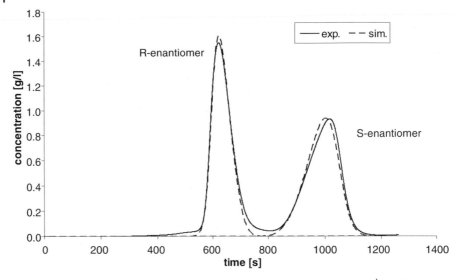

Fig. 6.31 Measured and simulated pulse experiment for the Tröger's base racemate ($\dot{V} = 1\,\text{ml min}^{-1}$, $V_{inj} = 1\,\text{ml}$, $c_{feed} = 2.2\,\text{g l}^{-1}$, $V_c = 7.9\,\text{ml}$; for additional data see Appendix B.2).

achieve a sufficient bed length. Even for industrial-scale plants with a column diameter of ca. 3 m the model approach is in good agreement with experimental data (Fig. 6.33). In this case the extra column dead volumes are especially important to account for.

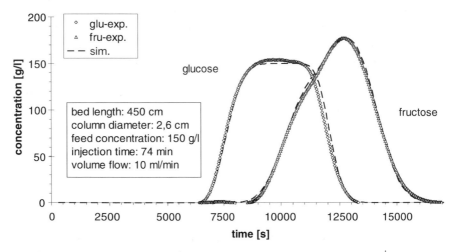

Fig. 6.32 Measured and simulated pulse experiment for the glucose–fructose mixture ($\dot{V} = 10\,\text{ml min}^{-1}$, $V_{inj} = 740\,\text{ml}$, $c_{feed} = 150\,\text{g l}^{-1}$, $V_c = 2.4\,\text{l}$; for additional data see Appendix B.3).

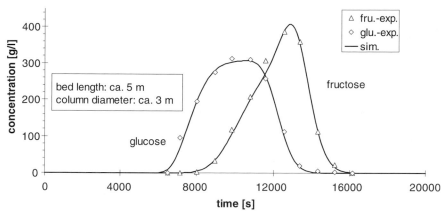

Fig. 6.33 Measured and simulated pulse experiment for the glucose–fructose mixture on an industrial plant.

6.7
Modeling of SMB Processes

6.7.1
Process Principle

The Simulated-Moving-Bed (SMB) process is an efficient way to realize a continuous chromatographic separation. In a more general view, the model-based design of this process is representative of a group of complex chromatographic operation modes described in Chapter 5. The common basis for the simulation of all these set-ups is a flowsheet connecting models for several chromatographic columns and additional plant peripherals. By supplying the necessary boundary conditions and process schedules many different kinds of continuous operations can be simulated. Importantly, simulating SMB processes is primarily a flowsheeting problem. Individual models, e.g. the column model, have already been validated by batch experiments.

In SMB operation the counter-current movement of liquid and solid phase is achieved by shifting the inlet and outlet streams in the direction of the fluid by valve switching (Chapter 5.3.4, Fig. 6.34). Figure 6.34 also shows the simplified concentration profile in the axial direction over all columns for a separation of two components. Due to the discrete shifting, the process reaches a cyclic steady state. Assuming identical columns, the profiles after each shifting time (t_{shift}) are identical, only shifted by one column. During the shifting or switching interval the profiles change with time, but generally the profiles are shown only at the end of one complete cycle. After each cycle the feed inlet stream enters once again at its initial position. The cycle time is equal to the number of columns multiplied by the shifting time.

The complex dynamics, cyclic operation and numerous influence parameters make a purely empirical design of SMB process difficult or even impossible. Since

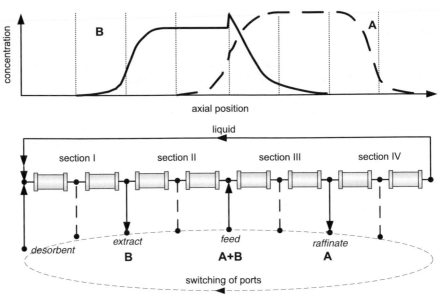

Fig. 6.34 Simplified axial concentration profile and flowsheet for an SMB process (standard configuration).

the introduction of this technology, models with different levels of details have been used to obtain the operating parameters (Ruthven and Ching, 1989, Barker and Ganestos, 1993, Storti et al., 1993b, Chu and Hashim, 1995, Strube, 1996, Zhong and Guichon, 1996 and Dünnebier, 2000).

One criterion for classifying SMB models is the type of isotherm considered. Models with linear isotherms, for which analytical solutions often exist (Ruthven and Ching, 1989, Storti et al., 1993b and Zhong and Guichon, 1996), are distinguished from models including complex isotherm equations (Strube 1996, Pais et al., 1998a, Migliorini et al., 2000 and Dünnebier, 2000). Another difference between models described in the literature is whether SMB or TMB models (True-Moving-Bed) are used to describe process behavior (Ruthven and Ching, 1989 and Barker and Ganestos, 1993). The SMB simulation approach takes into account the shifting of the inlet and outlet streams just as in the real plant and the process reaches a cyclic steady state (Hashimoto et al., 1983a, Storti et al., 1988, Strube, 1996, Zhong and Guichon, 1996, Pais et al., 1998a and Dünnebier, 2000). Since TMB models neglect this discrete dynamics they reach a true steady state and are simpler and faster to solve mathematically. Liapis et al. showed theoretically that the simulated converges to the true counter-current flow for an infinite number of columns and an infinitesimal shifting time (Liapis and Rippin, 1979). TMB and SMB models are compared in Section 6.7.3.

6.7.2
SMB Process Models

An SMB plant consists, for instance, of eight chromatographic columns connected by pipes (Fig. 6.35). The piping also includes valves for attaching external streams as well as measurement devices or pumps.

The overall model of an SMB process is developed by linking the models of individual chromatographic columns (Section 6.2). As with the chromatographic batch process, the plant set-up of Fig. 6.35 is converted into a simulation flowsheet. Figure 6.36 shows the SMB column model.

Mathematically, the SMB model is achieved by connecting the boundary conditions of each column model, including nodes represented by material balances of splitting or mixing models. These so-called node models (Ruthven and Ching, 1989) are given for a component i in the sections I–IV by:

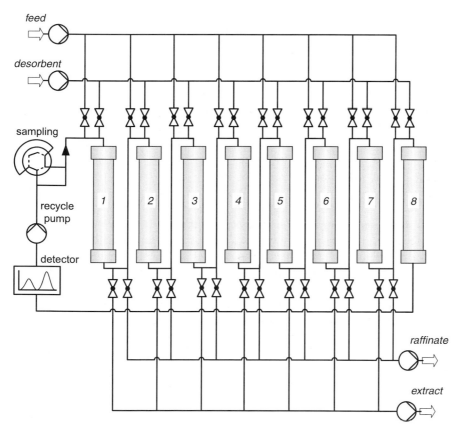

Fig. 6.35 Principle set-up of an SMB plant including detector systems in the recycle stream (8 column standard configuration).

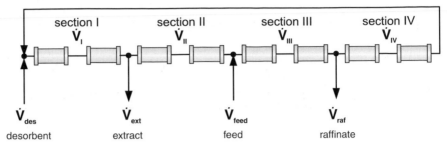

Fig. 6.36 Simulation flowsheet of the SMB process („SMB-column model").

Desorbent node:

$$\dot{V}_{des} = \dot{V}_I - \dot{V}_{IV}$$
$$c_{des,i}\,\dot{V}_{des} = c_{in,I,i}\,\dot{V}_I - c_{out,IV,i}\,\dot{V}_{IV}$$

(6.198)

Extract draw-off node:

$$\dot{V}_{ext} = \dot{V}_I - \dot{V}_{II}$$
$$c_{ext,i} = c_{out,I,i} = c_{in,II,i}$$

(6.199)

Feed node:

$$\dot{V}_{feed} = \dot{V}_{III} - \dot{V}_{II}$$
$$c_{feed,i}\,\dot{V}_{feed} = c_{in,III,i}\,\dot{V}_{III} - c_{out,II,i}\,\dot{V}_{II}$$

(6.200)

Raffinate draw-off node:

$$\dot{V}_{raf} = \dot{V}_{III} - \dot{V}_{IV}$$
$$c_{raf,i} = c_{out,III,i} = c_{in,IV,i}$$

(6.201)

Concentrations c_{in} are the inlet boundary conditions (Eq. 6.92 or 6.93) of the columns at the beginning of each section while c_{out} are the outlet concentrations calculated at the end of each section. Intermediate node balances consist of setting equal volume flows and assigning the outlet concentration to the inlet concentration of the subsequent column. Since SMB is a periodic process, the boundary conditions for every individual column are changed after a switching period t_{shift}. If SMB modifications such as VariCol, ModiCon, etc. (Chapter 5) are used the boundary conditions are modified accordingly.

Further extensions of the SMB model are necessary to account for the fluid dynamic effects of piping and other peripheral equipment such as measurement devices (Fig. 6.35), especially for plants with large recycle lines (Jupke, 2004). This is achieved by adding stirred tanks and pipe models (Section 6.3.2) to the simulation flowsheet, resulting in the extended SMB model given in Fig. 6.37. If the distribution of the dead volumes in the process is uneven, an asynchronous shifting (Chapter 5) of the inlet and outlet ports is required (Hotier and Nicoud, 1995 and Migliorini et

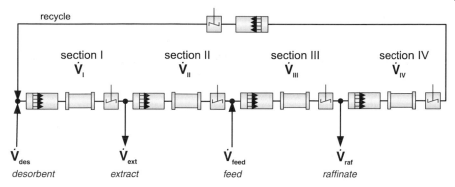

Fig. 6.37 Simulation flowsheet for the „Extended SMB model".

al., 1999). Sources for the dead volume between the columns are the connecting pipes as well as the switch valves. Dead volume in the recycle stream additionally includes the recycle pump and the measurement systems.

6.7.3
TMB Model

Because of the analogy between simulated and true counter-current flow, TMB models are also used to design SMB processes. As an example, the transport dispersive model for batch columns can be extended to a TMB model by adding an adsorbent volume flow $\dot{V}_{ads}$ (Fig. 6.38), which results in a convection term in the mass balance with the velocity u_{ads}. Dispersion in the adsorbent phase is neglected because the goal here is to describe a fictitious process and transfer the results to SMB operation. For the same reason, the mass transfer coefficient k_{eff} as well as the fluid dispersion D_{ax} are set equal to values that are valid for fixed beds.

The underlying node model of the TMB simulation is shown in Fig. 6.38.

The dynamic mass balance for component i and section j in the liquid phase is

$$\frac{\partial c_{j,i}}{\partial t} = -u_{int,j,TMB} \cdot \frac{\partial c_{j,i}}{\partial x} + D_{ax,j} \cdot \frac{\partial^2 c_{j,i}}{\partial x^2} - \frac{1-\varepsilon}{\varepsilon} k_{eff,j,i} \cdot \frac{3}{r_p}(c_{j,i} - c_{p,j,i}) \quad (6.202)$$

and in the adsorbent phase

$$\varepsilon_p \frac{\partial c_{p,j,i}}{\partial t} + (1-\varepsilon_p)\frac{\partial q_{j,i}}{\partial t} = +u_{ads}\left[\varepsilon_p \frac{\partial c_{p,j,i}}{\partial x} + (1-\varepsilon_p)\frac{\partial q_{j,i}}{\partial x}\right]$$
$$+k_{eff,j,i} \cdot \frac{3}{r_p}(c_{j,i} - c_{p,j,i}) \quad (6.203)$$

For process comparison, the stationary form of Eqs. 6.202 and 6.203 is normally used, in which the accumulation terms are zero. To transfer the results of TMB to

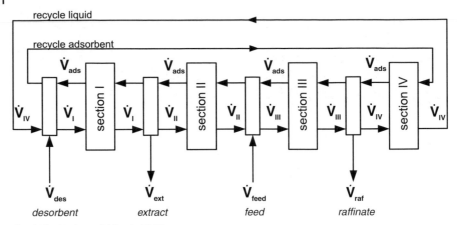

Fig. 6.38 Node model for the TMB process.

SMB, the interstitial velocity in the SMB process must be equal to the relative velocity of fluid and adsorbent in the TMB process:

$$u_{\mathrm{int,SMB}} = u_{\mathrm{int,TMB}} + u_{\mathrm{ads}} \tag{6.204}$$

u_{ads} is positive and the direction inverse to u_{int} is specified in Eq. 6.203 by a positive sign in front of the convection term.

Liquid phase velocities are related to the volume flows in each section while the adsorbent movement in the case of SMB is equal to the column volume moved per shifting time:

$$u_{\mathrm{int},j,\mathrm{SMB}} = \frac{\dot{V}_j}{\varepsilon A_{\mathrm{c}}} \tag{6.205}$$

$$u_{\mathrm{ads}} = \frac{\dot{V}_{\mathrm{ads}}}{A_c(1-\varepsilon)} = \frac{V_{\mathrm{c}}}{A_c\, t_{\mathrm{shift}}} = \frac{L_{\mathrm{c}}}{t_{\mathrm{shift}}} \tag{6.206}$$

6.7.4
Comparison between TMB and SMB model

Concentration profiles of SMB models with three or more columns per section show good agreement with those of TMB models (Ruthven et al., 1989, Lu and Ching, 1997 and Pais et al. 1998a). In this case, TMB models may be used to design SMB processes. Due to high investment costs SMB plants often have fewer columns. Therefore, SMB plant behavior differs considerably from the TMB process (Chapter 7, Chu and Hashim, 1995, Strube et al., 1998a and Pais et al., 1998a). Anyhow, even today (initial) process design is often based on true-moving bed assumptions (Storti et al., 1993b, Charton and Nicoud, 1995, Heuer et al., 1998, Ma and Wang, 1997 and Chapter 7).

Figure 6.39a gives an example of the difference in axial concentration profiles between TMB and SMB models, where the number of columns per SMB section is

a)

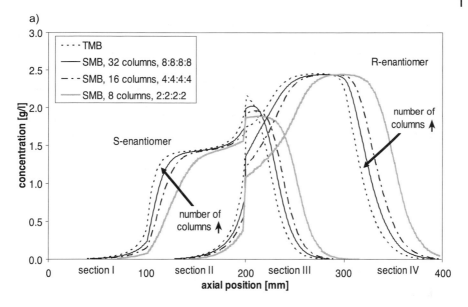

b)

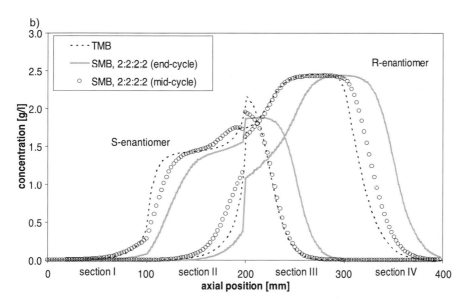

Fig. 6.39 Axial concentration profile for (a) TMB and SMB processes with different number of columns per section (end-cycle profiles); (b) TMB and SMB processes with 8 columns and profiles at end-cycle and mid-cycle (separation of EMD53986, equal operating parameters).

varied while the overall bed length is kept constant. The operating parameters are taken from an optimized TMB process with 99.9% purity of the product streams. Clearly, the end-cycle SMB profiles approach the TMB profile only for a high number of columns.

Some authors compare the mid-cycle profiles (Ma and Wang, 1997) with those of the TMB, in which case the observed deviation is different (Fig. 6.39b) – the difference between the SMB and the TMB profiles is smaller at the middle of the cycle than at the end. The use of end-cycle-profiles, however, permits a direct interpretation in terms of achievable purity and is thus related to the determination of operating conditions guaranteeing complete separation (Chapter 7).

6.7.5
Process and Operating Parameters

To simulate an SMB process, the model parameters of the individual columns (Section 6.5) and, if necessary, the dead volumes (Fig. 6.37) must be known.

It is advisable to use nearly identical columns (concerning bed length and packing structure), which is easily checked by comparing their outgoing signals from batch experiments with small injected amounts and determining the Henry coefficient. Comparison of the product peaks for the individual columns may also be used as a test for the packing. If strong deviations occur, the packing procedure must be repeated and checked for errors.

Model parameters should then be obtained only for one column, as these should be the same for all. Column parameters are determined by batch experiments or are known from previous tests. Finally, the dead volume inside an SMB plant (Fig. 6.37) can be determined from tracer pulse experiments by connecting the respective part of the plant directly to the pump and a detector.

For the operating parameters, it is necessary to specify five independent variables. A common method is to specify the four internal flow rates ($\dot{V}_{I-IV}$) and the switching time (SMB model) or solid flow (TMB model). Note that the four external flow rates have to fulfill the overall mass balance and supply only three independent specifications.

To achieve a separation, the operating parameters are selected on the basis of the effective migration velocity of the components in a counter-current bed (discussed in Chapter 7).

6.7.6
Experimental validation

6.7.6.1   Introduction
Ever since the development and application of mathematical models for the design of SMB processes, beginning in the 1980s, efforts have been made to validate these models by comparing measured and simulated data. SMB and TMB models of different complexity have been used for this task, for example the ideal and equilibrium

dispersive SMB model as well as TMB and SMB transport-dispersive models. A common approach is to compare the internal concentration profile with the simulation results. Another way is to use the product concentrations at extract and raffinate and their temporal evolution. Some characteristic points of the internal profiles can be obtained from taking samples (see sample valve in Fig. 6.35) and off-line analysis. Classical examples include those from applications in the petrochemical and sugar industries (Ruthven and Ching, 1989, Hashimoto et al., 1993a, Ching et al., 1993, Ma and Wang, 1997, Deckert and Arlt, 1997 and Beste et al., 2000). Example applications in enantioseparation are given in Pais et al. (1997a and 1997b), Heuer et al. (1998) and Kniep et al. (1998 and 1999). One disadvantage of taking samples is the normally low data density. Often only one sample per shifting interval is taken to reduce experimental effort and to limit the impact of withdrawn sample on the internal concentration profile. As the shifting interval may be up to several thousand seconds a detailed and meaningful model validation is difficult.

Chromatographic separations with product purities exceeding 99% and high separation costs require precise prediction of the optimal process design. This demands carefully validated models, especially in the separation sections of the SMB plant. Consequently, methods are needed to increase the number of measured data points.

This can be achieved by on-line analysis of the concentration profile. As depicted in Fig. 6.35, one or more detectors are placed in the recycle stream. They are positioned in front of the recycle pump, because some detectors are sensitive to high pressure. Due to the shifting of external streams the detector „travels" through every process section during one cycle (see below).

Yun and Guichon et al. (1997a and 1997b) placed one UV detector in the recycle stream to measure the fronts in sections I and IV for the separation of phenylethanol and phenylpropanol. The use of only one detector allows the measurement of the pure components in each regeneration section, but the concentration of the mixture in the separation section cannot be determined. Jupke et al. used a multi-detector set-up for binary separation, which gave the possibility of measuring the concentration profiles of all components in all sections individually (Jupke 2004, Epping, 2005, Mannschreck, 1992, Jupke et al., 2002 and Mihlbachler et al., 2002).

The general relation between the data obtained in the recycle stream, which can be used for model validation, and the axial profile, which can be used for process analysis, is given in Fig. 6.40. The set-up is identical to Fig. 6.35. In Fig. 6.40a, the simulated temporal profile measured behind column 8 during one complete cycle is shown. The concentration values after each shifting interval are given as symbols. In the ideal case they are identical to those of the samples taken for off-line analysis. As the process is in cyclic steady state, the axial profiles after each switching time are identical, only their position is shifted by one column. Thus, the symbols marked above can be translated into points of the axial profile at the end of one shifting period (Fig. 6.40b). Simulated curves are added for comparison. Although it is recommended to use the end-of-interval axial profile for process characterization (Chapter 7), the situation at mid-interval may be determined when taking the samples at these specific times. If no asymmetric distribution of dead volumes is present in the plant, the end-of-shifting-interval profiles are identical to those at the end of a cycle.

a)

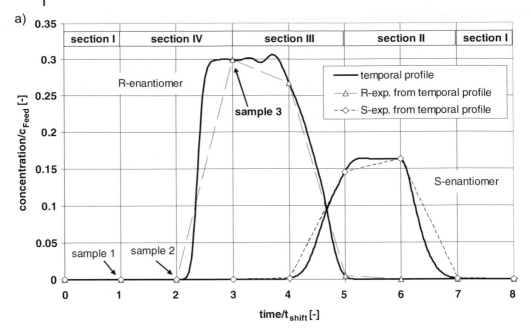

b)

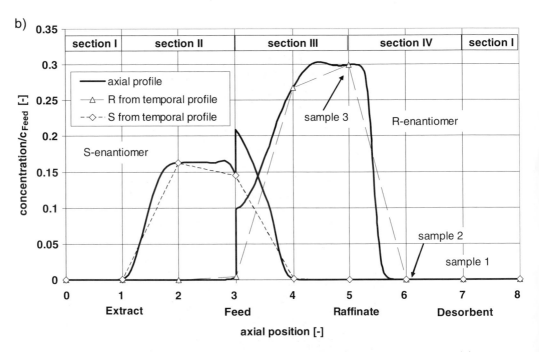

Fig. 6.40 Relationship between the temporal profile measured behind the eighth column (a) and the axial profile at the end of cycle just before the feed position is switched in front of the fifth column (b); symbols represent the identical points.

The axial position in Fig. 6.40b is normalized to the length of each column and, therefore, column 1 lies between the coordinates 0 and 1. The temporal positions 0, 1, 2...8 correspond to the axial position 8, 7...0. The feed is injected in front of column 5 at the start of the cycle and the last feed position at the end is in front of column 4 (Fig. 6.40). For clarity, no graphical shift is performed to have the feed position in front of column 5 as, for example, in Fig. 6.34.

Importantly, sampling gives only a limited number of points in the axial profile. The complete curve can not be reconstructed from any temporal measurement, as the concentration fronts inside the columns change from the beginning to the end of the shifting interval. However, increasing the number of experimental samples per interval of course allows a comparison of the simulated temporal profile on a broader base without on-line measurement.

Concerning model validation, the simulated complete axial profile used for process analysis can be trusted to represent the experimental state inside the plant if the temporal evolution of the concentration is properly predicted by the model.

6.7.6.2 **Results**

In the following some examples based on our own research are given for the validation of the SMB transport dispersive model, using an on-line detection system in the recycle stream. All flowsheet, column and plant models were implemented in the gPROMS (PSenterprise, UK) simulation tool and solved with OCFE methods (Section 6.4).

Fig. 6.41 Photograph of an SMB plant „Licosep Lab" from Novasep.

A more detailed discussion of the experiments and simulations for various operating points and conditions can be found in Jupke (2004) and Mihlbachler et al. (2002). Example results for the systems EMD53986, Tröger's Base and fructose–glucose are discussed. Pulse tests with the solutes proved that all columns for each SMB set-up behaved identically within small deviations. Most of the experiments were performed on a commercial plant „Licosep Lab" (Fig. 6.41) from Novasep (France).

In all cases good agreement between simulation and experiment was found.

Simulations were performed using the extended SMB model (Fig. 6.37), which included dead volumes and synchronous as well as asynchronous port switching. The latter accounts for the dead volume in the recycle stream, which is the dominant contribution (Hotier and Nicoud, 1995 and Migliorini et al., 1999). As mentioned in the previous section, the temporal concentration profiles are shown and the partitions of the time axis mark the switching time. The additional time added after the eighth period is the asynchronous switching time.

Figure 6.42 shows the simulated and measured concentration profiles for EMD53986. Good agreement can be observed for this system, which is characterized by its strong coupled and nonlinear adsorption behavior. Only slight deviation in the position of the fronts and the maximum height are observed. Especially, the steepness of the fronts is reproduced very well. The model can also predict the start-up of the SMB-plant (Fig. 6.43) correctly. Deviations in the first cycle are presumably due to the pressure fluctuations often encountered directly after start-up. Good agreement between simulation and experiment was observed for other operating conditions, too.

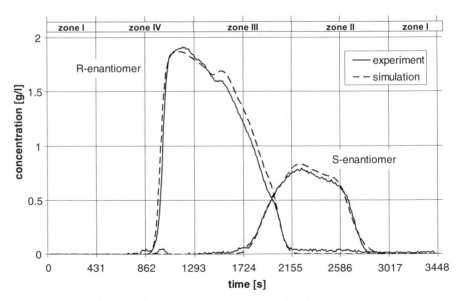

Fig. 6.42 Measured and simulated concentration profile in the SMB for EMD53986 (cycle 8, c_{feed} = $5 \, g \, l^{-1}$; for additional data see Appendix B.1. (Reproduced from Jupke et al., 2002.)

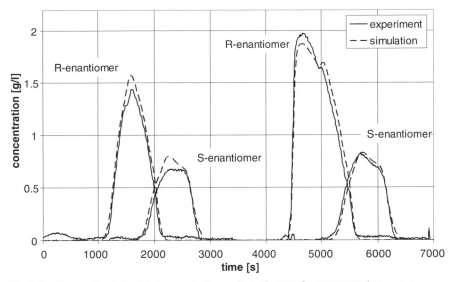

Fig. 6.43 Measured and simulated concentration profile in the SMB for EMD53986 during start-up (1. and 2. cycle; for other data see Fig. 6.42).

A racemic Tröger's base mixture was separated on an ICLC 16-10 plant (Prochrom) (Mihlbachler et al., 2001 and 2002 and Jupke, 2004). The measurement system consisted of polarimeter and UV detectors that had to be implemented in different parts of the product streams because the equipment does not have a recycle stream. All columns showed good homogeneity. Model validation was performed

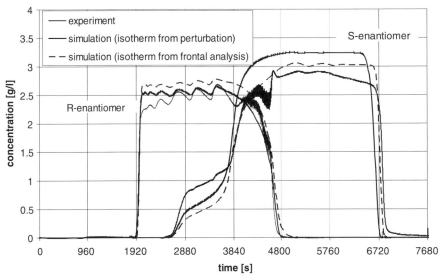

Fig. 6.44 Measured and simulated concentration profile in the SMB for Tröger's base (10th cycle, $c_{feed} = 7\,g\,l^{-1}$; for additional data see Appendix B.2). (Reproduced from Mihlbachler et al., 2002.)

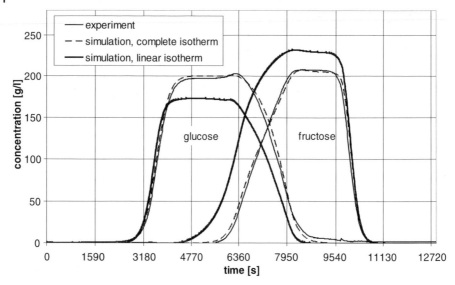

Fig. 6.45 Measured and simulated concentration profile in the SMB for fructose–glucose (10th cycle, $c_{feed} = 300\,g\,l^{-1}$; for additional data see Appendix B.3).

with the two different isotherms determined from frontal analysis (Eq. 6.187) and by the perturbation method (Eq. 6.186). The feed was pre-treated to remove impurities contained in the original substances, but a complete depletion was not possible. Figure 6.44 shows the resulting profiles.

Very good agreement is found for the *R*-enantiomer, while slight differences are encountered for the *S*-enantiomer. The positions of the fronts are well predicted, with only small deviations in the height. These slight differences can be attributed to impurities either affecting the isotherm determination or the value of the profile itself. The upwards curvature of the isotherm for the *S*-enantiomer for small concentrations causes a bulge to appear on the left front, which is qualitatively predicted by the simulations. The oscillations of the plateau concentrations are not observed for the other example systems. For example, for EMD53986 the higher mass transfer effects (lower number of stages compared with Tröger's base) effectively attenuate the oscillations.

The last example is the separation of fructose and glucose on the LicosepLab plant using a density sensor and a polarimeter for on-line detection (Jupke, 2004). After extensive experiments using different volume flows, feed concentration and temperatures, very good agreement with simulated data using the complete isotherm (Eq. 6.182) was found (Fig. 6.45). Figure 6.45 also shows the simulation results if only linear isotherms are assumed. This illustrates the importance of using complete isotherms with component interaction to obtain good agreement between experiment and simulation for industrial (high) feed concentrations. In sections of the process where only pure components are present, both isotherms lead to the same position of the front.

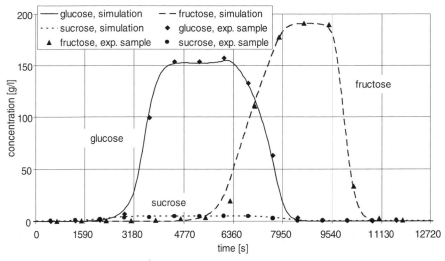

Fig. 6.46 Simulated and measured (sampled) concentration profile in the SMB for fructose–glucose ($c_{feed} = 300\,\mathrm{g\,l^{-1}}$) and sucrose ($c_{feed} = 18\,\mathrm{g\,l^{-1}}$) (8th cycle; for additional data see Appendix B.3).

Finally, results for the separation of an industrial feed mixture of fructose–glucose ($c_{feed} = 300\,\mathrm{g\,l^{-1}}$) containing 6% sucrose as an impurity are displayed in Fig. 6.46. Figure 6.47 shows an enlarged view of the sucrose profile. Once again, good agreement between experiment and simulation is found. The individual concentration profiles of this three-component mixture cannot be determined by two-detection systems. Therefore, additional samples (two per shifting interval) were taken and analyzed by HPLC.

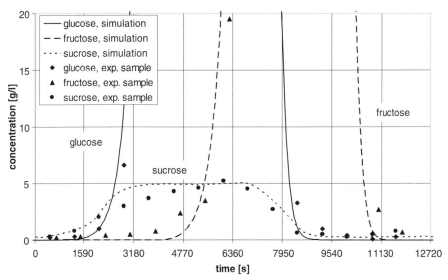

Fig. 6.47 Enlarged simulated and measured (sampled) concentration profile in the SMB for fructose–glucose–sucrose (other data see Fig. 6.46).**Table 6.1** Mass balance equations (general rate model).

Isotherm data for sucrose in this concentration range were determined from pulse experiments and represented by a linear isotherm. Because of the low adsorptivity, this impurity is collected at the raffinate together with the glucose.

7
Model Based Design and Optimization

A. Susanto, K. Wekenborg, A. Epping, A. Jupke

The necessity of optimizing chromatographic processes (and also other separation methods) in the pharmaceutical and biotechnological industries results from high separation costs, which represent a major part of the manufacturing cost. However, due to the numerous parameters and complex dynamic behavior, pure empirical design and optimization of the chromatographic processes are hardly possible. Therefore, mathematical models of the process are an essential basic requirement. While previous chapters deal with the optimal selection of a chromatographic system and process concepts, the present chapter deals with model-based design and optimization of a chromatographic plant, where the already selected chromatographic system and concept are applied. First, the basic principles of optimizing chromatographic processes will be explained. These include the introduction of commonly used objective functions and the degrees of freedom. Then, to reduce the complexity of the optimization and to ease scale-up of a plant, this chapter also emphasizes the application of dimensionless degrees of freedom. Examples for the scale-up of an optimized plant using these dimensionless parameters are also given.

Subsequently, methods for the model-based design and optimization of batch and SMB (Simulated Moving Bed) processes are introduced. For this purpose dynamic simulations of an experimentally validated model are used. When a model-based approach is not affordable, as for example in the design and optimization of complex SMB processes, a short-cut design and optimization method for SMB is introduced. Finally, the performances of both batch and SMB chromatography are compared.

Appendix B contains detailed information about the properties of the chromatographic systems used as examples in this chapter.

Preparative Chromatography. Edited by H. Schmidt-Traub
Copyright © 2005 WILEY-VCH Verlag GmbH & Co. KGaA, Weinheim
ISBN: 3-527-30643-9

7.1
Basic Principles

7.1.1
Objective Functions

Objective functions – such as yield, productivity or total cost – provide an evaluation of efficiency and quality of preparative chromatographic separations. Since values of the objective functions can be changed by altering operating and design parameters, they are dependent variables for any optimization problem. The definitions of each objective function are identical for every chromatographic process (e.g. batch or SMB) but different parameters must be applied for its calculation.

7.1.1.1 Characterization of Process Performance

The yield Y_i of a process is the ratio between the amount of a product (component i) that can be collected in the outlet and the amount that was introduced into the system through the feed stream within a batch cycle. Due to the continuous operation mode in an SMB plant, both collected and introduced amount of component i are calculated within a shifting interval.

$$Y_{i,\text{Batch}} = \frac{m_{\text{out},i}}{\dot{V}\, c_{\text{feed},i}\, t_{\text{inj}}} \tag{7.1}$$

$$Y_{i,\text{SMB}} = \frac{m_{\text{out},i}}{\dot{V}_{\text{feed}}\, c_{\text{feed},i}\, t_{\text{shift}}} \tag{7.2}$$

The amount of component i in the outlet stream can be calculated for a batch chromatographic process using Eq. 7.3, whereby $t_{1,i}$ and $t_{2,i}$ are the beginning and the end, respectively, of the fraction collection for pure component i.

$$m_{\text{out},i} = \dot{V} \int_{t_{1,i}}^{t_{2,i}} c_{i,\text{out}}(t)\, \mathrm{d}t \tag{7.3}$$

In comparison with batch chromatography t_1 and t_2 for an SMB plant are the beginning and the end, respectively, of a shifting interval. Furthermore, component i can be collected either from raffinate or extract stream.

$$m_{\text{out},i} = \dot{V}_{\text{rat/ext}} \int_{t_1}^{t_2} c_{i,\text{raf/ext}}(t)\, \mathrm{d}t \tag{7.4}$$

Using the purified amount of product i within a batch cycle, the average mass flow of purified product i (also called as production rate $\dot{m}_{\text{prod},i}$) in batch chromatography can be expressed as

$$\dot{m}_{\text{prod},i} = \frac{m_{\text{out},i}}{t_{\text{batch-cycle}}} \tag{7.5}$$

For an SMB plant the production rate must be determined within a shifting interval instead.

$$\dot{m}_{\text{prod},i} = \frac{m_{\text{out},i}}{t_{\text{shift}}} \tag{7.6}$$

The efficiency of a chromatographic separation can be expressed by comparing the purified amount of product i with the time taken and the resources necessary to produce it. This expression is referred to as productivity and can be applied in terms of:

• Volume specific productivity VSP_i (production rate over total solid adsorbent volume)
• Cross-section specific productivity ASP_i (production rate over free cross-section of the column)

$$ASP_{i,\text{Batch}} = \frac{\dot{m}_{\text{prod},i}}{A_c \, \varepsilon} \tag{7.7}$$

$$ASP_{i,\text{SMB}} = \frac{\dot{m}_{\text{prod},i}}{N_{\text{COL}} \, A_c \, \varepsilon} \tag{7.8}$$

$$VSP_{i,\text{Batch}} = \frac{\dot{m}_{\text{prod},i}}{V_{\text{solid}}} = \frac{\dot{m}_{\text{prod},i}}{V_c \, (1 - \varepsilon_t)} \tag{7.9}$$

$$VSP_{i,\text{SMB}} = \frac{\dot{m}_{\text{prod},i}}{V_{\text{solid,total}}} = \frac{\dot{m}_{\text{prod},i}}{N_{\text{COL}} \, V_c \, (1 - \varepsilon_t)} \tag{7.10}$$

Here, the number of columns N_{COL} in SMB has to be considered.

In many cases the cost of eluent (i.e. solvent) is not negligible and in some cases it even represents the greatest contribution to the total separation cost. Therefore, it is advisable in these cases to observe the eluent consumption, or better still, the efficiency of eluent-usage during the chromatographic separation. This can be characterized by specific eluent consumption EC_i, which means the amount of eluent required to purify a certain amount of product i.

$$EC_{i,\text{Batch}} = \frac{\dot{V}}{\dot{m}_{\text{prod},i}} \tag{7.11}$$

Since in an SMB plant the eluent is introduced into the system through the desorbent port as well as feed port, both flow rates have to be considered in the calculation.

$$EC_{i,\text{SMB}} = \frac{(\dot{V}_{\text{feed}} + \dot{V}_{\text{des}})}{\dot{m}_{\text{prod},i}} \tag{7.12}$$

Notably, eluent consumption is calculated by neglecting the influence of solute concentration since the change in liquid volume during the dissolution of solid components can generally be neglected.

Another objective function that is also often used as a boundary condition for a separation problem is the purity of a product i Pu_i. Product purity in a chromatographic process can be calculated by Eq. 7.13, where N_{COMP} is the number of components in the solution.

$$Pu_i = \frac{m_{out,i}}{\sum\limits_{i=1}^{N_{COMP}} m_{out,i}} \qquad (7.13)$$

7.1.1.2 Total Separation Cost

The importance of total separation cost as an objective function can be deduced from the work of Katti and Jagland (1998). This work gives a complete description of the contribution of different kinds of costs to the total separation costs for batch separation. Also, the influence of different physical properties of the separation system and the plant design is discussed. Katti and Jagland (1998) demonstrated that the cost structure of the separation problem changes with the scale of the separation. For smaller production amounts the contribution of capital, labor and maintenance costs to total cost is significantly higher than for bigger production rate. On this account, it is very important to consider the total separation cost, because only consideration of the separation cost will lead to an economical optimal process for both batch and SMB chromatography. Furthermore, selecting a „wrong" objective function will lead to a less economical process.

Yield, productivity, eluent consumption and purity are clear defined objective functions, while total separation cost and other objective functions related to economics depend on the individual company and are based on site-related parameters. Therefore, the calculation of total separation cost is complex due to various influencing parameters and cost structures. This chapter proposes the following cost functions, which can be easily adapted to specific conditions:

1. Separation problem independent costs (fixed costs)
   These can be considered as fixed costs and they are specific for each company.
   They can be divided into:
   - Annual operating cost ($C_{operating}$)
     Operating cost includes overhead costs as well as wages and maintenance cost.
   - Annual depreciation ($C_{depreciation}$)
     Annual depreciation is the allocation of investment cost over the depreciation years.
2. Separation problem dependent costs (variable costs)
   These costs are directly correlated to a given separation problem and the desired production rate. They are the annual costs for eluent, adsorbent and lost feed (also known as crude loss). All of these costs can be calculated by applying other objective functions that characterize the efficiency of the separation process (e.g. yield, productivity, etc.). Further requirements for the calculation of separation problem dependent costs are the production rate or the annual produced amount $\dot{m}_{prod,annual}$, prices for eluent f_{el} (€ per l eluent), for adsorbent f_{ads} (€ per kg adsor-

bent), and for feed f_{feed} (€ per kg feed). Additionally, to calculate the annual adsorbent cost C_{ads}, the lifetime of the adsorbent t_{life} has to be known or estimated.

$$C_{el} = EC_i \cdot \dot{m}_{prod,annual} \cdot f_{el} \tag{7.14}$$

$$C_{ads} = \frac{1}{VSP_i} \cdot \dot{m}_{prod,annual} \cdot \frac{\rho_{ads} f_{ads}}{t_{life}} \tag{7.15}$$

$$C_{crudeloss} = \frac{1 - Y_i}{Y_i} \cdot \dot{m}_{prod,annual} \cdot f_{feed} \tag{7.16}$$

The total annual separation cost, C_{total}, can then be determined by adding up the fixed costs as well as the variable costs.

$$C_{total} = C_{operating} + C_{depreciation} + C_{ads} + C_{el} + C_{crudeloss} \tag{7.17}$$

However, to be able to compare the total annual separation costs for different production scales, the production specific cost (cost per product amount) should be used instead:

$$C_{spec,total} = \frac{C_{total}}{\dot{m}_{prod,annual}} \tag{7.18}$$

7.1.2
Degrees of Freedom

7.1.2.1 Classification of Optimization Parameters
Models for a chromatographic column consist of many parameters that have already been classified into three groups in Chapter 6:

1. Operating parameters
   Alterable parameters during the operation of the plant, e.g.:
   - Flow rate
   - A number of operating parameters in batch chromatography result from the batch operation mode: switch times
   - To characterize the operation of an SMB plant precisely, these additional parameters are necessary: flow rate in each SMB section and shifting time
2. Design parameters
   Design parameters define the appearance of a plant and can not be changed during line operation, e.g.:
   - Column geometry (length and diameter)
   - Diameter of adsorbent particle
   - Maximum pressure drop
   - Additional degree of freedom in an SMB plant: column configuration (number of columns in each SMB section)

3. Model parameters

Model parameters are system inherent parameters that result from the choice of chromatographic system. They describe, for example, the following phenomena:

- Thermodynamics
- Fluid dynamics
- Dispersion effects
- Mass transfer resistance

A good process description based on validated models is needed to predict the position of the optimal state accurately. Since model parameters are determined experimentally on the chromatographic system, which is generally predefined before optimization starts, model parameters remain unchanged during the optimization. On this basis, all model-based optimization strategies in the following sections apply only to operating and design parameters.

7.1.2.2 Dimensionless Representation of Operating and Design Parameters

The chromatographic process contains many design and operating parameters. Hence, the optimization of all parameters requires a great amount of resources (e.g. faster computer, etc.) and more complex optimization algorithms. However, this can be avoided by summarizing the numerous design and operating parameters into a smaller number of dimensionless parameters, thus reducing the number of optimization parameters and the complexity of the optimization problem. This chapter deals with the dimensionless optimization parameters in a *transport-dispersive model*.

Dimensionless formulation of model equations for a chromatographic column can be found in Chapter 6. For clarity, the dimensionless parameters and the phenomena described by them will also be mentioned here:

1. Axial dispersion and convection: Bodenstein number

$$\text{Bo}_i = \frac{\text{convection}}{\text{dispersion}} = \frac{u_{\text{int}} L_c}{D_{\text{ax},i}} \tag{7.19}$$

2. Mass transfer and convection: modified Stanton number

$$\text{St}_{\text{eff},i} = \frac{\text{mass transfer}}{\text{convection}} = k_{\text{eff},i} \frac{6}{d_p} \frac{L_c}{u_{\text{int}}} \tag{7.20}$$

3. Adsorption:

For linear chromatography, adsorption behavior is determined by the dimensionless Henry-coefficient H_i.

$$H_i = \frac{Q_{\text{DL},i}}{C_{\text{p,DL},i}}$$

In most cases (nonlinear chromatography), however, the feed concentration plays a major role in the adsorption.

For example the Langmuir isotherm:

$$Q_{DL,i} = \frac{H_i \cdot C_{p,DL,i}}{1 + (c_{feed,i} \cdot b_i) \cdot C_{p,DL,i}}$$

Here the additional dimensionless number (product of $c_{feed,i}$ and b_i) must also be taken into account.

4. Process-related boundary conditions:
   - For batch processes: dimensionless injection time $t_{inj}/t_{0,int}$
     In this chapter another dimensionless number is used to represent the injection time, namely the loading factor LF_i, which was first introduced by Guiochon and Golshan-Shirazi (1989a) and is defined as

$$LF_i = \frac{\text{injected feed amount}}{\text{saturation capacity}} = \frac{c_{i,feed} \, \dot{V} \, t_{inj}}{q_{sat,i} \, V_c \, (1 - \varepsilon_t)} \tag{7.21}$$

The saturation concentration ($q_{sat,i}$) can be easily determined for all isotherm types with saturation concentration, for example the Langmuir type:

$$q_{sat,i} = \lim_{c_i \to \infty} \frac{H_i c_i}{1 + b_i c_i} = \frac{H_i}{b_i} \tag{7.22}$$

For other isotherm types with no saturation concentration (e.g. linear type) it is advisable to use the solid phase concentration at maximum feed concentration in the separation problem.

$$q_{sat,i} = q_{max,i} = q_i(c_{feed,max}) \tag{7.23}$$

The use of loading factor as degree of freedom is equivalent to the use of dimensionless injection time, as can be seen in Eq. 7.24.

$$LF_i = \frac{c_{i,feed}}{q_{sat,i}} \frac{\varepsilon}{1 - \varepsilon_t} \cdot \left(\frac{t_{inj}}{t_{0,int}} \right) \tag{7.24}$$

Since the loading factor is a more common parameter in practice, in this chapter it is preferred to the dimensionless injection time.

- For SMB processes: dimensionless shifting times $t_{shift}/t_{0,int,j}$

Once again, other dimensionless numbers are used to represent the shifting boundary conditions, namely the ratio m_j between net fluid flow in SMB section j and „simulated" solid flow, because it is an essential parameter for the design of counter-current chromatographic separation (Storti et al., 1993b).

$$m_j = \frac{\text{net fluid flow rate in section } j}{\text{simulated solid flow rate}} = \frac{\dot{V}_j - \dfrac{V_c \, \varepsilon_t}{t_{shift}}}{\dfrac{V_c \, (1 - \varepsilon_t)}{t_{shift}}} \tag{7.25}$$

In other words, the simplification of using the number of stages for process optimization is best applied if either mass transfer or dispersion dominates the peak broadening. Therefore, the optimization strategies discussed later in this chapter apply a validated transport dispersive model, which can flexibly consider mass transfer and/ or dispersion effect. Here, the number of stages is used as independent variable for the optimization criteria like productivity or eluent consumption. Another possible approach would be the use of simplified simulation model like equilibrium dispersive model (Seidel-Morgenstern, 1995).

Another important simplification, which makes the scale-up of chromatographic plants easier, is to always use the same feed concentration. Since higher feed concentration also results in higher productivity, feed concentration is fixed at the maximum allowed concentration, which depends on the chromatographic system and the solubility of the feed components. However, in SMB processes, feed concentration affects the size of the operating parameter region for a total separation and, therefore, should not be fixed at too high values in order not to hinder the ability to get „robust" operating parameters (Section 7.3). Thus, the remaining and applied degrees of freedom for the design and optimization methods for batch as well as SMB processes in this chapter are

- For batch processes: number of stages N_i and loading factor LF_i
- For SMB processes: total number of stages $N_{tot,i}$ and dimensionless flow ratios m_j

7.1.3
Scaling Up and Down

One of the advantages of the presented optimization method is the use of only a small number of optimization parameters, although nearly all design and operating parameters are considered by this method. Furthermore, chromatographic processes with identical dimensionless parameters will have similar behavior. In other words, the values of many objective functions (e.g. specific productivity, specific eluent consumption, etc.) for those processes will be the same and their dimensionless concentration profiles will also be identical. Therefore, no new optimization must be done after up scaling (or down scaling) the chromatographic plant, because the optimal dimensionless parameters from the old plant can be easily adapted to the new plant. This means:

- For batch processes:

$$N_{i,\text{new}} \geq N_{i,\text{old}} \tag{7.34}$$

$$LF_{i,\text{new}} = LF_{i,\text{old}} \tag{7.35}$$

- For SMB processes:

$$N_{\text{tot},i,\text{new}} \geq N_{\text{tot},i,\text{old}} \tag{7.36}$$

$$m_{j,\text{new}} = m_{j,\text{old}} \tag{7.37}$$

Since the adoption of the number of stages from the old chromatographic column into the new column can never be done exactly, the equal signs in Eqs. 7.34 and 7.36 are replaced with greater than equal signs. This should guarantee that the separation performance of the new column is equal to or better than the performance of the old column even if the concentration profiles and the values of objective functions are not exactly the same. The reason of this difference will be illustrated in the following subchapters with the aid of numbers of stages for two different chromatographic plants (an old, optimized plant and a new plant).

$$N_{i,\text{old}} = \frac{L_{c,\text{old}}}{\text{HETP}_{i,\text{old}}}$$

$$N_{i,\text{new}} = \frac{L_{c,\text{new}}}{\text{HETP}_{i,\text{new}}}$$

7.1.3.1 Influence of Different HETP Coefficients for Every Component

Two different plants will behave similarly only if the numbers of stages *for all components* are equal. If the equality condition is valid for the 1st component, then

$$N_{1,\text{new}} = N_{1,\text{old}}$$

$$\Leftrightarrow L_{c,\text{new}} = L_{c,\text{old}} \frac{\text{HETP}_{1,\text{new}}}{\text{HETP}_{1,\text{old}}}$$

Substituting this equation into the new number of stages for component 2 will result in

$$N_{2,\text{new}} = \frac{L_{c,\text{old}}}{\dfrac{\text{HETP}_{1,\text{old}}\,\text{HETP}_{2,\text{new}}}{\text{HETP}_{1,\text{new}}}}$$

The equality condition will therefore only be valid for the 2nd component, if

$$N_{2,\text{new}} = N_{2,\text{old}}$$

$$\Leftrightarrow \frac{\text{HETP}_{1,\text{old}}\,\text{HETP}_{2,\text{new}}}{\text{HETP}_{1,\text{new}}} = \text{HETP}_{2,\text{old}}$$

or as general expression

$$\frac{\text{HETP}_{i,\text{new}}}{\text{HETP}_{i,\text{old}}} = \frac{A_i + B_i u_{\text{int,new}}}{A_i + B_i u_{\text{int,old}}} = \text{const.} \tag{7.38}$$

Therefore, the equality condition for the number of stages is only fulfilled exactly by all components if the ratio between HETP of the old and new column are *equal for every component*. Due to the different HETP coefficients for each component the equality condition can only be exactly fulfilled if:

1. Both columns have the same length but different cross section areas

   Due to the identical column length the interstitial velocity in both columns must be kept constant in order to have the same column efficiency (i.e. same number of stages). Both columns have, however, different flow rates due to different cross section areas.

   $$u_{int,new} = u_{int,old}$$
   $$\Leftrightarrow \frac{HETP_{i,new}}{HETP_{i,old}} = 1 = \text{const.}$$

2. Dominant eddy diffusion

   For chromatographic system with negligible mass transfer resistance the eddy diffusion is the dominant effect. Thereby, HETP values are constant and independent of the interstitial velocity.

   $$\frac{HETP_{i,new}}{HETP_{i,old}} \approx 1 = \text{const.}$$

3. Dominant mass transfer resistance

   In this case the approximated HETP value is directly proportional to the interstitial velocity. Thus the HETP ratios between old and new column for all components are identical: it is always equal to the ratio of interstitial velocities.

   $$\frac{HETP_{i,new}}{HETP_{i,old}} \approx \frac{u_{int,new}}{u_{int,old}} = \text{const.}$$

Whenever those cases apply, the number of stages in the new and old column for other components will remain the same if the number of stages of one single component is not varied. Therefore, the column can be optimized by considering the loading factor and the number of stages of only one component (i.e. the reference component). Generally, however, the equality condition for all the number of stages can not be fulfilled exactly, so that a certain deviation has to be accepted. The magnitude of the deviation depends on the magnitude of the contributions to the HETP. The more dominant the influence of mass transfer resistance or eddy diffusion, the smaller the deviation between the concentration profiles. However, if the modified equality condition (Eqs. 7.34 and 7.36) is applied, the separation performance of the new column can be guaranteed to be equal to or better than that of the old column.

7.1.3.2 Influence of Feed Concentration

Actually, the definition of number of stages and HETP originates from linear chromatography though it can also be formulated for nonlinear chromatography. In nonlinear chromatography the number of stages and HETP are affected by the varying gradient of the adsorption isotherm; therefore, their value depends on the concentration range in the column. Due to the concentration dependency in the nonlinear case, the number of stages loses its original practical meaning as a measure of column efficiency.

For preparative chromatography, where we almost always have to deal with high feed concentrations and nonlinear adsorption isotherms, the following approach to the appropriate choice of feed concentration during model-based optimization is recommended:

1. Determination of HETP coefficients in the linear adsorption region (i.e. low feed concentration) as explained in Chapter 6.
2. Execution of model-based optimization using the same feed concentration that will be applied later in the preparative column. Generally, a high feed concentration is applied, which is close to the solubility limit of the feed components. Feed concentration is, therefore, not an optimization parameter and should not be varied during optimization. The number of stages, however, is calculated by applying the known HETP coefficients from the linear adsorption region.

Application of concentration independent HETP coefficients from the linear isotherm region in calculating the number of stages in a nonlinear region is only a formalism. However, since feed concentration is not varied during the optimization, the calculated number of stages can be nevertheless used further as it is a characteristic value for the specific chromatographic separation. Scale-up of the chromatographic column using an identical dimensionless number of stages will continue to produce identical concentration profiles and identical values of many objective functions (with relatively small deviation), provided the feed concentration is also constant during scale-up.

7.1.3.3 Examples for a Single Batch Chromatographic Column

To demonstrate the equality of concentration profiles, results from simulation of a transport-dispersive model with different design- and operating parameters but identical number of stages and loading factor are compared. Here it has to be kept in mind that these model calculations agree with experiments within measurement accuracy (Chapter 6).

The first exemplary problem is the separation of a racemic mixture EMD53986 (R:S-enantiomers = 1:1) on the chiral stationary phase Chiralpak AD from Daicel (20 μm particle diameter) (Tab. 7.1).

Please note that only the number of stages and loading factor of the R-enantiomer are taken into account, since R-enantiomer is used as the *reference component* in this separation problem. Furthermore, the number of stages is calculated using the

Table 7.1 Different design and operating parameters for a single column (EMD53986) $N_R = 25$; $LF_R = 1.25$; $c_{feed,R} = c_{feed,S} = 2.5\ g\ l^{-1}$.

	$\dot{V}$ (ml min^{-1})	u_{int} (cm s^{-1})	t_{inj} (s)	L_c (cm)	d_c (cm)	$St_{eff,R}$ (–)	Bo_R (–)
EMD-1	23.5	0.2247	208.0	10.8	2.5	21.63	3061.22
EMD-2	54.8	0.5237	206.6	25.0	2.5	21.48	7108.14
EMD-3	109.0	0.2605	207.7	12.5	5.0	21.59	3544.69

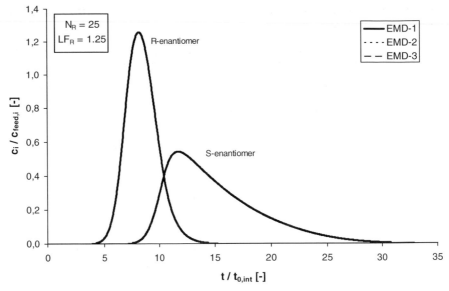

Figure 7.1 Equality of concentration profiles in a chromatographic column (EMD53986).

experimentally determined HETP coefficients (Eqs. 7.28 and 7.29). All applied model parameters (isotherms, mass transfer coefficients, etc.) for the simulation of this enantiomeric separation are already experimentally validated. The simulation results (i.e. simulated chromatograms) are presented in Fig. 7.1 in dimensionless form using the ratios $c_i/c_{feed,i}$ and $t/t_{0,int}$ instead of c_i and t, whereby $t_{0,int}$ is the mean residence time under non-adsorbing, non-penetrating conditions (i.e. the residence time of large tracer molecules, e.g. dextran).

$$t_{0,int} = \frac{L_c}{u_{int}} \tag{7.39}$$

As can be seen in Tab. 7.1, Bo_i far exceeds $St_{eff,i}$, clearly indicating that mass transfer resistance plays a major role in this separation problem. Therefore, the concentration profiles are completely identical despite the different Bo_i. This justifies simplification of the optimization method by using the number of stages instead of Bo_i and $St_{eff,i}$. Furthermore, the number of stages is calculated using the HETP coefficients determined in the linear isotherm region. However, even if the concentration range in the column no longer lies within the linear region of the adsorption isotherms, deviations of the concentration profiles are insignificant as long as the same feed concentration is used (Fig. 7.1).

Dimensionless degrees of freedom do not always transfer to other column designs as perfectly as shown in Fig. 7.1. In many cases some deviations in the concentration profiles have to be taken into account. These cases will now be demonstrated using a second exemplary separation problem: the chromatographic separation of 1:1 mixture of glucose and fructose on ion-exchange resin Amberlite CR 1320 Ca from Rohm & Haas (325 µm particle diameter) (Tab. 7.2).

Table 7.2 Different design and operating parameters for a single column (Glu/Fruc) $N_{Glu} = 100$; $LF_{Glu} = 0.5$; $c_{feed,Glu} = c_{feed,Fruc} = 300\,g\,l^{-1}$.

	$\dot{V}$ (ml min^{-1})	u_{int} (cm s^{-1})	t_{inj} (s)	L_c (cm)	d_c (cm)	$St_{eff,Glu}$ (–)	Bo_{Glu} (–)
FG-1	5.8	0.0490	328.1	56	2.6	14.76	950.15
FG-2	2.4	0.0201	428.0	30	2.6	19.25	505.52
FG-3	13.6	0.0313	367.7	40	5.0	16.54	676.07

Compared with the EMD53986 separation, Bo_i are smaller although they are still much greater than $St_{eff,i}$. This indicates that, although mass transfer effects still dominate, axial dispersion now plays a small but significant role. One reason why the axial dispersion coefficient grows is surely the bigger particle diameter at 325 µm. A certain deviation during the scale-up has therefore to be taken into account (Fig. 7.2).

Since the concentration profiles for this exemplary system are not perfectly identical, scale-up has to be done with caution. Especially, the „greater than or equal to" condition (Eq. 7.34 or 7.36) has to be fulfilled for all components. Therefore, in this case it is not sufficient to look only at the number of stages of one component but the number of stages of all components has to be taken into account. Table 7.3 and Fig. 7.3 show the resulting operating points and concentration profiles, if only the number of stages of glucose is considered during the scale-up. In this example the column „FG-2" is scaled up by changing its length from 30 to 100 cm.

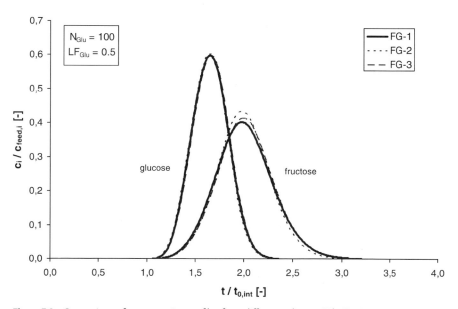

Figure 7.2 Comparison of concentration profiles from different columns (Glu/Fruc).

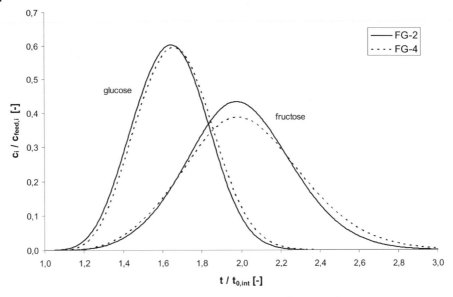

Figure 7.3 Comparison of concentration profiles with $N_{i,new}$ less than or equal to $N_{i,old}$.

Table 7.3 Different design and operating parameters with $N_{i,new}$ less than or equal to $N_{i,old}$.

	$\dot{V}$ (ml min^{-1})	t_{inj} (s)	L_c (cm)	d_c (cm)	N_{Glu} (−)	N_{Fruc}	LF_{Glu} (−)	LF_{Fruc}
FG-2	2.4	428.0	30	2.6	100.0	48.2	0.5	0.3
FG-4	11.5	293.4	100	2.6	100.0	38.8	0.5	0.3

Table 7.4 Different design and operating parameters with $N_{i,new}$ greater than or equal to $N_{i,old}$.

	$\dot{V}$ (ml min^{-1})	t_{inj} (s)	L_c (cm)	d_c (cm)	N_{Glu} (−)	N_{Fruc}	LF_{Glu} (−)	LF_{Fruc}
FG-2	2.4	428.0	30	2.6	100.0	48.2	0.5	0.3
FG-5	9.2	368.7	100	2.6	122.0	48.2	0.5	0.3

Table 7.3 shows that there are fewer stages of fructose in the new column („FG-4")
than in the old column („FG-2"). This causes a peak broadening of fructose, thus
worsening the separation of glucose and fructose (Fig. 7.3).

However, this can be avoided by taking fructose as reference component instead of
glucose. The resulting numbers of stages for both components in the new column
(„FG-5") are now greater or equal than in the old column (Tab. 7.4). The peak width
of glucose even decreases and, therefore, the worsening of the separation due to
scale-up is avoided (Fig. 7.4).

In summary the transport-dispersive model together with the simplified degrees of
freedom N_i and LF_i are very useful for optimization and scale-up of chromatographic
processes.

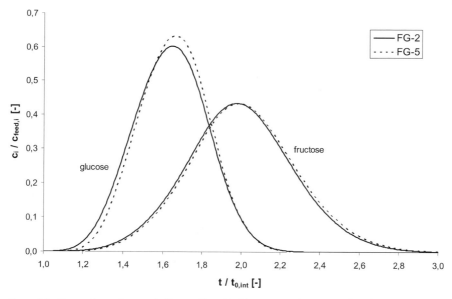

Figure 7.4 Comparison of concentration profiles with $N_{i,new}$ greater than or equal to $N_{i,old}$.

7.1.3.4 Examples for SMB Processes

An analogous procedure can also be applied to SMB processes (Tab. 7.5). The following compares the axial concentration profile of several 8-column SMB processes with different operating and design parameters but identical number of stages and dimensionless flow rate in each SMB section (m_j) (Tab. 7.6). For this purpose we use

Table 7.5 Identical dimensionless parameters for the SMB processes.

	SMB-1, -2 and -3
$N_{tot,R}$	66.5
m_1	34.0
m_{II}	6.9
m_{III}	10.9
m_{IV}	5.7

Table 7.6 Different operating and design parameters for SMB plant.

	SMB-1	SMB-2	SMB-3
N_{COL} (–)	8 (2 per section)	8 (2 per section)	8 (2 per section)
L_c (cm)	4.5	5.2	7.6
d_c (cm)	2.5	2.5	2.5
t_{shift} (s)	153.7	153.5	153.0
$\dot{V}_{des}$ (ml min^{-1})	68.3	79.1	115.9
$\dot{V}_{ext}$ (ml min^{-1})	65.4	75.7	111.0
$\dot{V}_{feed}$ (ml min^{-1})	9.7	11.2	16.4
$\dot{V}_{raf}$ (ml min^{-1})	12.6	14.5	21.3

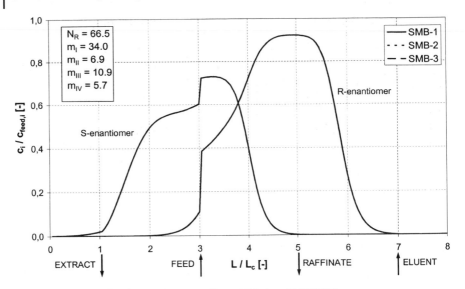

Figure 7.5 Equality of axial concentration profiles in SMB plant (EMD53986).

the same separation problem, a racemic mixture EMD53986, and the same feed concentration (i.e. $2.5\,g\,l^{-1}$) as above. Once again, R-enantiomer is used as reference component.

After the process simulation applying the experimentally validated model, all axial concentration profiles are plotted in a dimensionless diagram using $c_i/c_{feed,i}$ and L/L_c as axes. These profiles are taken at the end of a shifting interval from quasi-steady state. Figure 7.5 illustrates that all concentration profiles are completely identical."

Similar to batch processes, the transfer of number of stages is not always as perfect, as in the case of enantiomeric separation of EMD53986. If significant deviations between the concentration profiles are observed, another component has to be taken as reference component so that the total number of stages of every component in the new plant is greater or equal to the number of stages in the old plant (Eq. 7.36).

7.2
Batch chromatography

7.2.1
Fractionation Mode (Cut Strategy)

Within batch chromatography there exist many approaches for optimization. Basically, though, they can be classified into two approaches, which differ in their fractionation mode or cut strategy (Fig. 7.6).

In cut strategy II the products are collected consecutively. Or more precisely, the next fraction collection starts immediately after the previous fraction. High purity

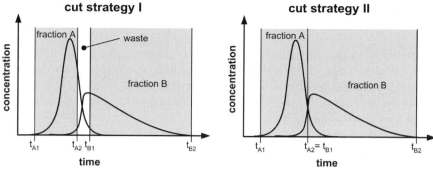

Figure 7.6 Cut strategies for a binary mixture.

demand can therefore only be satisfied by reducing the feed throughput. Furthermore, purity and yield are coupled and can not be varied independently. A high purity demand will consequently result in high yield.

Cut strategy I allows a waste fraction between the product fractions, which can be either discarded or processed further (e.g. by recycling the waste fraction or by application of other separation steps, such as crystallization, to purify it). Due to the introduction of a waste fraction the optimization problem gains additional degrees of freedom, i.e. times for the beginning and the end of waste collection. These additional degrees of freedom, however, will not increase the complexity of the optimization, because they are pinpointed automatically by the purity demand, which serves as a boundary condition for the optimization problem.

By use of cut strategy I the optimization problem can now be formulated with greater flexibility. For example, high purity demand can be satisfied either by reducing feed throughput or by increasing waste fraction, thus reducing the yield of separation. It is, however, essential in this cut strategy to consider the additional cost due to feed loss or cost for further treatment of waste fraction. Separation cost optimization with very high feed cost will often result in a very small waste fraction or even „base line separation", which can be interpreted as the marginal case for cut strategy I if t_{A2} equal to t_{B1} (Fig. 7.6). However, if feed cost is comparable with other costs, then it is more favorable to apply a waste fraction, thus reducing the yield but maximizing feed throughput at constant purity.

Since cut strategy II is only the marginal case of cut strategy I and since the application of cut strategy I provides more flexibility and often leads to a better global optimum, only *cut strategy I* will be used for model-based optimization in this chapter.

7.2.2
Design and Optimization Strategy for Batch Chromatographic Column

The main idea behind the design and optimization strategy for batch chromatographic columns is the equality of concentration profiles if their dimensionless parameters (i.e. number of stages and loading factor) are identical. As mentioned previously, column performance under this condition will be also identical (e.g. volume specific productivity, specific eluent consumption and yield at a defined purity). In other words, process performance will only vary if the number of stages and loading factor are varied. Therefore, it does not really matter which design or operating parameters are varied during the optimization, as long as this variation involves the variation of number of stages and loading factor. Based on this main idea a strategy can be developed that is suitable for the individual demand and boundary condition of the separation problem.

This chapter proposes a strategy that should be suitable for many separation problems. The following strategy should also be suitable for the optimization of operating parameters of an existing plant as well as the design of a new column:

- *Applied column model*: experimentally validated transport-dispersive model.
- *Applied cut strategy*: waste fraction is allowed (cut strategy I, Fig. 7.6)
- *Objective function*: to be able to compare total separation costs independently of production rate, the total separation cost per product amount $C_{spec,total}$ is used. Since the consideration of fixed costs in the cost structure is usually handled differently and is subject to company policies, it does not make sense to consider the fixed costs in this book. Therefore, the optimization is focused on variable costs only:

$$C_{spec,total} = C_{spec,el} + C_{spec,ads} + C_{spec,crudeloss}$$

$$C_{spec,total} = EC \cdot f_{el} + \frac{\rho_{ads}}{VSP\, t_{life}} f_{ads} + \frac{1-Y}{Y} f_{feed} \qquad (7.40)$$

- *Boundary conditions*: purity, maximum pressure drop Δp_{max}, desired production rate $\dot{m}_{prod}$. Since purity demand has to be fulfilled, the switch times for the fraction collection (t_{A1}, t_{A2}, t_{B1} and t_{B2}) can be determined and they are not subjects of optimization.
- The number of stages plays a major role in this optimization strategy and it can be easily calculated using the HETP equation (Eqs. 7.28 and 7.29). Therefore, experimental HETP coefficients for every feed component, or at least for the reference component in the separation problem, have to be acquired before running this optimization strategy.

The first step in this design strategy is to assume any initial values for the column geometry (L_c and d_c) (Fig. 7.7). If only the operating parameters of an existing chromatographic plant have to be optimized, L_c and d_c should be the length and diameter, respectively, of the current column. Using these initial values a series of dynamic simulations can be started to determine the values of following functions that characterize the column performance at different numbers of stages and loading factors:

- Volume specific productivity VSP
- Specific eluent consumption EC
- Yield at predefined purity Y

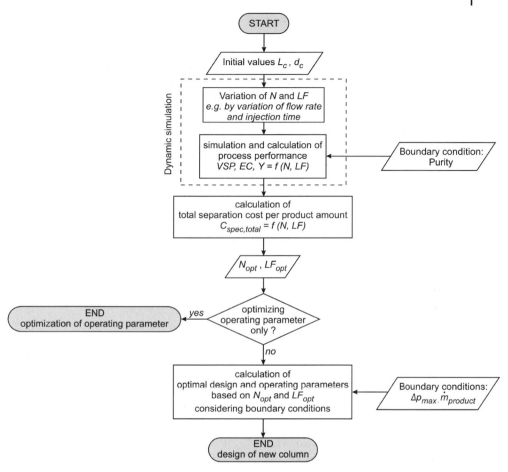

Figure 7.7 Design and optimization strategy for batch chromatographic column.

Provided prices for eluent, adsorbent and feed loss are known, the total separation cost per product amount $C_{spec,total}$ at each number of stages and loading factor can also be calculated (Eq. 7.40). As explained in Section 7.1.3.1, it is sufficient in most cases to take the number of stages and loading factor of a reference component into account. The number of stages and loading factor can be easily varied, e.g. by varying the injection time t_{inj} and flow rate $\dot{V}$, whereby the maximum pressure drop can be neglected for the time being. However, during optimization of an existing plant it has to be taken into account that a very small number of stages (i.e. very large flow rates) can not be realized due to the limiting pressure drop of the plant.

Afterwards, with the help of the results from dynamic simulation, the optimal number of stages N_{opt} and loading factor LF_{opt} can be determined where the value of objective function (i.e. total separation cost per product amount) is at its minimum. The applied total separation cost per product amount depends on productivity, eluent consumption and yield, which in turn only depend on the number of stages and loading factor. Therefore, any process with an identical number of stages and loading factor will also have the same total separation cost per product amount. Please consider that this is only valid for variable costs. Fixed costs should not be expressed as functions of number of stages and loading factor, because they only depend on instrumental effort, not process performance. Therefore, consideration of fixed costs in the total separation cost would require an additional optimization loop to optimize the instrumental effort. However, in practice fixed costs will hardly be an object of global optimization. In most cases there will be several options and the optimization task is reduced to several case studies.

When optimizing the operating parameters of an existing plant there is no need to determine the optimal column design with the help of N_{opt} and LF_{opt}. Optimization procedures can be terminated at this point because the operation of the batch column already reaches its optimum here. However, during the design of a new column one more step is necessary, to determine the optimal column geometry and operating parameters based on N_{opt}, LF_{opt} and the boundary conditions:

1. N_{opt} and δp_{max}

   Using optimal number of stages (N_{opt}), the unknown interstitial velocity, $u_{int,opt}$, and column length, $L_{c,opt}$, can be correlated as

$$N_{opt} = \frac{L_{c,opt}}{HETP(u_{int,opt})} = \frac{L_{c,opt}}{A + B u_{int,opt}}$$

   where A and B are the HETP coefficients of the reference component. At the same time, the maximum limit for pressure drop should not be violated. If the Darcy equation is applied, $L_{c,opt}$ and $u_{int,opt}$ can also be correlated as

$$\Delta p_{max} = \psi \cdot \frac{u_{int,opt} L_{c,opt} \eta_1}{d_p^2} \tag{7.41}$$

   It is a common misapprehension to assume that $u_{int,opt}$ for a column is always the maximum allowed interstitial velocity limited by the pressure drop for a certain column length. The truth is that for every objective function there are numerous optimal combinations of interstitial velocities and column lengths, which are represented by one optimal number of stages, N_{opt}. The problem is that this N_{opt} can not be realized if the selected column length is too large, so that the corresponding interstitial velocity exceeds the maximum velocity due to the limiting pressure drop. Therefore, the column length should be so selected that the optimal number of stages can be realized at the maximum or less pressure drop. Both unknown variables ($L_{c,opt}$ and $u_{int,opt}$) can therefore be calculated, considering both N_{opt} and δp_{max}.

$$L_{c,opt} = \frac{1}{2} \left(N_{opt} A + \sqrt{N_{opt}^2 A^2 + 4 N_{opt} B \frac{\Delta p_{max} d_p^2}{\psi \eta_1}} \right) \tag{7.42}$$

$$u_{\text{int,opt}} = \frac{\Delta p_{\max} d_p^2}{\psi \eta_1 L_{\text{c,opt}}} \tag{7.43}$$

If the effect of mass transfer resistance in the HETP equation is very dominant, then Eq. 7.42 can be simplified to

$$L_{\text{c,opt}} = \sqrt{N_{\text{opt}} B \frac{\Delta p_{\max} d_p^2}{\psi \eta_1}} \tag{7.44}$$

2. Desired production rate, $\dot{m}_{\text{product}}$

Since values that characterize column performance (i.e. yield, eluent consumption and productivity) at the optimal point are known, the required column diameter for a desired production rate can be easily determined using volume-specific productivity at the optimal point (Eq. 7.9).

$$d_{\text{c,opt}} = \sqrt{\text{VSP}_{\text{opt}} \frac{\dot{m}_{\text{product}}}{(1 - \varepsilon_t)} \cdot \frac{4}{\pi L_{\text{c,opt}}}} \tag{7.45}$$

The optimal flow rate can then be determined.

$$\dot{V}_{\text{opt}} = u_{\text{int,opt}} \varepsilon \frac{\pi}{4} d_{\text{c,opt}}^2 \tag{7.46}$$

3. Optimal loading factor, LF_{opt}

Using the optimal loading factor the last unknown operating parameter, the injection time, can be determined.

$$t_{\text{inj,opt}} = \frac{\text{LF}_{\text{opt}} q_{\text{sat}} \pi d_{\text{c,opt}}^2 (1 - \varepsilon_t)}{4 c_{\text{feed}} \dot{V}_{\text{opt}}} \tag{7.47}$$

7.2.3
Process Performance Depending on Number of Stages and Loading Factor

To improve the basic understanding of column performance in a batch chromatographic separation, the dependency of yield, eluent consumption, and productivity on the applied dimensionless parameters (i.e. number of stages and loading factor) will be investigated in here. Once again, the batch chromatographic separation of a racemic mixture EMD53986 on chiral stationary phase Chiralpak AD from Daicel (20 μm particle diameter) will be used as an exemplary chromatographic system. All the figures in this chapter result from the dynamic simulation of an experimentally validated column model.

Figure 7.8 illustrates the dependency of yield for 99% purity on the dimensionless parameters, showing that high yields can be realized at very low column loadings (which correspond to loading factors < 1). Here only a small amount of waste fraction is required. As the loading factor increases, the overlapping region in the concentration profile grows rapidly. This causes the amount of waste fraction to increase

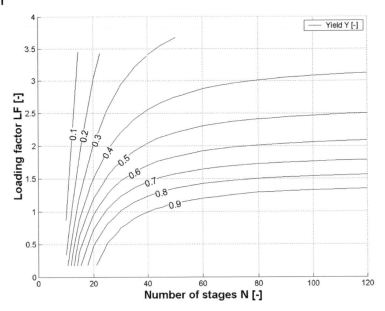

Figure 7.8 Dependency of yield for 99% purity on number of stages and loading factor.

in order to meet the purity demand, thus reducing the yield. However, the size of the overlapping concentration region, and consequently also the amount of waste fraction, can be reduced by increasing the number of stages, which also increases the peak resolution. Therefore, high yield at relatively higher loading factor can be realized in a column with a large number of stages.

Contrary to yield the productivity has a maximum in the investigated parameter region (Fig. 7.9). An increase of number of stages increases the yield due to better peak resolution. Separation becomes more efficient and a higher loading factor can be applied at constant yield, which, at first, results in a steep rise of production rate and volume specific productivity. But a higher number of stages also involves greater column length (i.e. bigger column) and/or reduction of interstitial velocity. The decrease of interstitial velocity increases the mean residence time of solutes in the column, thus increasing the batch cycle time. If the interstitial velocity falls below a certain point, the high batch cycle time will start to reduce production rate in spite of the higher loading factor. Therefore, a further increase in the number of stages will cause a smooth decrease in the volume-specific productivity, which is defined as production rate over required adsorbent volume.

The influence of loading factor on productivity is roughly similar to that of the number of stages. At first, the higher loading factor the higher production rate, and thus productivity also increases. However, as seen in Fig. 7.8, the yield already drops rapidly at low loading factors (LF at ca. 1). Therefore, a further increase of loading factor will cause the production rate and productivity to decrease despite higher feed throughput since a considerable amount is lost in the waste fraction.

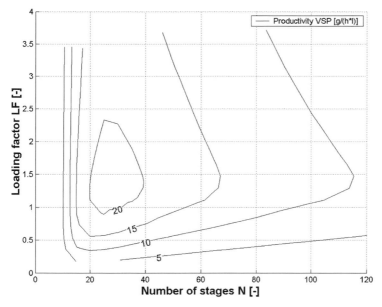

Figure 7.9 Dependency of volume specific productivity on number of stages and loading factor.

Figure 7.10 illustrates the *inverted* values of specific eluent consumption in dependency of dimensionless parameters. Higher values are therefore the more favorable values in Fig. 7.10 as they indicate a small eluent consumption. The dependency of

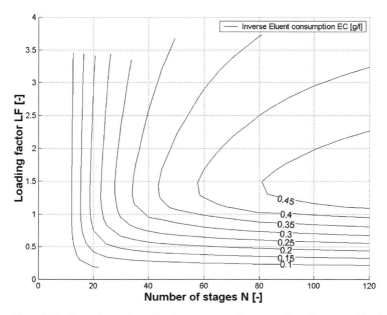

Figure 7.10 Dependency of specific eluent consumption on number of stages and loading factor.

specific eluent consumption on loading factor is similar to that of productivity, although the absolute eluent amount does not depend on the loading factor. However, in this chapter *specific* eluent consumption is used instead of just the amount of consumed eluent in order to compare the efficiency of eluent consumption. Since specific eluent consumption is defined as eluent per product amount (or as flow rate over production rate), a higher loading factor causes, at first, more favorable values but further increase in loading factor shifts the specific eluent consumption to less favorable values due to decreasing production rate caused by an increasing amount of waste fraction.

Contrary to productivity, however, a high number of stages does not result in less favorable values for specific eluent consumption. As mentioned before, a high number of stages applies in columns with relatively low interstitial velocity. A definite increase in the number of stages causes a decrease of flow rate (eluent flow rate) at the same degree, whereas the production rate drops slowly. Further increase of their number will, therefore, continue to improve the specific eluent consumption, at least in the parameter region investigated in this chapter.

A very important conclusion from the case study in this chapter is that yield, productivity and eluent consumption represent oppositional optimization goals during the design and optimization of batch chromatographic column. To demonstrate this, three particular examples within the N–LF diagram for a given column geometry will be compared. Each of these exemplary operating points represents one of the three extreme cases:

- Maximum productivity
- High yield
- Low eluent consumption

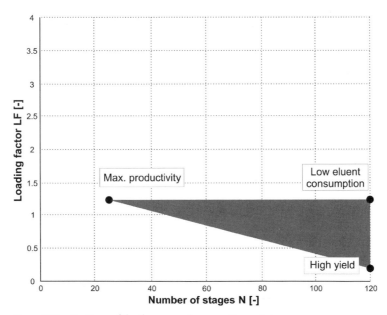

Figure 7.11 Positions of the three exemplary operating points.

Table 7.7 Objective functions at the exemplary operating points. Column geometry: d_c = 2.5 cm, L_c = 10.8 cm.

	$\dot{V}$ (ml min^{-1})	t_{inj} (s)	N_R (–)	LF_R (–)	VSP [g (l h)$^{-1}$]	EC (l g^{-1})	Y (–)
Max. productivity	23.5	208.0	25	1.25	**23.36**	4.07	0.60
Low eluent consumption	4.7	1045.2	120	1.25	9.89	**1.92**	0.94
High yield	4.7	167.2	120	0.20	1.80	10.53	**1.00**

The positions in N–LF diagram and the values of the objective functions for each case are shown in Fig. 7.11 and Tab. 7.7.

A closer look at the chromatogram at maximum productivity (Fig. 7.12) reveals a large overlapping area (i.e. large waste fraction) between the concentration peaks of the pure components. This confirms the fact that maximum productivity can be reached in the region with relatively higher loading factor, where the yield already drops significantly below 100% (i.e. 60%). Furthermore, although productivity is at its maximum there is still an optimization potential for the eluent consumption (Tab. 7.7).

High numbers of stages for a given column always represent low flow rate. Additionally, by increasing the number of stages the peak width will become smaller (Fig. 7.13), thus reducing the batch-cycle time but increasing the production rate. Both phenomena benefit the eluent consumption. Low eluent consumptions can therefore be reached in the region with very high number of stages although the productivity already decreases significantly below its maximum value (Tab. 7.7).

As can be seen in Tab. 7.7 and Fig. 7.14, a high yield can be reached by reducing the amount of feed introduced into the column (i.e. low loading factor) and by reduc-

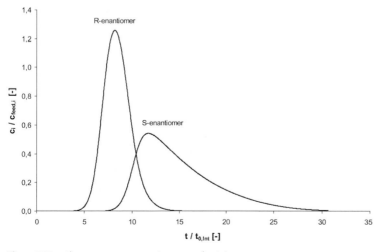

Figure 7.12 Chromatogram at maximum productivity.

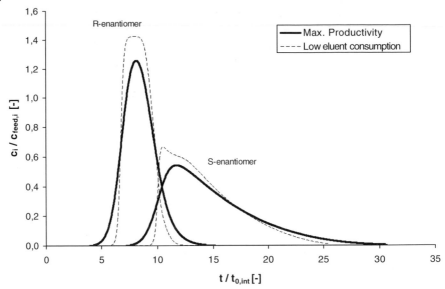

Figure 7.13 Comparison of chromatograms (max. productivity; low eluent consumption).

ing the flow rate (i.e. high number of stages). However, at very low loading factors the productivity of the column and the efficiency of eluent usage will deteriorate seriously. Therefore, low loading factors benefit only yield of separation and not utilization of adsorbent and eluent.

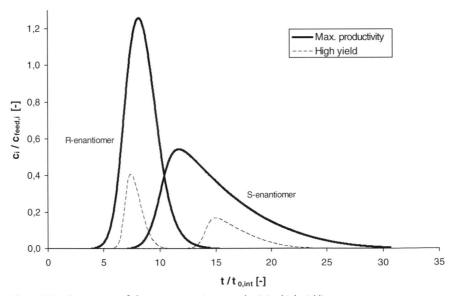

Figure 7.14 Comparison of chromatograms (max. productivity; high yield).

Figure 7.15 Triangle of oppositional optimization goals.

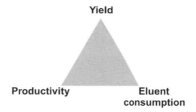

Yield

Productivity   Eluent
                               consumption

The only evaluation method that unifies the contributions of these optimization goals is the total separation cost. The position of the optimum after a cost optimization depends heavily on the magnitude of each contribution relative to each other. If the price of adsorbent or fixed costs (maintenance and capital cost) is a dominant factor, then the cost optimum will coincide with the maximum productivity. For separation problems with low solubility of the components and/or very high eluent price, the cost optimum is approximately equal to the minimum of eluent consumption (i.e. very high number of stages). In other words, only the optimization of total separation cost leads to the „real" (economically) optimum of the separation problem (Fig. 7.15).

7.2.4
Design and Optimization Strategies from Other Authors

The selection of chromatographic system is critical to process performance and process economy and can be regarded as the biggest potential for optimization. Therefore, most approaches to optimization of batch chromatography deal with the variation of the chromatographic system – thus varying parameters such as selectivity, saturation concentrations, etc., whereby design and operating parameters are somewhat left unnoticed. However, some authors also deal with the optimization of batch processes by variation of design and operating parameters. Table 7.8 gives an overview of articles with a conclusion of the applied optimization strategies, models, boundary conditions and objective functions. Furthermore, a conclusion for all publicized results from model-based optimization of batch chromatography until 1994 can be found in Guiochon et al. (1994b).

Guiochon and Golshan-Shirazi's (1989a) initial steps towards a model-based optimization of design and operating parameters are based on an analytical solution of the ideal chromatographic models by applying cut strategy I. Guiochon and Golshan-Shirazi (1989b) then extend this approach by considering a finite number of stages of a real column. For this purpose they apply in the optimization procedure an empirical correlation from Knox (1977) called the Knox plate-height equation, which is also based on an analytical solution of the ideal model. Later, Guiochon and Golshan-Shirazi (1991) discovered that the production rate of a column at a constant purity demand can be maximized by reducing the yield demand. But since all of these approaches are based on analytical solution of the ideal model with competitive Langmuir-isotherm, they can not be applied for other isotherm models due to the lack of an analytical solution.

Based on these facts, Felinger and Guiochon (1992a) maximized the production rate for different column lengths without yield demands using cut strategy I. Here they solved the simplified equations of an equilibrium-dispersive model for small concentrations and competitive Langmuir-isotherms numerically (Guiochon and Czok, 1990 and Guiochon et al., 1994b), whereby the physical dispersion in the column is approximated by a suitable numerical dispersion. However, all solute components are postulated to have the same numerical dispersion, while in a real column different dispersions for each component have to be assumed. With the help of this model they investigated separations with different selectivities and capacity factors (different isotherms). They examined different particle diameters d_p for each isotherm and concluded that, in every case, an optimal ratio d_p^2/L_c exists, for which the production rate is at its maximum. If this ratio can not be realized due to pressure drop limitations, the maximum production rate is located directly on the pressure drop limit, which is, consequently, not the global maximum. Soon after this, Felinger and Guiochon (1992b) published an analogous study for displacement chromatography.

Felinger and Guiochon (1993) performed the transition to an optimization based on the general equation of an equilibrium-dispersive model in a direct comparison between elution and displacement chromatography. Again, physical dispersion is approximated using numerical dispersion. In this study they also concluded that, for a given isotherm, there has to be an optimal ratio d_p^2/L_c, where maximum production rate is achieved. Furthermore, they investigate the influence of material dependent parameters, especially retention factor k', on the maximum achievable production rate and obtain an optimal retention factor for a given separation problem, which lies apparently below the common value. Thereby they assume that the retention factor can be varied independent of other material dependent parameters, such as selectivity, saturation load and solubility. This assumption is however not reliable because retention factor generally can not be varied without changing other material dependent parameters.

Subsequently, Felinger and Guiochon (1994) discovered that both production rate and eluent consumption must be considered to achieve the economical optimal separation condition. A cost optimization can thus only be done based on a hybrid objective function that considers both contributions.

Production rate was generally maximized in previous studies while tolerating a decrease in yield, thus neglecting the potentially very high costs resulting from feed loss. To solve this problem Felinger and Guiochon (1996) proposed the product of yield and production rate as a new objective function. More recent investigations of Felinger and Guiochon (1998) use this new objective function in dealing with the optimization and comparison of elution, displacement and non-isocratic chromatography. This investigation covers a wide range of retention factors, but they neglect the fact that the retention factor in real systems generally can not be varied independently of other material-dependent parameters.

In contrast to the work quoted so far, Strube and Dünnebier applied a different approach by optimizing elution chromatography using cut strategy II. Strube uses a transport-dispersive model for his simulation studies and optimized the productivity

Table 7.8 Quoted works about optimization of batch chromatographic separation.

Source	Cut strategy	Applied model	Suitable for isotherm type	Objective function	Boundary conditions	Degrees of freedom
Guiochon and Golshan-Shirazi (1989a)	I	ideal	Langmuir	prod.rate	purity, press.drop	design and oper. par.
Guiochon and Golshan-Shirazi (1989b)	I	ideal	Langmuir	prod.rate	purity, press.drop	design and oper. par.
Guiochon and Golshan-Shirazi (1990b)	II	ideal	Langmuir	prod.rate	purity, press.drop	design and oper. par.
Guiochon and Golshan-Shirazi (1991)	I	ideal	Langmuir	prod.rate	purity, press.drop, yield	design and oper. par.
Felinger and Guiochon (1992a)	I	equilibrium dispersive	Langmuir	prod.rate	purity, press.drop	design and oper. par.
Felinger and Guiochon (1992b)	I	equilibrium dispersive	Langmuir	prod.rate	purity, press.drop	design and oper. par.
Felinger and Guiochon (1993)	I	equilibrium dispersive	all	prod.rate	purity, press.drop, yield	design and oper. par.
Felinger and Guiochon (1994)	I	equilibrium dispersive	all	prod.rate, eluent cons.	purity, press.drop	design and oper. par.
Felinger and Guiochon (1996)	I	equilibrium dispersive	all	prod.rate * yield	purity, press.drop	design and oper. par.
Felinger and Guiochon (1998)	I	equilibrium dispersive	all	prod.rate * yield	purity, press.drop	design and oper. par.
Katti and Jagland (1998)	I	n.a.	n.a.	costs	purity, press.drop	design and oper. par.
Strube et al. (1998b)	II	transport dispersive	all	vol.spec.productivity	purity, press.drop, yield	operating par.
Strube (2000)	II	transport dispersive	all	vol.spec.productivity	purity, press.drop, yield	operating par.
Dünnebier (2000)	II	general rate	all	vol.spec.productivity	purity, press.drop, yield	operating par.
Epping (2005), Jupke et al. (2002)	I	transport dispersive	all	costs	purity, press.drop	design and oper. par.

of batch (and SMB) separations (Strube et al., 1998b and Strube, 2000). Dünnebier (2000) uses a general-rate model and implements yield and purity demand as boundary conditions. However, since the „touching band" strategy (cut strategy II) is used, yield and purity are not independent, thus restricting the maximum achievable productivity.

Katti and Jagland (1998) have performed a systematic analysis of different parameters that influence the separation cost. They investigate cost contributions of eluent, feed loss, packing materials, equipment and investment. Each cost contribution is also examined individually, depending on various free optimization parameters. Although the exact optimization strategy for an optimal operating point remains unclear, their work gives good insight into the general correlations and dependencies in a cost optimization and, therefore, represents a good basis for the analysis of cost structures and the economic efficiency of chromatographic processes (Tab. 7.8).

Based on this conclusion and Epping (2005) and Jupke et al. (2002) both optimized batch and SMB processes, regarding not only the common process performance (eluent consumption, productivity, etc.) but also the total separation cost, which is the summation of various cost contributions. Hereby, they applied an optimization strategy that utilizes dimensionless parameters. The applied optimization strategy in Epping (2005) and Jupke et al. (2002), its derivation and advantages are explained and demonstrated in Sections 7.2 and 7.3.

7.3
SMB Chromatography

Continuously operated chromatographic processes such as Simulated Moving Beds (SMB) are well established for the purification of hydrocarbons, fine chemicals and pharmaceuticals. They have proven ability to improve the process performance in terms of productivity, eluent consumption and product concentration, especially for larger production rates. These advantages, however, are achieved with higher process complexity with respect to operation and layout. A purely empirical optimization is rather difficult and, therefore, models with different levels of detail have been introduced to assist layout and optimization.

Starting with the simplest model, the True Moving Bed (TMB) model, it will be demonstrated how to determine first parameters for the operation of SMB processes. Based on these TMB short-cut methods, a more detailed optimization of operating parameters applying complete SMB models will be presented. Finally, a strategy to determine the optimal design of an SMB unit is illustrated, based on similar considerations to those already made for optimizing single column batch chromatography.

7.3.1
Optimization of Operating Parameters

The operating point of an SMB unit is characterized by the flow rates of the liquid phase in the four sections, $\dot{V}_j$, as well as the switching time, t_{shift}. Another degree of freedom is the concentration of the solute in the feed stream – but according to Charton (1995) the feed concentration should be at its maximum to achieve best productivity. For the following considerations feed concentrations as well as temperatures are kept constant.

During operation of an SMB plant the propagation of components to be separated are influenced by the internal fluid flow rates in the different sections as well as the switching time that simulates movement of the solid. By appropriate choice of operating parameters the movement of the less retained component is focused on the raffinate port while the more retained component is collected in the extract stream. Figure 7.16 shows an optimal axial profile of the liquid concentrations at the end of a switching interval after the process has reached a periodic steady state. The adsorption and desorption fronts of both components have to start or stop at given points to achieve complete separation at maximum productivity.

For complete separation the desorption fronts of the two components must not exceed points 1 and 2 respectively, which are located one column downstream the desorbent and extract port. Since the concentration profile displayed in Fig. 7.16 demonstrates the situation at the end of a switching interval, all ports will move one column downstream in the very next moment. In the case were the desorption front of component B does exceed point 2, the extract stream, meant to withdraw the more retained component A only, will be polluted with B after the ports have been switched. The same applies to point 1. If component A is shifted into section IV the adsorbent will transfer it to the raffinate port and the raffinate will be polluted. For the adsorption fronts, components A and B must not violate points 3 and 4, respec-

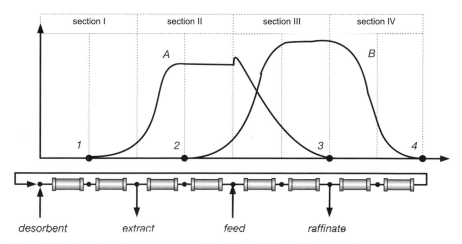

Figure 7.16 Optimal axial concentration profile at the end of a switching interval.

tively, at the end of the switching period. Otherwise, component A will pollute the raffinate, and component B will enter section I and pollute the extract.

To achieve 100% purity for both components it is necessary to fulfill the constraints mentioned before. In addition, it is favorable to push the fronts as far as possible towards points 1 to 4 to realize the highest throughput as well as the lowest eluent consumption.

If product purities lower than 100% are required, there are generally two possibilities to achieve this with a positive impact on other process parameters:

- A higher feed rate at constant eluent consumption can be realized when the adsorption front of component A exceeds point 3 and/or the desorption front of component B exceeds point 2.
- A lower eluent consumption at constant feed rate can be realized when the adsorption front of component B exceeds point 4 and/or the desorption front of component A oversteps point 1.

In these cases the raffinate will be polluted with the more strongly retained solute A and/or the extract will be polluted with the less retained component B. Depending on purity requirements and the objective function, any combination of these two cases can also be chosen.

Whenever dealing with the operation of SMB units one always has to consider the time the process needs to reach its periodic steady state. Depending on the chromatographic system, the process set-up and the columns geometry it might take 15 or more complete cycles to reach steady state.

7.3.1.1 Process Design Based on TMB Models (Short-cut Methods)

Due to the similarity between true moving (TMB) and simulated moving bed (SMB) processes (Chapter 6.7) the TMB approach is quite often used to estimate the operating parameters of SMB units. Operating parameters for TMB processes are the liquid flow rates in the four sections, $\dot{V}_{j,TMB}$, and the volumetric flow rate of the adsorbent, $\dot{V}_{ads}$. These TMB parameters can be transferred into SMB operating parameters ($\dot{V}_{j,SMB}$ and t_{shift}) following the relationships listed below.

$$\dot{V}_{ads} = \frac{(1 - \varepsilon) V_c}{t_{shift}} \tag{7.48}$$

$$\dot{V}_{j,SMB} = \dot{V}_{j,TMB} + \frac{V_c \varepsilon}{t_{shift}} \tag{7.49}$$

Ideal Model, Linear Isotherm

Ruthven and Ching (1989), Hashimoto et al. (1983a), and Nicoud (1992) have used analytical solutions of a TMB process under the assumption of the ideal model (neglecting axial dispersion and mass transfer phenomena) and linear isotherms. In each section *j* of the TMB process the fluid phase moves with the flow rate $\dot{V}_{j,TMB}$ in one direction and the solid phase with $\dot{V}_{ads}$ in the opposite direction. Because, in the

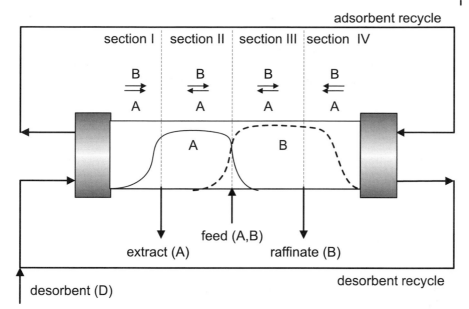

Figure 7.17 Directions of migration of two components in a TMB separation.

ideal model, equilibrium is assumed to be reached immediately, the mass flow $\dot{m}_{i,j}$ of component i depends on these two flow rates and the isotherm. Movement in the direction of the liquid flow will give a positive sign while migration with solid results in a negative one.

$$\dot{m}_{i,j} = (\dot{V}_{j,\text{TMB}} - \dot{V}_{\text{ads}}\varepsilon_\text{p})c_{i,j} - \dot{V}_{\text{ads}}(1 - \varepsilon_\text{p})q_{i,j} \tag{7.50}$$

Figure 7.17 depicts the directions of migration of two components for a TMB separation.

The parameter $\gamma_{i,j}$ indicates the direction of motion of the components in every section. It is defined as the ratio of the mass flow of component i dissolved in the liquid phase to the mass flow of component i adsorbed to the stationary phase.

$$\gamma_{i,j} = \frac{(\dot{V}_{j,\text{TMB}} - \dot{V}_{\text{ads}}\varepsilon_\text{p})c_{i,j}}{\dot{V}_{\text{ads}}(1 - \varepsilon_\text{p})q_{i,j}} = \frac{\dot{V}_{j,\text{TMB}} - \dot{V}_{\text{ads}}\varepsilon_\text{p}}{\dot{V}_{\text{ads}}(1 - \varepsilon_\text{p})H_i} \tag{7.51}$$

For linear isotherms the ratio between fluid concentration and loading of the adsorbent is constant and a function of the Henry coefficient H_i.

At $\gamma_{i,j} > 1$ components will move in the direction of the liquid phase while $\gamma_{i,j} < 1$ indicates migration upstream with the solid. Figure 7.17 shows the movement required for a complete separation of two components. In this case the following constraints have to be fulfilled.

Section I: $\gamma_{A,I} > 1,\ \gamma_{B,I} > 1$ (7.52)
Section II: $\gamma_{A,II} < 1,\ \gamma_{B,II} > 1$ (7.53)
Section III: $\gamma_{A,III} < 1,\ \gamma_{B,III} > 1$ (7.54)
Section IV: $\gamma_{A,IV} < 1,\ \gamma_{B,IV} < 1$ (7.55)

As the ideal model does not take any band broadening effects into account, safety margins have to be considered when applying this method for the layout of SMB separations.

Storti et al. (1993b) have presented a similar approach by introducing the dimensionless flow rate ratio m_j as the ratio between liquid and solid flow in every section.

$$m_j = \frac{\dot{V}_{j,\mathrm{TMB}} - \dot{V}_{\mathrm{ads}}\varepsilon_p}{\dot{V}_{\mathrm{ads}}(1 - \varepsilon_p)} \qquad (7.56)$$

This leads to the following constraints for complete separation of a two-component feed mixture.

Section I: $m_I \geq H_B,\ m_I \geq H_A$ (7.57)
Section II: $H_A \geq m_{II} \geq H_B$ (7.58)
Section III: $H_A \geq m_{III} \geq H_B$ (7.59)
Section IV: $m_{IV} \leq H_B,\ m_{IV} \leq H_A$ and   (7.60)
$H_B \leq m_{II} \leq m_{III} \leq H_A$ (7.61)

According to these constraints the dimensionless flow rate ratios can be chosen. To achieve the best performance with respect to productivity, however, the difference between m_{II} and m_{III} should be as high as possible, representing the highest possible feed flow rate. The flow rate of fresh desorbent and, therefore, the eluent consumption can be minimized by choosing m_I as low and m_{IV} as high as possible.

To visualize all possible operating points for sections II and III, the dimensionless flow rate ratio m_{III} is plotted versus m_{II} in the so-called „triangle" diagram. For linear isotherms all possible operating points that fulfill the constraints are within the triangle shown in Fig. 7.18.

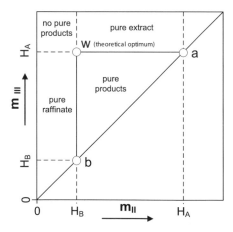

Figure 7.18 Operating or triangle diagram for linear isotherms.

Points a and b result from Henry coefficients. The theoretical optimum for process productivity is achieved at the point w, where the difference between m_{II} and m_{III} is maximal.

Ideal Model, Nonlinear Isotherm

Application of the method described so far becomes more complicated when the isotherms are no longer linear and the migration velocities of the components are strongly influenced by fluid concentrations. One approach for determining the operating parameters is the explicit solution of the ideal TMB process model (Chapter 6.7), as proposed by Storti, Mazzotti and Morbidelli. By introducing the dimensionless time

$$\tau = (t\,\dot{V}_{ads})/V_c$$

and the dimensionless axial position

$$z = x/L_c$$

as well as the flow rate ratio m_j, the balance for the TMB process under the assumption of ideal conditions is:

$$\frac{\partial}{\partial \tau}\left[\varepsilon_t c_{i,j} + (1-\varepsilon_t)q_{i,j}\right] + (1-\varepsilon_t)\frac{\partial}{\partial z}\left[m_j c_{i,j} - q_{i,j}\right] = 0 \tag{7.62}$$

This partial differential equation can be solved analytically for certain types of isotherms by applying the method of characteristics (Hellferich and Klein, 1970 and Rhee et al., 1970). Following this approach direct solutions are available for isotherms with constant selectivity, such as the multi-Langmuir or the modified competitive multi-Langmuir isotherm (Storti et al., 1993b, Mazzotti et al, 1994, Storti et al., 1995, Mazzotti et al., 1996b and Mazzotti et al., 1997b). For multi-bi-Langmuir isotherms with non-constant selectivities a numerical determination is necessary (Gentilini et al., 1998 and Migliorini et al., 2000).

As an example the procedure for determining the operating diagram is explained for the EMD53986 system (Appendix B). The adsorption equilibrium of this system can be described by the multi-Langmuir isotherm (Eq. 7.63).

$$q_i = \frac{H_i c_i}{1 + \sum b_i c_i} \qquad i = A, B \tag{7.63}$$

Based on this isotherm, the nulls ω_G and ω_F ($\omega_G > \omega_F > 0$) have to be determined for the following quadratic equation.

$$(1 + a_A c_{feed,A} + a_B c_{feed,B})\omega^2 - \left[H_A(1 + a_B c_{feed,B}) + H_B(1 + a_A c_{feed,A})\right]\omega - H_A H_B = 0 \tag{7.64}$$

With these data, the isotherm parameters for the EMD53986 and the feed concentration, which is 2.5 g l^{-1} for each component, a plot as shown in Fig. 7.19 can be created. The different points and lines are defined as follows.

$$\text{line } wf: \left[H_A - \omega_G(1 + b_A c_{feed,A})\right] m_{II} + b_A c_{feed,A}\,\omega_G\, m_{III} = \omega_G(H_A - \omega_G) \tag{7.65}$$

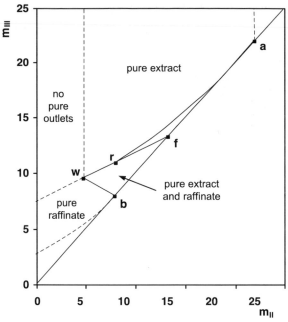

Figure 7.19 Operating or triangle diagram for nonlinear isotherms. Here: EMD53986 ($c_{feed} = 2.5\,g\,l^{-1}$).

line wb: $\left[H_A - H_A(1 + b_A\,c_{feed,A})\right] m_{II} + b_A\,c_{feed,A}\,H_B\,m_{III} = H_B(H_A - H_B)$

$$(7.66)$$

curve ra: $m_{III} = m_{II} + \dfrac{\left(\sqrt{H_A} - \sqrt{m_{II}}\right)^2}{b_A c_{feed,A}}$

$$(7.67)$$

point a: (H_A, H_A), point b: (H_B, H_B), point f: (ω_G, ω_G)

point w: $\left(\dfrac{H_B \omega_G}{H_A}, \dfrac{\omega_G\left[\omega_F(H_A - H_B) + H_B(H_B - \omega_F)\right]}{H_B(H_A - \omega_F)}\right)$

$$(7.68)$$

Due to the strong nonlinearity of the isotherms, combined with the competitive adsorption behavior, the triangle is completely different to that for linear isotherms.

Since the slope of such isotherms depends strongly on the fluid concentration, the feed concentration has a remarkable influence on the shape of the separation region. Figure 7.20 illustrates the correlation between feed concentration and the operating diagram.

To complete the set of possible operating parameters for a four-section SMB unit, values for the dimensionless flow rate ratios m_I and m_{IV} have to be determined as well. In sections I and IV things are a little less complicated, since only the adsorption of single components is involved.

Because the solid phase is regenerated in section I the more strongly retained component A has to be desorbed by the fluid flow. This can be guaranteed for all Langmuir-type isotherms by the following condition.

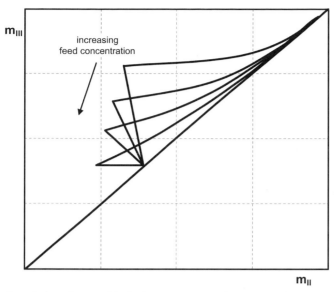

Figure 7.20 Influence of the feed concentration on the operating diagram.

$$m_I \geq m_{I,\min} = H_A \tag{7.69}$$

In section IV the less retained component B has to be adsorbed and carried towards the raffinate port in order to regenerate the liquid phase. For the EMD53986 system with its multi-Langmuir isotherm, the corresponding constraint on the dimensionless flow rate ratio m_{IV} is

$$\frac{-\varepsilon_p}{1 - \varepsilon_p} < m_{IV} \leq m_{IV,\max} = \frac{1}{2} \left[H_B + m_{III} + b_B c_{\text{feed},B}(m_{III} - m_{II}) \right. \tag{7.70}$$

$$\left. - \sqrt{\left[H_B + m_{III} + b_B c_{\text{feed},B}(m_{III} - m_{II}) \right]^2 - 4 H_B m_{III}} \right]$$

The maximal m_{IV} is seen to be a function of flow rate ratios m_{II} and m_{III} as well as the feed concentration.

Table 7.9 lists the theoretical optimal values of all flow rate ratios for the separation of EMD53986 at a feed concentration of $2.5 \, \text{g} \, l^{-1}$. These values indicate the set of operating parameters were the productivity is at its maximum (point w), since the difference between m_{II} and m_{III} is biggest. In addition, eluent consumption reaches its minimum because the difference between m_I and m_{IV} is smallest.

Table 7.9 Operating parameters for EMD53986 obtained by the TMB short-cut method.

$m_{I,\min}$	19.75
m_{II} (at point w)	5.29
m_{III} (at point w)	10.57
$m_{IV,\max}$	6.28

The procedure described so far requires a detailed knowledge of the adsorption equilibrium. Naturally, the more accurately the isotherm parameters have been determined, the more reliable are the obtained operating parameters m_j. However, as discussed in Chapter 6.5, experimental determination of isotherm parameters might consume quite a lot of time and substance.

Therefore, the basic correlations for situations present in the different sections of the SMB process, combined with a simplified short-cut approach, will be presented. Without knowing the adsorption equilibria, an operating point can be generated with a minimum number of experiments. All considerations made in the following are valid only for Langmuir-type isotherms.

The retention time for a concentration c_i^+ of the disperse desorption front can be described as a function of the isotherms derivative (Chapter 6.2). This implies that the maximum time for desorption or the minimum migration velocity is given by the derivative of the isotherms at lowest concentration ($c_i^+ \to 0$), which is equal to the Henry coefficient.

$$t_{R,i}(c^+) = t_0 \left(1 + \frac{1 - \varepsilon_t}{\varepsilon_t} \cdot \left. \frac{\partial q_i}{\partial c_i} \right|_{c^+ \to 0} \right) = t_0 \left(1 + \frac{1 - \varepsilon_t}{\varepsilon_t} H_A \right) \tag{7.71}$$

For an adsorption shock front, the time for the front's breakthrough under Langmuir behavior can be calculated according Eq. 7.72, where the isotherm's secant at feed concentration is relevant. The smaller the ratio q_i/c_i the higher the velocity of the shock front.

$$t_{R,i,\text{shock}}(c^+) = t_0 \left(1 + \frac{1 - \varepsilon_t}{\varepsilon_t} \cdot \left. \frac{q_i}{c_i} \right|_{c^+ = c_{\text{feed}}} \right) \tag{7.72}$$

These correlations can be transferred to continuous SMB or TMB chromatography where disperse as well as shock fronts are also present. The dimensionless flow rate ratios m_j can then be described as function of either the initial slope or the secant of the isotherm, depending on the situation in every zone.

Section I:
In section I the more strongly retained component A has to desorb while no B is present. The minimum m_I is therefore the initial slope or the Henry coefficient of the isotherm.

$$m_{I,\text{min}} = \left. \frac{dq_A}{dc_A} \right|_{c_A \to 0;\ c_B = 0} = H_A \tag{7.73}$$

As described in Chapter 6.5 the Henry coefficient can be determined by single pulse experiment at low concentrations of component A.

Section II:
In section II the minimum value for the corresponding flow rate ratio m_{II} is fixed to the maximal possible initial isotherm slope of the less retained component, which

has to be desorbed here. The highest initial slope for component B is reached if no A is around. Of course, A is always present in this zone (Fig. 7.17). Since it is quite difficult to predict the concentration of A in this section its presence causes a further decrease of the initial slope, and the worst case estimation of m_{II} is

$$m_{II,min} = \left. \frac{dq_B}{dc_B} \right|_{c_A=0;\ c_B \to 0} = H_B \tag{7.74}$$

Again the initial slope of the isotherm has to be determined by a pulse experiment of B.

Section III:
The third section of an SMB process is characterized by the adsorption of more strongly retained component A. To achieve complete separation one has to ensure that the shock front of A does not exceed the raffinate port. The highest possible velocity of that front occurs when the q_A/c_A is smallest. This is the case when both components are present at their highest possible concentration. Again, it must be pointed out that it is difficult to determine the exact concentrations, but the feed concentrations are a good first guess, as well as the worst case estimation.

$$m_{III,max} = \left. \frac{q_A}{c_A} \right|_{c_A=c_{feed};\ c_B=c_{feed}} \tag{7.75}$$

Experimentally, this parameter is determined by a breakthrough experiment where both components are fed at feed concentration to the chromatographic column.

Section IV:
In the last section the eluent is regenerated by complete adsorption of component B. Again the velocity of a shock front is limiting. In this case the highest possible velocity is determined by the lowest slope of the secant to the isotherm of component B. Since no A should be present the minimum slope is reached when B appears at feed concentration. Therefore, the correlation for this section is as follows.

$$m_{IV,max} = \left. \frac{q_B}{c_B} \right|_{c_A=0;\ c_B=c_{feed}} \tag{7.76}$$

This parameter can be extracted from a breakthrough experiment with pure B.

Following this procedure a very first initial guess for the four dimensionless operating parameters can be obtained with a minimum number of experiments. The resulting parameters for the EMD53986 system are listed in Tab. 7.10.

Table 7.10 Operating parameters for EMD53986 obtained by the experimental short-cut method.

$m_{I,min}$	19.75
$m_{II,min}$	7.88
$m_{III,max}$	10.15
$m_{IV,max}$	6.53

Comparison of these results with the operating point obtained by the first short cut method for nonlinear isotherms shows that, especially for the separation sections II and III, a point (m_{II}, m_{III}) has been found that is located within the area of complete separation (Fig. 7.19).

Again it must be pointed out that this procedure for determining the operating diagram is based on strong simplifications and represents something like a worst case scenario. Nevertheless, this can be done very quickly, leading to quite safe operating parameters. More detailed shortcut methods that also need only rather small experimental effort have been published by Mallmann et al. (1998) and Migliorini et al. (2002). The advantage of these approaches is a more detailed estimation of the amount of components entering sections II and III. This can be achieved by hodographic analysis of breakthrough experiments.

Beside the description of the TMB process based on the ideal model, other model approaches can be used to follow the same goal of determining reasonable operating parameter for an SMB unit. One possibility is to utilize stage models as introduced in Chapter 6.2 for the batch column and extend these to the TMB set-up (Ruthven and Ching, 1989 and Pröll and Küsters, 1998). In addition to determining operating parameters, TMB stage models have also been used to optimize the design of SMB plant in terms of, for example, column geometry (Charton and Nicoud, 1995 and Biressi et al., 2000). Heuer et al. (1998) have applied a TMB equilibrium-dispersive model for the determination of operating parameters. In this approach effects such as axial dispersion and mass transfer are lumped together in an apparent dispersion coefficient. Hashimoto et al. (1983a and 1993a) have used a TMB transport model to assist the layout.

The advantages of all approaches based on the stationary TMB process are the quite simple models and, in some cases, the possibility of direct analytical solutions. Without the need of dynamic simulations, much information with respect to the operation of SMB unit can be gathered. Nevertheless, these methods described so far can only provide the basis for a more or less proper guess of the operating parameters since the dynamics due to port switching as well as mass transfer and axial dispersion are neglected. In addition to this of course the quality of the isotherm plays an important role. Incomplete (e.g. neglecting competitive adsorption behavior) or wrong parameter determination influence significantly the quality of the operating point. Therefore, safety margins have always to be considered. Further optimization can be done afterwards by applying a more detailed model as presented below. The necessity of extra operating parameter optimization is shown below (Fig. 7.25) were it can be seen that the theoretical optimum does not coincide with the real optimum of an SMB process.

7.3.1.2 Process Design Based on SMB Models

Due to the simplifying assumptions made for the TMB approaches, accurate design and optimization of SMB processes is not possible. Several approaches based on SMB models have been suggested to improve the prediction and optimization of the SMB operation. Zhong and Guiochon (1996) have presented an analytical solution for an ideal SMB model and linear isotherms. The results of this ideal model are compared with concentration profiles obtained by an equilibrium-dispersive model as well as experimental data. Strube et al. (1997a and 1997b) used a transport-dispersive model in a systematic case study to optimize operating parameters. The strategy has been successfully tested on an industrial-scale enantioseparation (Strube et al., 1999). A similar approach has been published by Beste et al. (2000) to optimize the separation of sugars. A more detailed model, a general rate model, was used in Dünnebier and Klatt (1999) as well as in Dünnebier (2000a and 2000b) to optimize SMB operating parameters.

This section presents a systematic strategy for the optimization of operating parameters, based on the transport-dispersive SMB model (Chapter 6.7). A first approximation is performed by applying the triangle theory described above. Notably, the basic strategy of this procedure is not limited to a model-based design but can also be applied for the experimental optimization of a running separation process.

Operating Parameter Estimation
Knowing the isotherms, an initial guess of operating parameters can be performed by applying the short-cut methods described previously. With the triangle theory, one obtains values for the four dimensionless operating parameters m_j. Since simplifying assumptions have been made a safety margin of at least 10% should always be considered. The four dimensionless parameters m_j have then to be transferred to a „real" SMB process where the number of independent operating parameters is five (four flow rates and one switching time). Thus, for a given plant set-up, one out of these five parameters has to be fixed first. One possibility is to set the flow rate in section I to the maximum allowable pressure drop Δp_{max}, assuming that this flow rate is predominant in all sections. If Darcy's correlation for the description of the pressure drop is used, the flow rate in section I can be calculated by Eq. 7.77.

$$\dot{V}_I = \frac{\Delta p_{max} A_c d_p^2}{\psi N_c L_c \eta_f} \tag{7.77}$$

Here N_C stands for the total number of columns in the SMB plant. This procedure leads to higher pressure drops than in the real plant, since section I is the part of the process where, in general, the highest flow rate occurs. As demonstrated later, is not necessary to fix the flow rate in section I to the maximum allowed pressure drop to achieve optimal process performance. However, for a given flow rate in section I and all dimensionless flow rate ratios m_j, the switching time can be calculated according to Eq. 7.78.

$$t_{shift} = \frac{[m_I(1 - \varepsilon_t) + \varepsilon_t] V_c}{\dot{V}_I} \tag{7.78}$$

The missing internal flow rates $\dot{V}_j$ are calculated by Eq. 7.79.

$$\dot{V}_j = \frac{\left[m_j(1 - \varepsilon_t) + \varepsilon_t\right] V_c}{t_{shift}} \tag{7.79}$$

The flow rates of all external streams such as desorbent, feed, extract and raffinate are determined by the following balances for the port nodes.

Desorbent node: $\dot{V}_{des} = \dot{V}_I - \dot{V}_{IV}$ (7.80)

Extract node: $\dot{V}_{ext} = \dot{V}_I - \dot{V}_{II}$ (7.81)

Feed node: $\dot{V}_{feed} = \dot{V}_{III} - \dot{V}_{II}$ (7.82)

Raffinate node: $\dot{V}_{raf} = \dot{V}_{III} - \dot{V}_{IV}$ (7.83)

Applying this strategy, a first set of operating parameters is generated and it is possible to either run the SMB separation or make further optimizations using a more detailed model. Since the determination of these operating parameters is based on the TMB model with all its simplifying assumptions (neglecting all non-idealities such as axial dispersion, mass transfer and port switching) the estimated operating point should always be handled with care.

Optimization Strategy for Linear Isotherms

Figure 7.21 shows the internal axial concentration profiles at the end of a switching interval for a system of linear isotherms and no competitive interaction of the two components. After switching all ports downstream in the direction of the liquid flow, the extract will be polluted with component B because the desorption front of B violates point 2 (Fig. 7.16).

To improve processes performance and achieve a complete separation of the two components, the desorption front of component B has to be pushed in the direction of the feed port. Since the isotherms of both components are linear in this example and

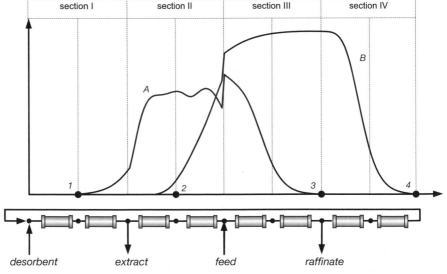

Figure 7.21 Axial concentration profile with pollution of the extract (system with linear isotherms).

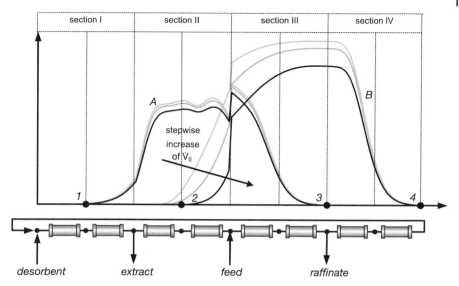

Figure 7.22 Axial concentration profiles for different flow rates in section II (system with linear isotherms).

the isotherm of one component is not influenced by the presence of the other, the front can be moved in the desired direction by simply increasing the internal flow rate in section II, $\dot{V}_{II}$. This can be realized by decreasing the flow rates of extract and feed by the same amount. The effect of a stepwise decrease of these external flow rates with the associated increase of $\dot{V}_{II}$ is depicted in Fig. 7.22.

Only the desorption front of component B is influenced by the increased flow rate in section II – all other fronts are not influenced and remain at their initial position. The decrease in feed flow rate has an impact on the total height of the concentration plateaus, especially in section III. This procedure of shifting the fronts is also applicable for all other sections of the SMB process, to optimize the internal concentration profile according to Fig. 7.16 and improve the process performance with respect to purities, productivity and eluent consumption.

Optimization Strategy for Nonlinear Isotherms
Process optimization gets more difficult when the isotherms are no longer linear and the equilibrium of one component is strongly influenced by the presence of the other, as is the case for the EMD53986. Again an initial guess based on the TMB model has been made for a given plant set-up. The resulting internal concentration profile is plotted in Fig. 7.23. Again, the desorption front of component B violates the optimization criterion for point 2, leading to pollution of the extract stream.

Applying the same strategy as introduced for linear isotherms, an internal concentration profile as shown in Fig. 7.24 is obtained. Again only the flow rate in section II has been increased by decreasing both extract and feed flow rate. The desorption front of component B is again pushed in the correct direction. However, in this case a change of flow rate in section II does also effect the situation in other sections of

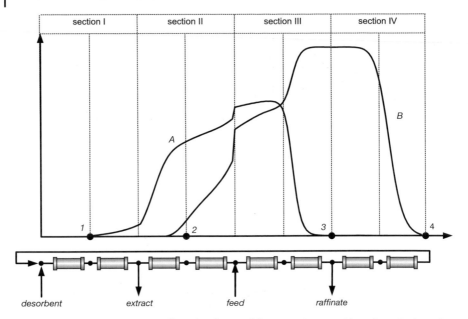

Figure 7.23 Axial concentration profile with pollution of the extract (system with nonlinear isotherms).

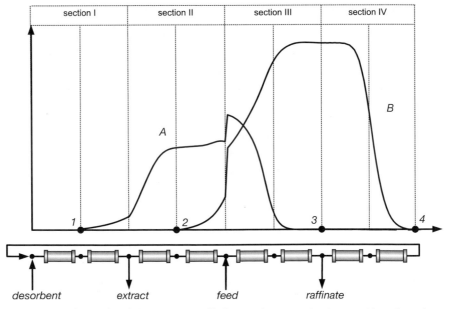

Figure 7.24 Suboptimal axial concentration profile for complete separation (system with nonlinear isotherms).

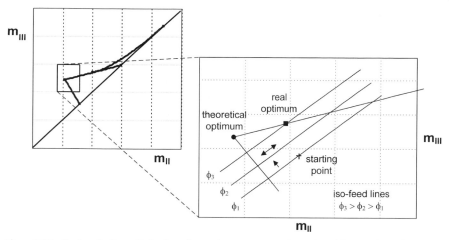

Figure 7.25 Strategy for the optimization of operating parameters. (Reproduced from Jupke et al., 2002.)

the process. In this example the adsorption front of component A has moved away from the raffinate port where, actually, the stopping point 3 (Fig. 7.16) is located. Even though the two products can now be withdrawn with maximum purity, the whole process is not operated at its optimum with respect to productivity.

To achieve this goal a new optimization strategy has to be applied. The most difficult step is to adjust the flow-rate ratios in sections II and III, especially for strongly nonlinear isotherms with competitive interaction. Therefore, the m_{II}–m_{III} plane is divided into segments by parallel lines to the diagonal. Each of these lines represents combinations of m_{II}–m_{III} values of constant feed flow – the so-called iso-feed lines (Fig. 7.25).

If for instance, as in our example, the purity of the extract stream is too low while the adsorption front of component A is still far from stopping point 3, the operating point has to be varied along one iso-feed line. In this case, m_{II} and m_{III} and, therefore, also the flow rates in sections II and III are increased by the same amount. This can be done by decreasing the extract flow rate and increasing the raffinate flow rate while the feed rate is held constant. When both the extract and the raffinate purity are higher than required, or the concentration profile can still be optimized according to Fig. 7.16, the total feed flow can be increased by jumping to the next iso-feed line closer to the vertex of the operating triangle. Otherwise, if both purities are too low the feed flow has to be decreased. When the flow rates in sections II and III have been optimized to achieve the required purities and the highest productivity, sections I and IV have to be checked and optimized separately.

After adjusting the flow rates in all sections, process conditions for complete separation with highest productivity and lowest solvent consumption have been found. The resulting internal concentration profile (Fig. 7.26) exhibits all the constraints of an optimal concentration profile (Fig 7.16).

From this optimization with a detailed SMB model another very important aspect can be observed. Due to the simplifying assumption made to determine the operat-

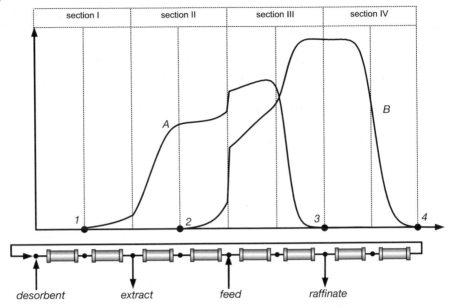

Figure 7.26 Optimized axial concentration profile (system with nonlinear isotherms).

ing diagram (Section 7.3.1.1), the „real" optimum of the SMB process does not coincide with the predicted theoretical optimum, as it can be seen from Fig. 7.25.

The mentioned optimizations of operating parameters can be performed for a given plant set-up, where the total number of columns, their distribution over the different sections and their geometry are given before. With such information and all operating parameters the specific process parameters such as productivity $VSP_{i,SMB}$ (Eq. 7.10) and eluent consumption $EC_{i,SMB}$ (Eq. 7.12) can be calculated. But since these parameters are only valid for one given set of design parameters a strategy to optimize these will be presented next.

7.3.2
Optimization of Design Parameters

Needless to say that, beside the operating parameters, the design of an SMB plant has a significant influence on the process performance in terms of various objective functions like the volume and cross section specific productivity ($VSP_{i,SMB}$ and $ASP_{i,SMB}$) as well as the eluent consumption $EC_{i,SMB}$. Important design parameters for SMB processes are

- Column geometry L_c and A_c
- Total number of columns N_c
- Distribution of columns over different sections j ($N_{c,j}$)
- Maximum pressure drop

Few published papers deal with the optimization of design parameters. Charton and Nicoud (1995) and Nicoud (1998) have presented a strategy based on a TMB stage model to optimize the operation as well as the design of SMB processes. The basic influence of the number of stages and the particle diameter on productivity and eluent consumption are illustrated and compared with batch elution chromatography. Based on this work, Biressi et al. (2000) presented a method to optimize operating parameter and investigate the influence of the particle diameter as well as the feed concentration on process performance. Ludemann-Hombourger et al. (2000a) have also analyzed the influence of particle diameter on the productivity.

In this contribution the focus will be on the optimization of column geometry for a fixed column distribution first, while the influence of the total number of columns as well as their distribution will then be shown for some examples. A more detailed description of this strategy has been published by Jupke (2004) and Jupke et al. (2002).

The basic idea for the optimization of process performance is, similar to batch chromatography, the summarization of all influencing parameters into dimensionless parameters. Relevant dimensionless parameters for an SMB processes are the total number of plates (N_{tot}) in the system as well as the four dimensionless flow rate ratios m_j. Therefore, a single set of N_{tot} and m_j represents not only one unique SMB configuration but a series of different SMB configurations with similar behavior. As demonstrated before, nearly identical concentration profiles (Fig. 7.5) can be observed for different plant layouts if these parameters are kept constant. Also, since only the dimensionless parameters are used as optimization parameters, violation of the maximum pressure drop can be neglected during optimization. The maximum pressure drop is considered only at the end of the optimization in order to determine the „real" SMB configuration based on optimal values for the dimensionless parameters $N_{tot,opt}$ and $m_{j,opt}$.

Figure 7.27 illustrates the design procedure for a new SMB plant. For an initial column geometry and a fixed column distribution (e.g. 2/2/2/2) the optimal operating parameters, namely the flow rate ratios m_j, can be obtained following the strategy described in the previous section. From these parameters the present dimensionless design parameter N_{tot} can be calculated, as well as the corresponding objective functions. Since the initial values for the column geometry, especially the column length, do not consequently represent the optimal design parameters, other plate numbers have to be examined. This can be done by choosing a different column length from the initial one but, to realize a different total number of theoretical plates in the system, the flow rate in section I, as it had to be chosen before, should be held constant. Another possibility is to fix the column length to a constant value but to change the flow rate in section I. For the new set-up again an optimization of operating parameters has to be performed to fulfill purity requirements etc.

The algorithm in Fig. 7.27 is characterized by two optimization loops. The inner loop describes the operating parameter optimization based on the strategies introduced before. When for one set-up (e.g. one column length) the optimized operating parameters have been found, a new set up (e.g. another column length) has to be chosen, represented by the outer design parameter optimization loop. This proce-

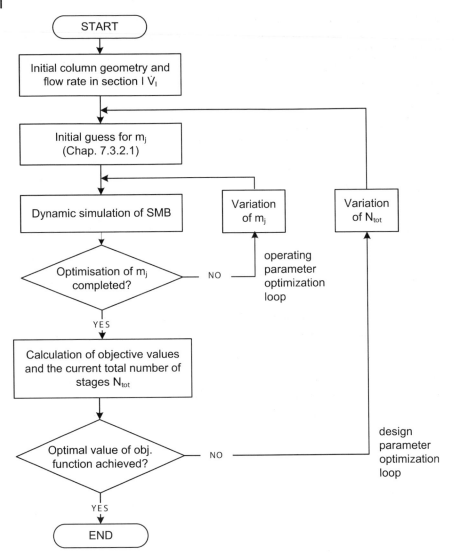

Figure 7.27 Algorithm for a complete SMB design optimisation.

dure is repeated until the optimum for the objective function (e.g. VSP, EC or costs) has been found.

Results for the systematic optimization of the volume-specific productivity (VSP) for the separation of the EMD53986 system are summarized in Tab. 7.11 and displayed in Fig. 7.28."

In this example, starting with a given set-up, the flow rate in section I was held constant while the column length has been changed to realize different plate numbers. Again it must be pointed out that the pressure drop is not considered during the optimization.

Table 7.11 Optimization of plate number for EMD53986.

Constant parameters

N_c	8 (2/2/2/2)
d_c (cm)	2.5
$\dot{V}_1$ (ml min^{-1})	1.472
$c_{feed,i}$ (g l^{-1})	2.5

Parameters varied during optimization

Step	1	2	3	4	5	6	7
L_c (cm)	20	10	6	5	4.5	4.26	4.0
N_{tot} (–)	231.3	124.6	83.3	71.9	65.6	62.3	57.1
t_{shift} (s)	500.7	275.5	187.7	163.4	151.2	145.1	136.3
m_I (–)	24.30	27.00	31.00	32.50	33.50	33.99	34.00
m_{II} (–)	5.55	5.89	6.30	6.57	6.90	7.15	8.00
m_{III} (–)	10.63	10.71	10.80	10.84	10.95	11.00	11.00
m_{IV} (–)	6.32	6.20	5.90	5.75	5.70	5.70	5.70
VSP (g h^{-1} l^{-1})	11.42	19.70	26.98	29.41	30.13	29.85	24.77
EC (l g^{-1})	1.81	2.12	2.63	2.91	3.15	3.34	4.17

According to Tab. 7.11 a maximum VSP is reached for a total number of plates of approximately 65 and the corresponding flow rate ratios m_j. This is more than a two-fold increase over the productivity at a plate number of 231. But, in addition to the development of productivity, the specific eluent consumption (EC) has increased. As reported by Jupke (2002) analogue correlations can be found for different particle diameter. With decreasing particle diameter an increase in terms of productivity can be observed, while the total number of stages remains nearly constant.

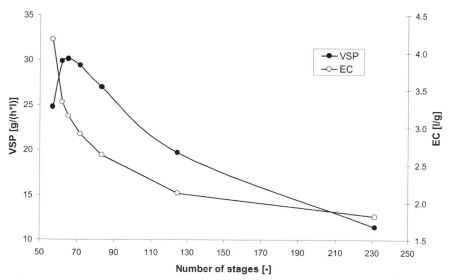

Figure 7.28 Influence of the plate-number on the specific productivity and specific eluent consumption.

The separation problem has been transferred completely to the dimensionless space, where the best performance in terms of productivity VSP, eluent consumption (EC) etc. can be achieved by one set of five parameters (N_{tot}, m_j). The next step is to transfer these parameters to an SMB process by considering a given production rate $\dot{m}_{prod,i}$, purity requirements and pressure drop limitations.

For the desired production rate, $\dot{m}_{prod,i}$, together with the volume-specific productivity ($VSP_{i,SMB}$), as it had been introduced before (Eq. 7.10), the optimal volume of one column $V_{c,opt}$ can be calculated according to Eq. 7.84.

$$V_{c,opt} = A_{c,opt}\, L_{c,opt} = \frac{\dot{m}_{prod,i}}{(1 - \varepsilon_t)\, N_c\, VSP_{i,SMB}} \tag{7.84}$$

This will lead to an infinite number of possible A_c/L_c ratios and, therefore, the column length has to be chosen according to the following equations.

The first equation represents the total number of theoretical plates of one component in the system, which has to be held constant to enable a proper scaling of the process.

$$N_{tot,i} = \sum_{j=I}^{IV} \left(\frac{N_{c,j}\, L_{c,opt}}{A_i + B_i\, u_{int,j}} \right) \tag{7.85}$$

But, in analogy to the procedure in batch chromatography, one has to ensure that for both components the total number of stages of the new process is equal to or greater than for the old one. This can be done, as introduced for the single column before, by choosing the correct reference component.

In addition to the plate number the dimensionless flow rate ratios should remain constant.

$$m_{j,opt} = \frac{u_{int,j}\, A_{c,opt} - \dfrac{V_{c,opt}\, \varepsilon}{t_{shift}}}{\dfrac{V_{c,opt}(1 - \varepsilon)}{t_{shift}}} \tag{7.86}$$

Also, as a final constraint, pressure drop limitations have to be considered, which are described by Eq. 7.87.

$$\Delta p_{max} = \frac{\psi\, \eta_l}{d_p^2} \cdot \sum_{j=I}^{IV} (u_{int,j}\, N_{c,j}\, L_{c,opt}) \tag{7.87}$$

This procedure leads to one $L_{c,opt}/A_{c,opt}$ ratio that exactly fulfils the maximum pressure drop condition for the given feed stream. Of course, smaller L_c/A_c ratios are also applicable to reach the same optimal performance of the SMB plant but with a lower pressure drop. Notably, a lower pressure drop will decrease the investment cost of a chromatographic unit. However, the smaller that ratio becomes, and with that the lower the pressure drop is, the shorter the columns will be. Consequently, this leads to difficulty in ensuring a proper fluid distribution inside the column.

Finally, it is worth mentioning that, in the presented strategy, beside the total number of stages, the operating parameters in terms of the flow rate ratio were also held constant. Therefore, no further optimizations of the operating parameters have to be done.

Beside the column's geometry, of course, the total number of columns and their distribution over the unit are design parameters that influence the process performance. To compare different process configurations with the initial 2/2/2/2 set-up, a complete optimization of the column geometry based on optimization of the number of stages has been performed for different configurations. Productivity again was the objective function to be optimized in these considerations. The results for selected configurations (2/3/2/2, 2/2/3/2, 2/3/3/2 and 3/3/3/3) are shown, relative to the data for the initial eight column set-up, in Fig. 7.29.

This example indicates that the column configuration can be used to improve the process performance. For every configuration it is not enough just to change the number of columns and optimize the operating parameters, because every new configuration shows the best results in terms of productivity at its own optimal total number of stages.

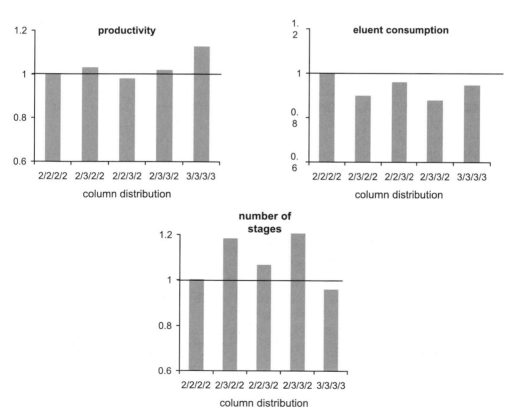

Figure 7.29 Influence of the number of columns and column distribution on the specific productivity and the specific eluent consumption.

Process configurations with a reduced number of columns, such as 1/2/2/2, 2/2/2/1 and 1/2/2/1, are not considered. Since the number of columns is reduced in the sections meant to regenerate the solid (section I) as well as the fluid phase (section IV), a large increase in eluent consumption will be the result. Nevertheless, these configurations should also be taken into account when fresh eluent is not the major block in the cost structure.

7.4
Comparison of Batch and SMB Chromatography

Although chromatographic processes are often considered as too expensive, intensive research and development during the last decade has helped to improve the acceptance and industrial applicability of chromatographic processes. Many attempts have been made to lower separation costs by developing new optimization strategies and even new process concepts. Since the total separation cost is mainly dominated by the cost of stationary phase (i.e. adsorbent) and mobile phase (i.e. eluent), investigations are therefore focused on achieving higher productivity and lower eluent consumption. Here, the two most common process concepts, batch and SMB chromatography, are compared. Productivity and eluent consumption will also be the main focus of this comparison.

Comparison of process concepts between batch and SMB chromatography reveals at first a crucial drawback of SMB. While one can have any number of product fractions in batch processes, the number of product outlets in SMB is limited to two (i.e. extract and raffinate), which still makes it a powerful separation tool for binary mixtures like enantiomers. For multi-component systems, however, only one pure component can be collected. On the other hand, SMB processes benefit from the simulated counter-current movement between the stationary and mobile phase, resulting in better utilization of adsorbent and eluent. SMB processes are therefore generally known to have much higher productivity and lower eluent consumption than batch processes. This has been excellently demonstrated in Nicoud (1998). Nicoud (1998) pointed out among other things, that the achievable productivity of SMB is about 300% higher than batch chromatography at a selectivity of 1.8. Eluent consumption of SMB is also 50% lower than batch. Furthermore, these advantages of SMB grow even larger at lower selectivity.

The following comparison between batch and SMB chromatography once again uses the enantiomeric separation of EMD53986 as an exemplary chromatographic system. Similar to Nicoud (1998), volume specific productivity VSP and specific eluent consumption (EC) for both processes are compared in dependency of the number of stages. The loading factor (LF) (for batch processes) and the flow ratios m_j (for SMB processes) at each number of stages are optimized regarding the current objective functions. Please note that the product purities are kept constant at 99.9%. In addition, since no waste fraction is allowed (i.e. base-line separation) in this comparison, yields are also constant at ca. 100%. Figures 7.30 and 7.31 illustrate the results of this comparison.

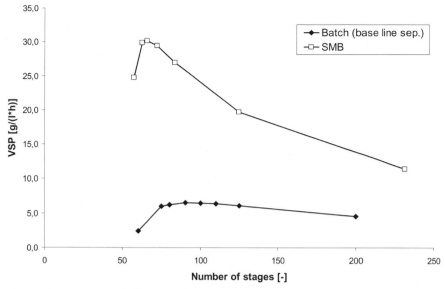

Figure 7.30 Comparison of volume specific productivity at 99.9% purity.

The first insights that can be gathered from Figs. 7.30 and 7.31 are that the maximum productivity of SMB is about five times greater. Although the absolute minima of eluent consumption are not visible in Fig. 7.31, it is still obvious that the achievable value for eluent consumption in SMB processes is lower than in batch processes. Further advantages of SMB over batch chromatography appear if we take a

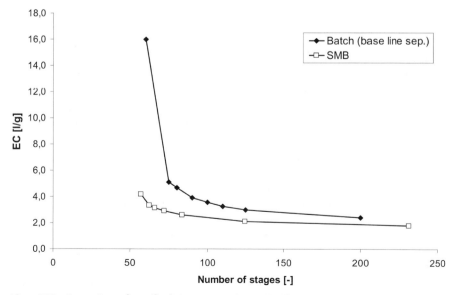

Figure 7.31 Comparison of specific eluent consumption at 99.9% purity.

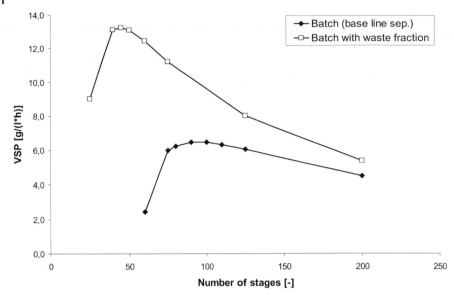

Figure 7.32   Comparison of volume-specific productivity (VSP) between two batch modes.

closer look at the number of stages. Figure 7.30 shows that the maximum productivity in SMB can be achieved with relatively fewer stages than in batch process, while Fig. 7.31 shows a steep rise of eluent consumption in batch process at a low number of stages. Therefore, as in Nicoud (1998), greater efficiency of adsorbent and eluent usage in SMB processes can also be confirmed for this separation problem.

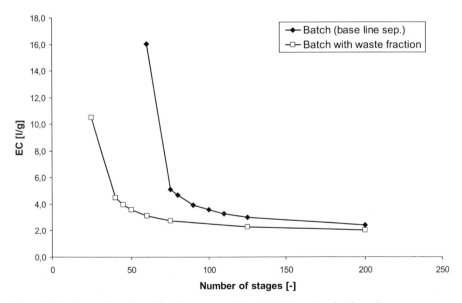

Figure 7.33   Comparison of specific eluent consumption (EC) between two batch modes.

However, the performance of batch chromatography with regard to productivity and eluent consumption can be improved significantly by allowing a waste fraction between the product fractions (cut strategy I in Fig. 7.6). As seen in Fig. 7.32, not only can the maximum productivity be doubled but also the number of stages required to achieve the maximum productivity can be reduced significantly. Furthermore, Fig. 7.33 shows lower eluent consumption for batch processes with a waste fraction, especially at a small number of stages.

The main reason for these phenomena is that a highly efficient column is necessary to separate only a small amount of feed at high purity demand during base-line separation (i.e. high number of stages and small loading factor). As an alternative to base-line separation, high purity demand can also be achieved by decreasing the product collection time, thus creating a gap (i.e. waste fraction) between the product fractions. In other words, high purity demand can also be satisfied by decreasing the separation yield. Therefore, by allowing a waste fraction the separation can be accomplished at lower efficiency and higher „loadability" of the column (i.e. low number of stages and high loading factor). This expands the optimization region and allows the batch chromatographic process to reach more favorable operating points, as illustrated in Section 7.2.3. Additionally, as can be seen and calculated from the isotherms of EMD53986, the selectivity factor at current feed concentration ($2.5\,\mathrm{g\,l^{-1}}$) is 1.16 if the competitive effect between the enantiomers can be neglected, but it increases to 1.53 if the competitive effect is considered. This so-called displacement effect is fully utilized to increase the separation efficiency by not performing a base-line-separation and allowing an overlapping concentration region during the separation.

Theoretically, it might also be possible to improve SMB performance by decreasing the separation yield in a similar way to batch processes with a waste fraction. This can be accomplished by interrupting the otherwise continuous product collection at the extract and/or raffinate port. However, the applicability of this method has never been investigated.

The quintessence of this comparison is that the application of SMB results in higher productivity and lower eluent consumption, thus reducing adsorbent and eluent costs significantly and making it a very attractive process concept in preparative- and industrial-scale chromatography for binary separations. The performance of batch processes, however, is not as bad as generally predicted. Efficiency of adsorbent and eluent usage in batch processes can be increased significantly by allowing waste fractions. Although one should not forget that waste fractions always cause additional costs (e.g. cost due to feed loss or cost for further treatment of waste fractions), it might be worth considering whether the increase of productivity and the decrease of eluent consumption can still reduce the total separation cost.

Please note that batch and SMB chromatography are compared here regarding the productivity and eluent consumption and not regarding the total separation cost. This is because productivity, eluent consumption, yield and purity are clearly defined functions, while the total separation cost and other objective functions related to economics depend on the individual company. Therefore, the magnitude of economic benefits in the application of SMB in comparison to batch chromatography depends on the applied cost structure. Jupke et al. (2002) have provided a detailed comparison between batch and SMB chromatography regarding the total separation cost.

8
Chromatographic Reactors

T. Borren and J. Fricke

This chapter introduces chromatographic reactors that combine the chromatographic separation, discussed in the earlier chapters, and a chemical reaction. The reaction type A $\rightleftharpoons$ B + C is considered to explain the operating principles of the different chromatographic reactors. Section 8.2 overviews the different kinds of reactors and the types of reactions investigated.

Rigorous modeling is, so far, the only effective tool to describe the behavior of chromatographic reactors in detail. Therefore, the modeling approach presented in Chapter 6 is extended for chromatographic reactors. Section 8.3 presents this approach, as well as an overview of further models.

Section 8.4 covers the operation and design of chromatographic reactors for reactions of the type A $\rightleftharpoons$ B + C. Rules for the choice of process conditions for integrated chromatographic processes as well as the operating conditions of batch wise and continuous operation are given.

Finally, Section 8.5 discusses the design of chromatographic reactors, using as examples the esterification of β-phenethyl alcohol with acetic acid and the isomerization of glucose.

8.1
Introduction

In standard chemical processes a reaction unit is followed by one or more separation steps. This enables a straightforward design of each unit operation and offers a high degree of freedom for process design. However, with equilibrium limited reactions the yield is limited and can only be enhanced by additional recycle streams. This increases the dimensions of all apparatuses and the investment costs, especially for low conversion rates.

Another possibility for enhancing reaction yield and separation efficiency is to integrate both unit operations within a single piece of equipment. Here the operating parameters of both the reaction and the separation must match, and they also

Preparative Chromatography. Edited by H. Schmidt-Traub
Copyright © 2005 WILEY-VCH Verlag GmbH & Co. KGaA, Weinheim
ISBN: 3-527-30643-9

have to fulfill restrictions due to equipment design. Therefore, integrated processes often suffer due to a reduced window of possible operating points. Examples of integrated processes are reactive distillation, reactive extraction, reactive absorption, reactive membrane separation, reactive crystallization and reactive adsorption (Kulprathipanja, 2002).

A chromatographic reactor combines a chemical or biochemical reaction with a chromatographic separation. Within the integrated process, homogeneous or heterogeneous reactions can occur. The latter requires a solid catalyst or an immobilized enzyme, which are mixed with the chromatographic packing material. In special cases such as esterifications the adsorbent can act as catalyst as well (Mazotti et al., 1996a). When different materials for catalyst and adsorbent are needed, establishing a reliable packing is a major problem. Due to different particle sizes and densities a heterogeneous distribution of catalyst within the bed can occur. One possible way of overcoming these difficulties is to partition the bed with alternating layers of adsorbent and catalyst (Meurer et al., 1997). For a homogeneously catalyzed reaction, the catalyst has to be either recovered or separated after the chromatographic reactor.

Since the late 1950s, analytical chromatographic reactors have been applied to studies of reactions (Coca et al., 1993), determination of reaction kinetics (Bassett and Habgood, 1960), characterization of stationary phases (Jeng and Langer, 1989), or examination of interactions between mobile and stationary phase (Coca et al., 1989).

The main focus of recent developments has been on preparative-scale applications. Carr and Dandekar (2002) have recently reviewed gas-phase chromatographic reactors. Application of liquid-phase chromatographic reactors will be discussed in the following.

8.2
Types of Chromatographic Reactors

8.2.1
Chromatographic Batch Reactor

A reaction of type A $\rightleftarrows$ B + C is used in Fig. 8.1 to explain the principles of chromatographic batch reactors (BR). In the given example the reactant A has intermediate adsorption behavior and products B and C are the more strongly and weakly adsorbed components. Reactant A is injected as a sharp pulse into the column. During its propagation along the column, A reacts to give B and C. Owing to their different adsorption behavior all components propagate with different velocities and are separated. Therefore, the restriction of an equilibrium-limited reaction can be overcome to convert the reactant totally. Additionally, no separation unit following the chromatographic reactor is necessary to gain the high purity products. In cases of several products an appropriate number of fractions can also be collected at the outlet of the reactor. Due to batch operation, only a small part of the bed is used for reaction and separation. Therefore, the operation requires large amounts of desorbent

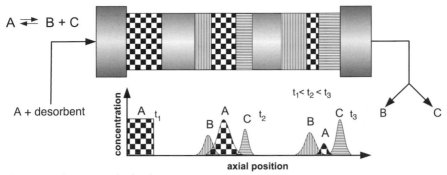

Figure 8.1 Chromatographic batch reactor.

and results in a high product dilution. Furthermore, the productivity of the process is usually rather low.

To separate the products and the reactant the chromatographic system (Chapter 4) has to provide reasonable selectivity for all components. For instance, for the reaction $A \rightleftarrows B + C$ it is favorable if the products have the highest and lowest affinity while the retention of the reactant has to be intermediate, as assumed in Fig. 8.1. For multiple reactants a chromatographic system with a separation factor close to one in respect of the reactants should be chosen. In all other cases the reactants separate and complete conversion is not possible, so that special operation modes have to be applied (Section 8.4).

Chromatographic batch reactors are employed to prepare instable reagents on the laboratory scale (Coca et al., 1993) and for the production of fine chemicals. These applications include the racemic resolution of amino acid esters (Kalbé et al., 1989), acid-catalyzed sucrose inversion (Lauer, 1980), production of dextran (Zafar and Barker, 1988) and saccharification of starch to maltose (Sarmidi and Barker, 1993a). Sardin et al. (1993) employed batch chromatographic reactors for different esterification reactions such as the esterification of acetic acid with ethanol and the transesterification of methylacetate. Falk and Seidel-Morgenstern (2002) have investigated the hydrolysis of methyl formate.

8.2.2
Continuous Annular Reactors

The continuous annular chromatographic reactor (CACR) realizes a continuous operation. The general set-up of an annular chromatograph has been explained in Chapter 5.3.2.

The annulus is packed with adsorbent and catalyst and rotates continuously. Considering again the reaction $A \rightleftarrows B + C$ a fluid element entering the annular gap is propelled axially by the eluent and in a tangential direction by the rotation. Reactant A is injected into the annulus continuously at a certain point at the top of the bed while eluent enters along the remaining circumference. Reaction occurs and both

$$A \rightleftharpoons B + C$$

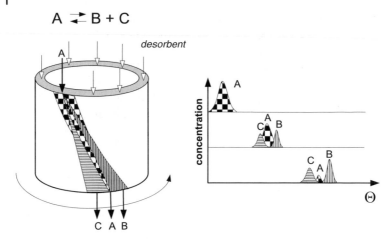

Figure 8.2 Rotating annular reactor.

reactant and products describe helical paths through the bed (Fig. 8.2). The more strongly adsorbed component B is transported further in tangential direction and has therefore a longer residence time than component C. Products as well as non-converted reactant can be collected at different outlet ports at the bottom of the annulus.

Using the annular reactor, any number of products can be collected at the outlet. The annular reactor can be considered as a certain number of parallel batch reactors. Therefore, the characteristics of operation are quite similar to those of a batch reactor. However, the packing of the annular gap with adsorbent or a mixture of adsorbent and catalyst, as well as the sealing, restricts the volume flow and therefore the application.

The rotating annular reactor was first described by Martin (1949). An application of this concept has been investigated by Carr (1993), considering the hydrolysis of methyl formate. Other reactions performed using the rotating cylindrical annulus are the saccharification of starch to maltose (Sarmidi and Barker, 1993a), enzyme-catalyzed inversion of sucrose (Sarmidi and Barker, 1993b) and the homogeneously catalyzed redox reaction between iridium and iron (Herbsthofer et al., 2002).

8.2.3
Counter-current Chromatographic Reactors

8.2.3.1 True Moving Bed Reactor
A counter-current flow of solid and fluid offers the advantage of continuous operation. The easiest concept of such a reactor is the true moving bed reactor (TMBR). It is a direct adoption of the true moving bed (TMB) process explained in Chapter 5.3.4.

Figure 8.3 illustrates the basic process principles for the reaction type $A \rightleftharpoons B + C$. Sections I and IV have the same purpose as in the TMB process – they regenerate the

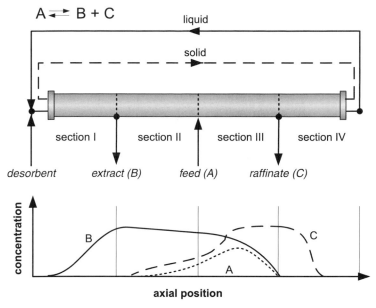

Figure 8.3 Chromatographic true moving bed reactor.

adsorbent and the mobile phase respectively. In sections II and III, reaction and separation occur simultaneously. The stronger adsorbed product is carried with the solid stream to the extract port and the less strongly retained product is moved with the mobile phase towards the raffinate port. If total conversion of reactant A is achieved, products B and C can be withdrawn in high purity.

In contrast to chromatographic batch and annular reactors, at most two pure products can be withdrawn. To obtain an optimal process, sections II and III have to be carefully designed with respect to total conversion. After the start up period the TMBR reaches steady state.

The counter-current moving bed reactor was first mentioned by Takeuchi et al. (1978). However, so far only gas-phase investigations are known. The lack of applications of the TMBR are for the same reasons as with the TMB process, namely difficulties with particle movement, back mixing of the solid, and abrasion of the particles.

8.2.3.2 Simulated Moving Bed Reactor

The simulated moving bed reactor (SMBR) based on the simulated moving bed (SMB) process is a practical alternative for implementing counter-current continuous reactors. Counter-current movement of the phases is simulated by sequentially switching the inlet and outlet ports located between the columns in direction of the liquid flow (Fig. 8.4). As with the SMB process, two different concepts are known to realize the counter-current flow. One is based on switching the ports and the other on the movement of columns. However, both require elaborate process control concepts to realize the movement. Owing to the periodical changes of the set-up the pro-

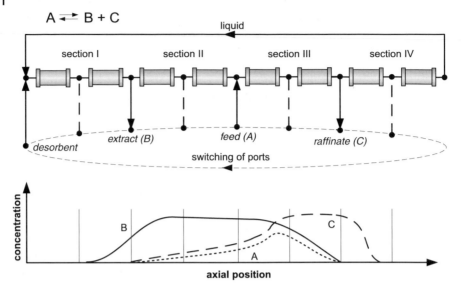

Figure 8.4 Chromatographic simulated moving bed reactor.

cess reaches, after start up, a cyclic steady state. At most, two pure products can be withdrawn and in the case of partial reactant conversion further product purification is necessary.

The design of SMBR processes has to take into account the requirements of different types of reactions. Therefore, different types of flowsheets and operating modes can be chosen. As with a semi-continuous operation, a process without section IV (regeneration of the solvent), a five-section process or a four section process without recycle of the eluent may be advantageous. Relevant design criteria for the SMBR are discussed in Section 8.4.

The chromatographic SMB reactor has been examined for various reaction systems, with the main focus on reactions of the type A + B $\rightleftarrows$ C + D. Examples are esterifications of acetic acid with methanol (Lode et al., 2003b), ethanol (Mazotti et al., 1996a) and β-phenethyl alcohol (Kawase et al., 1996) as well as the production of bisphenol A (Kawase et al., 1999). The same reaction type can also be found for various hydrocarbons, such as the transfer reaction of sucrose with lactose to lactosucrose (Kawase et al., 2001) and the hydrolysis of lactose (Shieh and Barker, 1996). Barker et al. (1992) focused on reactions of the type A $\rightleftarrows$ B + C, such as enzyme-catalyzed sucrose inversion and the production of dextran. Also, reactions of the type A $\rightleftarrows$ B have been investigated, e.g. isomerization of glucose to fructose by Fricke (2005) as well as Tuomi and Engell (2004). Michel et al. (2003) have examined the application of electrochemical SMB reactors for consecutive reactions and used as an example the production of arabinose.

Recently developed improved SMB processes, e.g. VariCol (Ludemann-Hombourger et al., 2000b), PowerFeed (Zhang et al., 2003) and ModiCon (Schramm et al., 2003), can also be used in reactive chromatography. Application of the VariCol concept leads directly to the same advantage as in the separation case, namely higher productivity. The other concepts may offer advantages concerning the driving force for the reaction, but no investigations have been published so far.

8.2.3.3 Hashimoto Process

Chromatographic SMB reactors are not well suited for reactions of the type A $\rightleftarrows$ B. Due to the homogeneous mixture of adsorbent and catalyst it is only possible to slightly overcome the chemical equilibrium and total conversion is not possible. However, specially designed processes can overcome this limitation.

Hashimoto et al. (1983b) proposed a reactive counter-current chromatographic process for the production of higher fructose syrup. This process, also known as the Hashimoto process, can be applied for these reactions. It is based on a partial deintegration of reaction and separation. Columns and reactors are alternately arranged in one or several sections of the SMB process (Fig. 8.5). Characteristic for this process is that the reactors are stationary in the assigned section. Reactors are shifted together with the inlet and outlet nodes in the direction of the liquid flow. For the example shown in Fig. 8.5, after shifting the nodes the reactors will be located in front of the columns numbered 6 and 7. Movement of the reactors is also responsible for the characteristic axial concentration profile within the Hashimoto process. By shifting the reactors their liquid content is moved from the inlet to the outlet of the column. At the reactor outlet a much higher concentration level exists than at the inlet of the following column. Therefore, a discontinuity in the axial concentration profile arises, which is balanced over the switching interval by adsorption and hydrodynamic effects. Owing to the reactors placed in section III it is often not possible to withdraw a raffinate stream with high purity.

The Hashimoto process design separates the functionalities separation and reaction. This concept offers the possibility to exchange adsorbent and, for example,

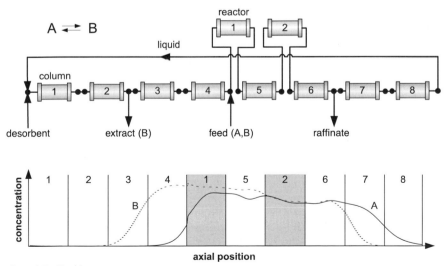

Figure 8.5 Hashimoto process.

immobilized enzyme after different operating times. Also, different operating conditions, e.g. temperatures, can be applied for reaction as well as separation.

So far, the concept has been used for the isomerization of glucose to fructose (Hashimoto et al., 1983 and Borren and Schmidt-Traub, 2004), but further applications such as thermal racemizations are conceivable. An application for reactive systems with more than two components is also possible if special catalysts or operating conditions are required. However, an even more complex design, as well as control of the process, has to be taken into account.

8.3
Modeling of Chromatographic Reactors

8.3.1
Models of Chromatographic Batch Reactors

Modeling approaches for chromatographic reactors are similar to models used to describe chromatographic separations (discussed in Chapter 6 in detail). In the mass balances, homogeneous or heterogeneous reactions also have to be taken into account to extend the derived models for chromatographic reactors. Owing to the high dilution the heat of reaction can be neglected for all reactions.

A common modeling approach for chromatographic batch reactors is the equilibrium-transport dispersive model (Chapter 6.2.4.1). Therefore, only the equations for this approach are discussed here. The differential mass balance (Fig. 8.6) takes into account axial dispersion as well as mass transfer between fluid and both solid phases.

Mass transfer kinetics are given by simple linear driving force models. The fraction of catalyst within the whole fixed bed is described by the factor X_{cat}. Assuming

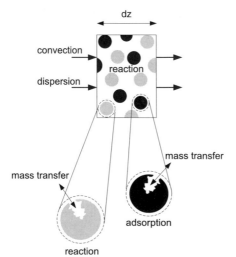

Figure 8.6 Model of the chromatographic batch reactor.

reaction occurs homogeneously in the liquid phase and heterogeneously at the surface of the catalyst, the differential mass balance for component i can be written for the liquid phase as:

$$\frac{\partial c_i}{\partial t} = -u_{int}\frac{\partial c_i}{\partial x} + D_{ax}\frac{\partial^2 c_i}{\partial x^2} - (1 - X_{cat})\, k_{eff,ads,i}\,\frac{6}{d_{p,ads}}\frac{1-\varepsilon}{\varepsilon}(c_i - c_{p,ads,i})$$
$$- X_{cat}\, k_{eff,cat,i}\,\frac{6}{d_{p,cat}}\frac{1-\varepsilon}{\varepsilon}(c_i - c_{p,cat,i}) + v_i\, r_{hom,i} \tag{8.1}$$

and for both solid phases:

$$\frac{\partial q_{ads,i}}{\partial t} = k_{eff,ads,i}\,\frac{6}{d_{p,ads}}(c_i - c_{p,ads,i}) \tag{8.2}$$

$$\frac{\partial q_{cat,i}}{\partial t} = k_{eff,cat,i}\,\frac{6}{d_{p,cat}}(c_i - c_{p,cat,i}) + v_i\, r_{het,i} \tag{8.3}$$

Adsorption equilibrium can be described by any type of isotherm and gives a correlation between the concentrations in the adsorbed and liquid phases:

$$q_i = f(c_{p,i}, \ldots, c_{p,n}) \tag{8.4}$$

Boundary conditions proposed by Danckwerts (1953) are used to solve the differential equation:

$$u_{int}\, c_{in,i} = u_{int}\, c_i\Big|_{x=0} - D_{ax}\frac{\partial c_i}{\partial x}\Big|_{x=0} \tag{8.5}$$

$$\frac{\partial c_i}{\partial x}\Big|_{x=L_c} = 0 \tag{8.6}$$

The correlation proposed by Chung and Wen (1968) can be applied to calculate the axial dispersion coefficient:

$$Pe = \frac{\dot{V}d_p}{A_c \varepsilon D_{ax}} \tag{8.7a}$$

$$Pe = \frac{0.2}{\varepsilon} + \frac{0.011}{\varepsilon}(\varepsilon Re_p)^{0.4} \tag{8.7b}$$

$$Re_p = \frac{\rho_l \dot{V}d_p}{A \varepsilon \eta_l} \tag{8.7c}$$

8.3.2
Models of Continuous Annular Chromatographic Reactors

For separation in an annular chromatograph, different modeling approaches have been developed. Similar to column models, theoretical plate models, equilibrium stage models and models based on a differential mass balance equation can be distinguished. The differential mass balance model is often applied to describe continuous annular reactors. Different approaches have been developed by Carr (1993), Sarmidi and Barker (1993b) and Herbsthofer et al. (2002). The general model has to take into account angular convection of the stationary phase, convection of the mobile phase in angular and axial directions, angular and axial dispersion, mass transfer and reaction. The general balance equation for the solute can be written as (Sarmidi and Barker, 1993a):

$$\frac{\partial c_i}{\partial t} = -\omega \frac{\partial c_i}{\partial \Theta} - u_{\text{int}} \frac{\partial c_i}{\partial x} + D_{\text{an}} \frac{\partial^2 c_i}{\partial \Theta^2} + D_{\text{ax}} \frac{\partial^2 c_i}{\partial x^2}$$
$$- (1 - X_{\text{cat}}) k_{\text{eff,ads},i} \frac{6}{d_{\text{p,ads}}} \frac{1 - \varepsilon}{\varepsilon} (c_i - c_{\text{p,ads},i}) \qquad (8.8)$$
$$- X_{\text{cat}} k_{\text{eff,cat},i} \frac{6}{d_{\text{p,cat}}} \frac{1 - \varepsilon}{\varepsilon} (c_i - c_{\text{p,cat},i}) + \upsilon_i r_{\text{hom},i}$$

and for the solid phases:

$$\frac{\partial q_{\text{ads},i}}{\partial t} = -\omega \frac{\partial q_{\text{ads},i}}{\partial \Theta} + k_{\text{eff,ads},i} \frac{6}{d_{\text{p,ads}}} (c_i - c_{\text{p,ads},i}) \qquad (8.9)$$

$$\frac{\partial q_{\text{cat},i}}{\partial t} = -\omega \frac{\partial q_{\text{cat},i}}{\partial \Theta} + k_{\text{eff,cat},i} \frac{6}{d_{\text{p,cat}}} (c_i - c_{\text{p,cat},i}) + \upsilon_i r_{\text{het},i} \qquad (8.10)$$

In continuous operation the process reaches steady state. Therefore, a steady state mass balance is sufficient to describe the process behavior. Transport by angular dispersion is often neglected.

8.3.3
Models of Counter-Current Chromatographic Reactors

Two different modeling approaches are used for simulated moving bed reactors. The first approach combines the model of several batch columns with the mass balances for the external inlet and outlet streams. By periodically changing the boundary conditions the transient behavior of the process is taken into account. The model is based on the SMB model introduced in Chapter 6 and is, therefore, referred to as the SMBR model. The second approach assumes a true counter-current flow of the solid and the liquid phase like the TMBR. Therefore, this approach is called the TMBR model.

Table 8.1 overviews the different approaches that have been applied for continuous chromatographic reactors in the liquid phase.

Table 8.1 Modeling approaches for continuous chromatographic reactors.

Approach	Reaction	Adsorption isotherm	Remark	Ref.
SMBR – general rate	$A \rightleftarrows B$	Nonlinear	Optimization-based control	Tuomi and Engell (2004)
SMBR – general rate	$A + B \rightleftarrows C + D$	Miscellaneous	Model-based optimization	Dünnebier et al. (2000a)
SMBR – equilibrium transport disversive	$A \rightleftarrows B$	Nonlinear	Model-based design	Borren and Schmidt-Traub (2004)
SMBR, TMBR – equilibrium transport dispersive	$A \rightarrow B + C$ $A \rightleftarrows B + C$	Linear	Model-based design	Fricke and Schmidt-Traub (2003)
SMBR – equilibrium transport dispersive	$A + B \rightleftarrows C + D$	Langmuir	Validated for synthesis of methyl acetate	Lode et al. (2001)
SMBR – equilibrium transport dispersive	$A \rightarrow B + C$ $A \rightleftarrows B + C$	Miscellaneous	Model-based analysis	Meurer et al. (1997)
SMBR – equilibrium transport	$A + B \rightleftarrows C + D$	Linear	Parallel reactions	Kawase et al. 2001
SMBR – equilibrium transport	$A + 2B \rightleftarrows C + D$	Langmuir		Kawase et al. (1999)
SMBR – equilibrium transport	$A + B \rightleftarrows C + D$	Langmuir	Validated for production of β-phenethyl acetate	Kawase et al. (1996)
SMBR – ideal stage	$A + B \rightleftarrows C + D$	Linear	Validated for hydrolysis of lactose	Shieh and Barker (1996)
TMBR, SMBR – equilibrium transport	$A \rightleftarrows B$	Linear	Validated for glucose isomerization	Hashimoto et al. (1993b)
TMBR, SMBR – ideal stage	$A + B \rightleftarrows C + D$	Linear		Pai et al. (1996)
TMBR – ideal	$A \rightarrow B$	Miscellaneous	Analytic solution	Altshuller (1983)
TMBR – miscellaneous	$A \rightleftarrows B$	Linear	Analytic solution	Cho et al. (1982)
CACR – equilibrium transport	$A + B \rightleftarrows C + D$	Linear	Validated for redox reaction of iridium	Herbsthofer et al. (2002)
CACR – equilibrium dispersive	$A + B \rightleftarrows C + D$	Linear	Validated for sucrose inversion	Sarmidi and Barker (1993b)

8.3.3.1 SMBR Model

By combining the node balance and a reactive column model as described in Section 8.3.1, an SMBR model can be created. Each section consists of one or more columns. The affiliation of the single columns and the boundary conditions are changed periodically. The flowsheet of the SMBR process is similar to that for the TMBR process given in Fig. 8.7. Please note that the origin of the axis system is located in the feed port. Internal flow rates are related to the external streams by balances of the inlet- and outlet nodes. A mass balance for each node is also necessary to calculate the concentration at the inlet and outlet of each single section.

Desorbent node:

$$\dot{V}_{\mathrm{des}} = \dot{V}_{\mathrm{I}} - \dot{V}_{\mathrm{IV}} \tag{8.11}$$

$$c_{i,\mathrm{des}}\,\dot{V}_{\mathrm{des}} = c_{i,\mathrm{I}}(L_{\mathrm{sec}})\,\dot{V}_{\mathrm{I}} - c_{i,\mathrm{IV}}(L_{\mathrm{sec}})\,\dot{V}_{\mathrm{IV}} \tag{8.12}$$

Extract draw-off node:

$$\dot{V}_{\mathrm{ext}} = \dot{V}_{\mathrm{I}} - \dot{V}_{\mathrm{II}} \tag{8.13}$$

$$c_{i,\mathrm{ext}} = c_{i,\mathrm{I}}(0) = c_{i,\mathrm{II}}(L_{\mathrm{sec}}) \tag{8.14}$$

Feed node:

$$\dot{V}_{\mathrm{feed}} = \dot{V}_{\mathrm{III}} - \dot{V}_{\mathrm{II}} \tag{8.15}$$

$$c_{i,\mathrm{feed}}\,\dot{V}_{\mathrm{feed}} = c_{i,\mathrm{III}}(0)\,\dot{V}_{\mathrm{III}} - c_{i,\mathrm{II}}(0)\,\dot{V}_{\mathrm{II}} \tag{8.16}$$

Raffinate draw-off node:

$$\dot{V}_{\mathrm{raf}} = \dot{V}_{\mathrm{III}} - \dot{V}_{\mathrm{IV}} \tag{8.17}$$

$$c_{i,\mathrm{raf}} = c_{i,\mathrm{III}}(L_{\mathrm{sec}}) = c_{i,\mathrm{IV}}(0) \tag{8.18}$$

8.3.3.2 Ideal TMBR Model

The ideal TMBR model is based on a differential mass balance of each section (Fig. 8.7). Only convection and homogeneous or quasi-heterogeneous reaction is taken into account. Axial dispersion and mass transfer resistances are neglected.

The following differential equations can be retrieved for component i in section j in the stationary state:

$$+\dot{V}_{l,j}\frac{\mathrm{d}c_i}{\mathrm{d}x} - \dot{V}_{\mathrm{solid},j}\frac{\mathrm{d}q_i}{\mathrm{d}x} - \upsilon_{\mathrm{het},i}\,A_{\mathrm{c}}(1-\varepsilon)r_{i,\mathrm{het}} - \upsilon_{\mathrm{hom},i}\,A_{\mathrm{c}}\,\varepsilon\,r_{i,\mathrm{hom}} = 0 \tag{8.19a}$$
$$(j = \mathrm{I},\,\mathrm{II})$$

$$-\dot{V}_{l,j}\frac{\mathrm{d}c_i}{\mathrm{d}x} + \dot{V}_{\mathrm{solid},j}\frac{\mathrm{d}q_i}{\mathrm{d}x} - \upsilon_{\mathrm{het},i}\,A_{\mathrm{c}}(1-\varepsilon)r_{i,\mathrm{het}} - \upsilon_{\mathrm{hom},i}\,A_{\mathrm{c}}\,\varepsilon\,r_{i,\mathrm{hom}} = 0 \tag{8.19b}$$
$$(j = \mathrm{III},\,\mathrm{IV})$$

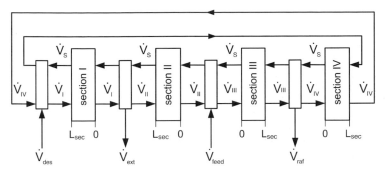

Figure 8.7 Ideal TMBR model. (Reproduced from Fricke and Schmidt-Traub, 2003.)

To apply the TMBR model for a SMBR process the flow rates have to be transferred. The solid flow rate can be calculated by taking into account the column length L_c and cross section A_c as well as the shifting interval, t_{shift}:

$$\dot{V}_{solid} = \frac{(1-\varepsilon)A_c L_c}{t_{shift}} \tag{8.20}$$

Internal flow rates are related by:

$$\dot{V}_j^{TMB} = \dot{V}_j^{SMB} - \frac{\varepsilon \dot{V}_{solid}}{(1-\varepsilon)} \tag{8.21}$$

The total mass balance equations, describing the relationship between the internal and external flow rates, are the same for both the TMBR and SMBR models (Eqs. 8.11, 8.13, 8.15 and 8.17). Component balances, however, have to take into account the solid streams.

Desorbent node:

$$c_{i,des} \dot{V}_{des} = c_{i,I}(L_{sec}) \dot{V}_I - c_{i,IV}(L_{sec}) \dot{V}_{IV} + \left[q_{i,IV}(L_{sec}) - q_{i,I}(L_{sec})\right] \dot{V}_{solid} \tag{8.22}$$

Extract draw-off node:

$$c_{i,feed} \dot{V}_{feed} = c_{i,III}(0) \dot{V}_{III} - c_{i,II}(0) \dot{V}_{II} + \left[q_{i,II}(0) - q_{i,III}(0)\right] \dot{V}_{solid} \tag{8.23}$$

Feed node:

$$c_{i,feed} \dot{V}_{feed} = c_{i,III}(0) \dot{V}_{III} - c_{i,II}(0) \dot{V}_{II} + \left[q_{i,II}(0) - q_{i,III}(0)\right] \dot{V}_{solid} \tag{8.24}$$

Raffinate draw-off node:

$$c_{i,raf} \dot{V}_{raf} = c_{i,III}(L_{sec}) \dot{V}_{III} - c_{i,IV}(0) \dot{V}_{IV} + \left[q_{i,IV}(0) + q_{i,III}(L_{sec})\right] \dot{V}_{solid} \tag{8.25}$$

Notably, the idealized TMBR model is often extended by taking into account axial dispersion and mass transfer resistances to achieve a more accurate description of the process behavior.

8.3.3.3 Comparison of Modeling Approaches

With small interactions between different components and a sufficiently large number of columns in each section a TMB model can be used to approximate the mean concentration profile of a periodic steady state SMB separation process.

Fricke and Schmidt-Traub (2003) as well as Lode (2002) used the idealized TMB approach to derive analytical solutions and used these results to develop short-cut methods for SMB separations.

Strube (1996) has shown that an increasing interaction of components or a decreasing number of columns per section results in significant deviations of both the calculated concentration profiles and purities. Model based optimal design requires a correct description of the dynamic behavior of the SMB process. Therefore, Dünnebier et al. (2000a) recommend the use of the detailed SMB model. These considerations are also valid for SMBR processes. Additionally, Lode et al. (2003a) have shown that the residence time calculated with the TMBR model differs from that in the SMBR model and, in consequence, different conversion rates are calculated.

Owing to the effect of simplifications of the TMBR model described above it is strongly recommended to use SMBR models for the final design of SMBR processes. Nevertheless, TMBR models are useful in determining initial guesses of the operating conditions of SMBR processes.

8.3.4
Determination of the Model Parameters

The determination of model parameters for chromatographic systems has been explained in detail in Chapter 6.5. The same methods can be applied for reactive chromatographic systems, except that the interaction parameters of adsorption isotherms can be determined with high accuracy only for non-reacting components (Mazotti et al., 1996a).

Additionally, the parameters of the reaction rate equation have to be determined. Levenspiel (1999) has described different methods that can be applied. For homogeneous-catalyzed reactions the experiments can be done in a stirred tank reactor. If the reaction is catalyzed heterogeneously the use of a fixed bed reactor is recommended. In this case, the mass transfer resistance and the hydrodynamic parameters of the packed bed also have to be taken into account.

8.4
Operation and Design of Chromatographic Reactors

8.4.1
Choice of the Process Conditions for Chromatographic Reactors

The first step in designing a chromatographic reactor is the choice of suitable process conditions. These are the pressure and temperature. In liquid-phase applications the pressure or the maximum pressure drop has no influence on either the separation or the reaction. The maximum pressure is theoretically only limited by the destruction of the adsorbent or the apparatuses. However, often, the maximum pressure is not applied because the corresponding volume flow is not suitable for an integrated process. Due to the high volume flow the residence time is reduced and, therefore, lower conversion rates are achieved.

Higher temperatures are favorable for most reactions due to higher reaction rates. But higher temperatures also enhance desorption and hence reduce the separation factor. Therefore, a compromise between reaction rate and separation efficiency has to be found in each case. Matsen et al. (1965) recommend to select the process conditions in a way that the reaction rate is fast and therefore the separation is the limiting step within the chromatographic reactor.

8.4.2
Operating Conditions of Batch Reactors

After selecting the chromatographic system the operation mode of the batch reactor has to be chosen. High productivities require a high throughput. Therefore, pulsed operation is used (Fig. 8.8). Reactants are supposed to be injected as a rectangle pulse of period t_{cycle} and duration t_{inj}. These parameters are strongly affected by the reaction kinetics, reaction stoichiometry and adsorption isotherm.

Based on the maximum of the conversion, Sardin et al. (1993) derived simple algebraic equations to calculate the optimum pulse period. These equations are based on the following assumptions:

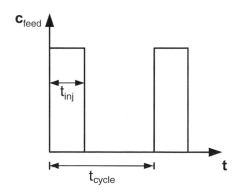

Figure 8.8 Feed concentration in pulsed operation mode. (Reproduced from Sardin et al., 1993.)

- Linear adsorption
- Plug flow
- Reaction not affecting the shape of the reactant pulse

Therefore, an optimization of the operating conditions applying of these assumptions results in a maximization of the contact time between reactants. Nevertheless, they can be applied as an initial guess of the final operating conditions.

For a reaction of type A $\rightleftharpoons$ B + C, conversion can especially be improved if reactant A is the medium-retained component and products B and C can be separated. In this case the optimal pulse period is equal to the elution time and can be calculated by:

$$t_{cycle} = t_0 \frac{1-\varepsilon}{\varepsilon}(H_B - H_C) + t_{inj} \tag{8.26}$$

When both products are eluted before or after the reactant, high conversions are only possible for small feed concentrations. Therefore, these cases are not of interest for preparative scale reactive separations.

For a reaction of type A + B $\rightleftharpoons$ C + D it is important that products C and D are separated to inhibit the reverse reaction, and that the reactants A and B have the same adsorption behavior and are thus not separated. This ideal situation almost never occurs. Several cases have to be distinguished.

Where A and B have similar retention times or one of the reactants can be used as eluent the problem can be reduced to the simpler reaction A $\rightleftharpoons$ C + D and Eq. 8.26 can be applied.

When component B is the stronger retained reactant the contact of the two components can be intensified by successive injection (Fig. 8.9). The less strongly retained component A is injected at the end of the injection of B but leaves the column together with the early injected portion of B if no reaction occurs. Maximum conversion is achieved when the faster propagating component leaves exactly with the first amount of the slower propagating component. Therefore, the optimal injection times for batch reactors are related by:

$$t_{inj,B} = t_0 \frac{1-\varepsilon}{\varepsilon}(H_B - H_A) + t_{inj,A} \tag{8.27}$$

Owing to the different adsorption behavior of the products, different cases can be distinguished in respect to the optimal cycle time. If the products elute on both sides of the reactant A or the retention of one of the products is similar to reactant A the cycle time can be calculated by:

$$t_{cycle} = t_0 \frac{1-\varepsilon}{\varepsilon}(H_D - H_C) + t_{inj,A} \tag{8.28}$$

If both products elute before or after the reactants a sufficient separation is not possible. Therefore, these cases are of minor practical interest.

For the consecutive reaction A + B $\rightarrow$ C and C + B $\rightarrow$ D with C as desired product, the yield is governed by the reaction rate of the second reaction and the separation of B and C. Therefore, the concentration of reactant A has to exceed that of B in the

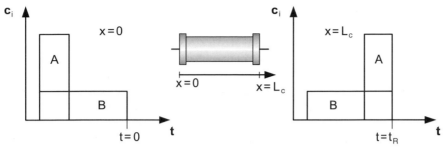

Figure 8.9 Propagation of the reactants under separation and no reaction. (Reproduced from Sardin et al., 1993.)

reaction zone in order to enhance the yield of product C. Assuming an equimolar feed, component A has to be injected as a sharp pulse either before or after the second reactant B. The injection time $t_{inj,B}$ can be estimated using Eq. 8.27.

8.4.3
Operating Conditions of SMB Reactors

8.4.3.1 Deduction of the Design Criteria

Successful operation of SMB chromatographic reactors depends on the correct choice of operating conditions, particularly of the internal flow rates in each section and the switching time. As explained in Chapter 7.3 the triangle theory developed by Storti et al. (1993b) is often used to design SMB separation processes. Considering SMB reactors this method can also be used to determine operating conditions that have to be met to separate the products. Theories developed by Lode (2002) as well as Fricke (2005) take into account the chromatographic separation as well as different kinds of reaction. Based on their theories it is possible to calculate an initial guess of the residence time necessary to achieve a given conversion rate in the chromatographic SMBR process.

As an example the reaction type $A \rightleftarrows B + C$ is considered here. Results for other processes can be taken from the literature mentioned above.

The direction of propagation of a component within the TMBR depends on the dimensionless flow rates in each section of the process. Assuming a linear isotherm a component propagates with the fluid if the dimensionless flow rate is higher than the Henry coefficient. If the flow rate is smaller than the Henry coefficient a component is transported in the direction of the solid flow. Therefore, Lode et al. (2001) subdivided the separation region into the three regions shown in Fig. 8.10.

If the operating point of the TMBR is located in region T_3, reactant A is carried with the liquid stream towards the raffinate outlet and can not enter section II. The opposite situation occurs with an operating point within region T_2. Reactant A is carried with the solid and cannot enter section III of the process. Region T_1 is characterized by the propagation of the reactant into both sections next to the feed port.

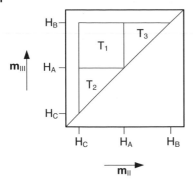

To calculate the conversion rate X of the reactant, the amount withdrawn by the product streams has to be known. As illustrated above, whether component A can reach the extract and/or the raffinate outlet depends on the chosen operating conditions. The ideal TMBR model and the resulting flow rates are shown in Fig. 8.7. Taking into account the external streams and dimensionless flow rates, X can be calculated by Eq. 8.29:

$$
\begin{aligned}
X &= 1 - \frac{\dot{V}_{\text{ext}}\, c_{\text{A,ext}} + \dot{V}_{\text{raf}}\, c_{\text{A,raf}}}{\dot{V}_{\text{feed}}\, c_{\text{A,feed}}} \\
&= 1 - \frac{(H_{\text{A}} - m_{\text{II}})\, c_{\text{A,II}}(L_{\text{sec}}) + (m_{\text{III}} - H_{\text{A}})\, c_{\text{A,III}}(L_{\text{sec}})}{(H_{\text{A}} - m_{\text{II}})\, c_{\text{A,II}}(0) + (m_{\text{III}} - H_{\text{A}})\, c_{\text{A,III}}(0)}
\end{aligned}
\tag{8.29}
$$

If the operating point is located in region T_2 or T_3 the term representing section III or II, respectively, is equal to zero.

The differential equation can be transferred into its dimensionless form by dividing Eq. 8.19 by the feed concentration c_{feed} and the solid flow rate $\dot{V}_{\text{solid}}$ and multiplying by the section length L_{sec}. Assuming a heterogeneously catalyzed reversible reaction as well as linear adsorption isotherms, the following dimensionless differential equations describe the behavior of the TMB reactor:

$$
(m_j - H_i) \frac{\mathrm{d} C_{\text{DL},i,j}}{\mathrm{d} Z} + v_i\, k_{\text{reac}} \frac{A_{\text{sec}} L_{\text{sec}}(1 - \varepsilon)}{\dot{V}_{\text{solid}}} \left(C_{\text{DL,A},j} - \frac{C_{\text{DL,B},j} C_{\text{DL,C},j}}{K_{\text{EQ}}} \right) = 0
$$
$$
(j = \text{I, II})
\tag{8.30a}
$$

$$
(H_i - m_j) \frac{\mathrm{d} C_{\text{DL},i,j}}{\mathrm{d} Z} + v_i\, k_{\text{reac}} \frac{A_{\text{sec}} L_{\text{sec}}(1 - \varepsilon)}{\dot{V}_{\text{solid}}} \left(C_{\text{DL,A},j} - \frac{C_{\text{DL,B},j} C_{\text{DL,C},j}}{K_{\text{EQ}}} \right) = 0
$$
$$
(j = \text{III, IV})
\tag{8.30b}
$$

As the conversion rate only depends on the concentration at the inlet and outlet of sections II and III it can be calculated if these concentrations are known. Differential equations describing the mass balances in section III are:

$$(H_A - m_{III}) \frac{dC_{DL,A,III}}{dZ} - Da_{solid} \left(C_{DL,A,III} - \frac{C_{DL,B,III} \, C_{DLC,III}}{K_{EQ}} \right) = 0 \qquad (8.31)$$

$$(H_B - m_{III}) \frac{dC_{DL,B,III}}{dZ} + Da_{solid} \left(C_{DL,A,III} - \frac{C_{DL,B,III} \, C_{DL,C,III}}{K_{EQ}} \right) = 0 \qquad (8.32)$$

$$(H_C - m_{III}) \frac{dC_{DL,C,III}}{dZ} + Da_{solid} \left(C_{DL,A,III} - \frac{C_{DL,B,III} \, C_{DL,C,III}}{K_{EQ}} \right) = 0 \qquad (8.33)$$

with the Damkoehler number Da_{solid} referenced by the solid stream:

$$Da_{solid} = \frac{k_{reac} \, A_{sec} \, L_{sec} \, (1 - \varepsilon)}{\dot{V}_{solid}} \qquad (8.34)$$

Direct integration is not possible as the three equations depend on each other. Fish and Carr (1989) suggested eliminating the reaction term by addition of Eq. 8.31 and Eq. 8.32 or 8.33. After integration and introducing a net flow rate ($W_{i,III}$) for components B and C, Eqs. 8.32 and 8.33 are transformed into Eq. 8.35:

$$W_{i,III} = (H_A - m_{III}) C_{DL,A,III} + (H_i - m_{III}) C_{DL,i,III} \quad (i = B, C) \qquad (8.35)$$

The net flow rate is constant for each component and does not depend on the position within the section. Based on this equation the concentration of components B and C can be calculated as a function of the reactant concentration and the operating parameters:

$$C_{DL,i,III} = \frac{-W_{i,III}}{m_{III} - H_i} - \frac{m_{III} - H_A}{m_{III} - H_i} C_{DL,A,III}$$
$$= K_{i,III} - M_{i,III} \, C_{DL,A,III} \qquad (8.36)$$

A similar procedure yields a corresponding equation for section II:

$$C_{DL,i,II} = \frac{-W_{i,II}}{H_i - m_{II}} - \frac{m_{II} - H_A}{H_i - m_{II}} C_{DL,A,II}$$
$$= K_{i,II} - M_{i,II} \, C_{DL,A,II} \qquad (8.37)$$

After inserting Eqs. 8.36 and 8.37 into the differential equation for the reactant (Eq. 8.31), integration leads to the following expression for the dimensionless outlet concentration of reactant A in section j:

$$C_{DL,A,j}(L_{sec}) = \frac{-A_{2,j}}{2A_{1,j}} + \sqrt{-\Delta_j} \frac{B_j + e^{-D_j}}{B_j - e^{-D_j}} \quad \text{for } \Delta_j < 0 \qquad (8.38a)$$

$$\frac{2}{\sqrt{\Delta_j}} \left[\arctan \frac{2 \cdot A_{1,j} \cdot C_{DL,A,j}(L_{sec}) + A_{2,j}}{\sqrt{\Delta_j}} - \arctan \frac{2 \cdot A_{1,j} \cdot C_{CL,A,j}(0) + A_{2,j}}{\sqrt{\Delta_j}} \right] = E_j$$

$$\text{for } \Delta_j > 0 \qquad (8.38b)$$

Definitions of the variables used here are summarized in Tab. 8.2.

Table 8.2 Variables used in Eq. 8.38.

		Section II	Section III
$K_{i,j}$		$\dfrac{-W_{i,\mathrm{II}}}{H_i - m_{\mathrm{II}}}$	$\dfrac{-W_{i,\mathrm{III}}}{m_{\mathrm{III}} - H_i}$
$M_{i,j}$		$\dfrac{H_A - m_{\mathrm{II}}}{H_i - m_{\mathrm{II}}}$	$\dfrac{H_A - m_{\mathrm{III}}}{m_{\mathrm{III}} - H_i}$
D_j		$\dfrac{\mathrm{Da_{solid}}\sqrt{-\Delta_{\mathrm{II}}}}{K_{\mathrm{EQ}}(H_A - m_{\mathrm{II}})}$	$\dfrac{\mathrm{Da_{solid}}\sqrt{-\Delta_{\mathrm{III}}}}{K_{\mathrm{EQ}}(m_{\mathrm{III}} - H_A)}$
E_j		$-\dfrac{\mathrm{Da_{solid}}}{K_{\mathrm{EQ}}(H_A - m_{\mathrm{II}})}$	$-\dfrac{\mathrm{Da_{solid}}}{K_{\mathrm{EQ}}(m_{\mathrm{III}} - H_A)}$
$A_{1,j}$		$-M_{B,j}\,M_{C,j}$	
$A_{2,j}$		$K_{\mathrm{EQ}} + K_{B,j}\,M_{C,j} - K_{C,j}\,M_{B,j}$	
$A_{3,j}$		$-K_{B,j}\,K_{C,j}$	
Δ_j		$N\,A_{1,j}\,A_{3,j} - A_{2,j}^2$	
B_j		$\dfrac{\left[2A_{1,j}\,C_{\mathrm{DL,A,II}}(0) + A_{2,j} + \sqrt{-\Delta_j}\right]}{\left[2A_{1,j}\,C_{\mathrm{DL,A,II}}(0) + A_{2,j} - \sqrt{-\Delta_j}\right]}$	

To evaluate Eqs. 8.38, the constant net flow rate ($W_{i,j}$) has to be specified for the different separation regions. If the operating point is within region T_3 and the separation condition inside section III is fulfilled, the stronger adsorbed component B cannot reach the raffinate node, i.e. $C_{B,\mathrm{III}}(L_{\mathrm{sec}})$ is equal to zero. In this case Eq. 8.35 can be reduced and the net flow rate of component B is:

$$W_{B,\mathrm{III}} = (-m_{\mathrm{III}} + H_A)C_{\mathrm{DL,A,III}}(L_{\mathrm{sec}}) \tag{8.39}$$

The net flow rate of component C can be derived by considering the inlet of section III. As long as the operating point is within the separation region, component C propagates towards the raffinate node. Therefore, its concentration at the inlet of section III is zero and $W_{C,\mathrm{III}}$ becomes:

$$W_{C,\mathrm{III}} = (-m_{\mathrm{III}} + H_A)C_{\mathrm{DL,A,III}}(0) \tag{8.40}$$

Now the concentration at the outlet of section III can be determined by introducing the net flow rates (Eqs. 8.39 and 8.40) into Eqs. 8.38. Based on this concentration the conversion rate of the TMBR can be calculated. Notably, the outlet concentration has to be evaluated numerically, as B_{III} depends on $C_{\mathrm{DL,A,III}}$. For a given limit of the conversion rate the flow rate in section III is determined so as to guarantee that the conversion calculated by Eq. 8.29 corresponds to the required conversion.

A balance for the inlet of section II and an operating point within region T_2 results in Eq. 8.41 for $W_{B,\mathrm{II}}$:

$$W_{C,\mathrm{II}} = (m_{\mathrm{II}} - H_A)C_{\mathrm{DL,A,II}}(L_{\mathrm{sec}}) \tag{8.41}$$

while $W_{C,II}$ can be calculated for the outlet of section II by Eq. 8.42:

$$W_{C,II} = (m_{II} - H_A)C_{DL,A,II}(L_{sec})$$ (8.42)

Therefore, the same procedure as in region T_3 can be used to calculate the concentration profile.

For an operating point within region T_1 the mass balances of the inlet of sections II (Eq. 8.41) and III (Eq. 8.40) are no longer valid. The concentration of component C at the inlet of section III cannot be neglected and the following expression has to be used to calculate $W_{C,III}$:

$$W_{C,III} = (-m_{III} + H_A)C_{DL,A,III}(0) + (-m_{III} + H_C)C_{DL,C,III}(0)$$ (8.43)

The concentration of component C at the outlet of section II, $C_{DL,C,II}(0)$, can be calculated using the net flow rate in this section, $W_{C,II}$. If this concentration is known the inlet concentration of section III can be determined by the mass balance of the feed node:

$$C_{DL,C,III}(0) = \frac{(m_{II} - H_C)}{m_{III} - H_C} \left[K_{C,II}\, C_{DL,C,II}(0) \right]$$ (8.44)

In the same way the concentration of component B at the inlet of section II can be determined:

$$C_{DL,B,II}(0) = \frac{(m_{III} - H_B)}{m_{II} - H_B} \left[K_{B,III}\, C_{DL,B,III}(0) \right]$$ (8.45)

After the transformations described above, concentrations $C_{DL,A,II}(L_{sec})$ and $C_{DL,A,III}(L_{sec})$ can be calculated numerically by solving Eqs. 8.38, and the conversion can be determined using Eq. 8.29. Finally, the flow rates in sections II and III are chosen so as to meet the specified conversion rate.

8.4.3.2 Application of Design Criteria

To examine the effect of reaction kinetics on the reaction region the derived design criteria are applied for the reversible solid-phase reaction $A \rightleftharpoons B + C$. A linear adsorption isotherm of the components is assumed, with Henry coefficients of 0.4 (reactant), 0.2 and 0.6 (products) respectively. A process with an equal number of columns in sections II and III is considered. The conversion that has to be reached is set to 99.99%.

Figure 8.11 shows the influence of the equilibrium constant on the separation region for a constant Damkoehler number ($Da_{solid} = 2$) in respect of the solid flow rate. Process performance can be evaluated by the distance of the operating point towards the diagonal. An increasing distance is equal to an increasing feed flow rate and indicates, therefore, a higher productivity of the process. Operating conditions that allow the desired conversion are limited by a triangular area. With large equilibrium constants the limits of the separation and reaction region are almost equal. With a decreasing equilibrium constant the triangle of the desired conversion becomes smaller, which can be explained by two different effects. The decrease in

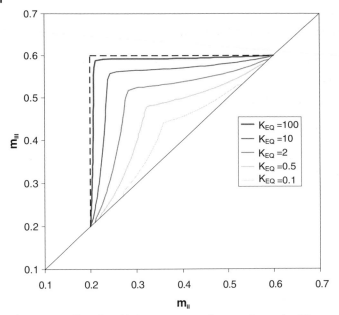

Figure 8.11 Effect of equilibrium constant on the separation region ($Da_{solid} = 2$).

the dimensionless flow rates is equivalent to lower liquid flow rates in the sections. Therefore, the residence time of the components increases and, for simultaneous separation, a higher reaction yield can be achieved. The driving force for the separation for a linear isotherm is proportional to the difference between the Henry coefficient H_i and the dimensionless flow rate m_j. Therefore, a smaller dimensionless flow rate enhances the separation efficiency of the integrated reactor.

For an equilibrium constant of 0.1 only a small area within the operating diagram ensures the desired conversion of 99.99 %. Nevertheless, compared with a conventional reactor with the equilibrium conversion limited to 27 % a significant improvement in conversion can be achieved within the integrated process. However, productivity is reduced significantly compared with the conventional process. Therefore, the consecutive process, where a reactor is followed by a separation, is always an alternative that has to be considered in the case of low equilibrium constants.

Lode (2002) and Fricke and Schmidt-Traub (2003) have discussed the influence of further parameters, e.g. Damkoehler number. Although, notably, the TMBR model only approximates an SMBR process, it can be applied to discuss the different influences on the SMBR.

8.5
Design Examples

Here the process design of chromatographic reactors is discussed in detail for two different reaction systems. All results have been derived by applying the transport-dispersive model as given by Eq. 8.1.

8.5.1
Esterification of β-Phenethyl Alcohol with Acetic Acid

β-Phenethyl acetate production from β-phenethyl alcohol and acetic acid is an example of an heterogeneously catalyzed esterification reaction:

$$CH_3COOH + C_6H_5CH_2CH_2OH \rightleftarrows C_6H_5CH_2CH_2COOCH_3 + H_2O$$

The desired product, β-phenethyl acetate, is used for the production of scents and perfume.

The reaction is catalyzed by an acidic ion-exchanger resin, which can also be used as adsorbent. Equation 8.46 can be applied to delineate the reaction at the surface of the resin:

$$r_i = v_i \, k_{\mathrm{reac}} \frac{c_{\mathrm{acid}} \, c_{\mathrm{alcohol}} - \dfrac{c_{\mathrm{ester}} \, c_{\mathrm{water}}}{K_{\mathrm{eq}}}}{(1 + k_{\mathrm{alcohol}} \, c_{\mathrm{alcohol}} + k_{\mathrm{water}} \, c_{\mathrm{water}})^2} \tag{8.46}$$

A Langmuir isotherm describes the adsorption. All model parameters are summarized in Appendix B.4.

Esterifications are typical applications for chromatographic reactors. Unless one of the reactants can be used as eluent, which is favorable for the equilibrium reaction and reduces the reaction stoichiometry to A $\rightleftarrows$ C + D, separation of the reactants can be inhibited to a certain extent by the choice of operating conditions.

First, the design of a chromatographic batch reactor for the production of β-phenethyl acetate is considered. Due to different adsorption of reactants, they are separated during their propagation within the column. To maximize the time for an overlap of both components, the more strongly adsorbed reactant is injected before the less strongly adsorbed one. Owing to the different propagation velocity the weaker adsorbed reactant passes the peak of the stronger adsorbed one. Figure 8.12 shows the dependence of the conversion rate on the difference between the injection times for two different temperatures. Reactor performance can be improved by injecting the reactants separately. Additionally, the effect of the separation factor of reactants becomes obvious. The separation factor of the reactants increases when the esterification reaction is performed at lower temperatures. If a batch chromatographic reactor is used for the production of β-phenethyl acetate at 50 °C, the time difference for the injection of the stronger adsorbed component and the weaker adsorbed component has to be increased. Please note that the decrease in conversion rate is not only caused by the increased separation factor of the reactants. In most cases the influence of temperature on the reaction rate will be predominant.

For SMB reactors, separation of the reactants cannot be compensated by different injection times. Therefore, a modified SMB reactor concept is recommended in the case of large separation factors of the reactants. Five sections are used in order to get two different feed nodes for the reactants (Fig. 8.13). The feed next to the extract node is for the less strongly adsorbed reactant. The more strongly adsorbed reactant is fed at the second feed node close to the raffinate node. The additional fifth section between the feed nodes is called the reaction section. In this section the flow

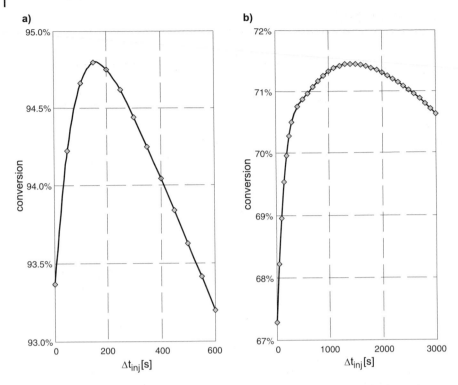

Figure 8.12 Effect of the difference in injection time of the reactants on conversion for the production of β-phenethyl acetate at (a) 85 and (b) 50 °C.

rate has to be adjusted so that the reactants propagate in different directions. The more strongly adsorbed reactant propagates towards the extract node whereas the less strongly adsorbed component propagates towards the raffinate node. This procedure increases the contact between both reactants and intensifies the reaction. If the process is properly designed the reaction takes place in the reaction zone and none of the reactants reach one of the product nodes.

The 4-section-SMBR and the 5-section-SMBR have been designed for esterification reactions to asses these concepts. Process parameters have been chosen so as to optimize the conversion of reactants. Table 8.3 summarizes the results for two different separation factors of the educts. For a small separation factor the conversion cannot be increased within the 5-section SMBR. Although a smaller conversion is reached, the corresponding purity of the withdrawn ester is greater than that in a 4-section SMBR.

If a reaction with a large separation factor is chosen, the 5-section SMBR shows superior performance to the 4-section SMBR. When a 4-section SMBR is employed the purity of the acetate does not exceed 61%. By using the 5-section process a purity of the ester up to 89% and higher conversion rates can be reached. This demon-

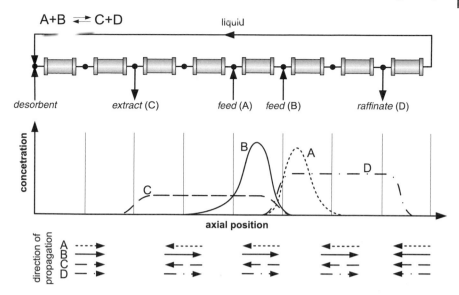

Figure 8.13 Scheme of the five-section SMB reactor.

Table 8.3 Comparison of four- and five-section process for different separation factors.

Separation factor α	Number of sections	Conversion X (%)	Ester purity Pu (%)
1.7	4	92.9	88.5
	5	87.8	99.0
3.1	4	65.8	60.8
	5	70.3	89.6

strates that the 5-section SMBR can significantly increase the process performance if significant separation of the reactants occur.

8.5.2
Isomerization of Glucose

Glucose–fructose syrup is principally used in the food industry. To increase the sweetness of the syrup the fructose content is raised by isomerization. Practically available syrups contain 42 or 55 wt.% (dry base) fructose (Schenck, 2003).

A batch reactor containing immobilized glucose isomerase is used to convert glucose into glucose–fructose syrup. The reaction shall yield a product with a minimum of 42 wt.% fructose in the dry substance. Production of 55 wt.% syrup is accomplished by employing chromatography to separate the 42 wt.% syrup into glucose- and fructose-rich fractions. The glucose is recycled to the isomerization step. Finally, the fructose-rich stream, containing between 80 and 95 wt.% fructose, is blended with 42 wt.% syrup to produce the 55 wt.% syrup.

The reaction is catalyzed by an immobilized enzyme and can be described quasi-homogeneously by the following rate equation:

$$r_i = v_i \, k_{reac} \left[\frac{K_{eq}}{1 + K_{eq}} \left(c_{glu} + \frac{c_{fru}}{K_{eq}} \right) \right] \tag{8.47}$$

The SMB separation uses an ion-exchanger resin whose non linear adsorption isotherm is described by Eq. 8.48:

$$q_i = A_i \, c_i + B_i \, c_i^2 + D_i \, c_i \, c_{j \neq i} \tag{8.48}$$

All model parameters are summarized in Appendix B.3.

Chromatographic reactors offer the opportunity to enhance the productivity of these chromatographic industrial-scale processes further.

Villermaux (1981) has investigated the applicability of the chromatographic batch reactor for isomerizations. The simulation studies show that the conversion within the integrated batch reactor can not exceed the conversion in a fixed bed reactor without separation. Therefore, the integrated batch reactor does not offer any advantages for reactions of type A $\rightleftarrows$ B compared with a sequential process.

However, a continuous counter-current chromatographic reactor offers the opportunity to overcome the chemical equilibrium. Different degrees of integration are possible within the chromatographic reactor (Borren and Schmidt-Traub, 2004). An SMBR with a homogeneous mixture of adsorbent and immobilized enzyme integrates reaction and separation totally. As an alternative Hashimoto et al. (1993) suggested a 3-section process without raffinate withdrawal, which causes a breakthrough of glucose into section I.

The so-called Hashimoto process partially deintegrates reaction and separation by using separate reactors and separation columns. Due to the distributed functionalities a breakthrough of glucose is not advantageous. Therefore, a process design with four sections, including a raffinate port, was chosen. Glucose withdrawn with the raffinate stream is reconcentrated, e.g. in a membrane module. In a following process step glucose is converted again into glucose–fructose syrup and recycled back into the process together with the initial feed stream.

A further alternative is the totally deintegrated sequential process, where the reactor is followed by an SMB separation and the glucose is recycled to the reactor. Figure 8.14 shows the three continuous concepts considered here.

If the flow rate ratios m_j in sections I and IV are chosen so that total desorption occurs in section I and total adsorption occurs in section IV these flow rate ratios have no further impact on the separation efficiency of the process. Therefore, the same appropriate flow rates in sections I and IV are chosen for all concepts. Then the operating conditions depend only on the flow rates in sections II and III.

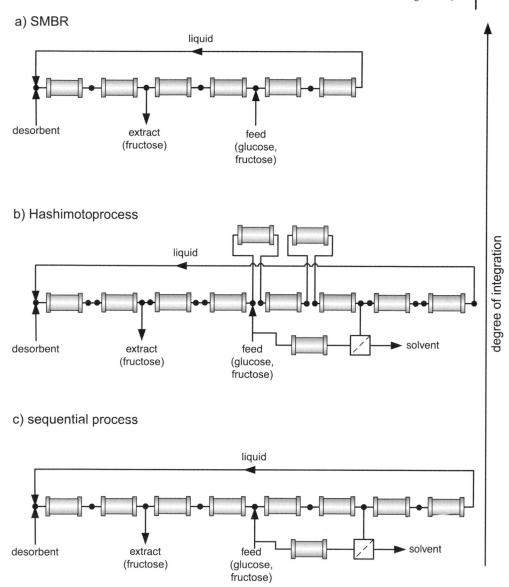

Figure 8.14 Reactor concepts for the production of fructose syrup.

Owing to the practical conditions for the production of fructose syrup no purity constraints exist. The purity of the extract in respect of fructose is:

$$\mathrm{Pu_{fru}} = \frac{m_{\mathrm{fru}}}{m_{\mathrm{fru}} + m_{\mathrm{glu}}} \tag{8.49}$$

while productivity in respect of the produced fructose is defined as:

$$\mathrm{Pr_{fru}} = \frac{\dot{m}_{\mathrm{fru}}}{m_{\mathrm{ads}}} \tag{8.50}$$

Another objective function for process optimization is the eluent consumption, which is not considered here because of the fixed flow rates in the regeneration sections. A further promising alternative for process optimization is the VariCol process proposed by Ludemann-Hombourger et al. (2000b) (Chapter 5.3.5.1). This process is based on an asynchronous movement of the in- and outlet valves and offers, therefore, the opportunity to apparently decrease the number of columns, with constant purity and the same amount of product. This concept can be applied for all of the processes considered here.

Product purities and productivities for all three processes have been determined by rigorous simulation. Figure 8.15 shows the dependency of the maximum achievable productivity on the purity for each concept. For low fructose purities, both four-section processes reach a maximum productivity, which is caused by the additional external recycle stream.

Figure 8.15 indicates operating areas where one of the concepts is advantageous. The three-section SMBR with a homogeneous mixture of adsorbent and catalyst is favorable for purities < 63%. In the three-section processes a breakthrough of glucose occurs in section I and reduces the driving force for the back reaction of fructose

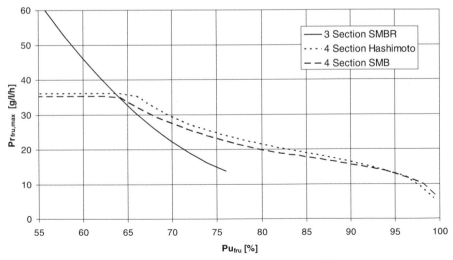

Figure 8.15 Comparison of different reactor concepts for the production of fructose syrup.

to glucose. Productivity decreases rapidly with increasing purity demands. Compared with the other process concepts, high purity production is not advantageous.

Both four-section processes achieve similar productivities. For fructose syrup purities of up to 95 % the productivity of the Hashimoto process is slightly higher. If purities higher than 95 % are required the non-integrated sequential process is slightly favorable. This concept also allows the highest purities of fructose syrup of all.

The integrated reactors can overcome the chemical equilibrium of the glucose isomerization. With the sequential process, recycling of the glucose withdrawn with the raffinate is necessary to achieve total conversion. As long as high purities are not required, integration of reaction and separation within a single piece of equipment increases the productivity of the process. Nevertheless, the increased complexity of plant and operation in the case of the Hashimoto process, as well as the deactivation of the immobilized enzyme within the SMBR, has to be taken into account.

9
Advanced Control of Simulated Moving Bed Processes

A. Toumi and S. Engell

This chapter gives an introduction to the concept of model predictive control and an overview of the concepts proposed for the control of simulated moving bed processes. Thereafter the benefits of a model-based optimizing control strategy for the example of a 6-column reactive SMB plant of pharmaceutical scale are presented.

9.1
Introduction

As discussed in several previous chapters, the use of detailed process models in the design and in the choice of the operating parameters of chromatographic separation processes may lead to considerable improvements. However, optimal settings of the operating parameters do not guarantee an optimal operation of the real plant. This is due to non-idealities of the column packings and the peripherals, the effects of external disturbances, changes of plant behavior over time, and to the inevitable discrepancies between the model and the real system. Usually, not even the constraints on process and product parameters (e.g. desired product purities) are met at the real plant if the operation parameters obtained from off-line optimization are applied.

Any real process is, therefore, operated using some sort of feedback control. Feedback control means that (some of) the degrees of freedom of the plant are modified during the operation based upon observation of some measurable variables. These measurements may be available quasi-continuously or with a more or less large sampling period (e.g. results of a laboratory analysis), and accordingly the operation parameters (termed inputs in feedback control terminology) are modified in either a quasi-continuous fashion or intermittently. In chromatographic separations, feedback control is most often realized by the operators, who change the operating parameters until the specifications are met.

Automatic feedback control is the continuous or repetitive modification of some operating parameters based on measurement data. Due to the dynamic nature of the process, the control algorithm (or feedback law) is usually also dynamic and has to be designed carefully to avoid instability. Standard feedback laws are proportional (P), integral (I) and differentiating (D) control and combinations thereof, e.g. PI control where the controlled input depends on the instantaneous control error and on its integral, guaranteeing steady-state accuracy and a more or less quick response to sudden disturbances. The design of automatic feedback controllers becomes nontrivial when several actuated variables affect the parameters (outputs) of interest in interacting fashion, and when the plant exhibits nonlinear behavior and/or complex dynamics.

9.2
Model Predictive Control

During the last two decades, model-predictive control (MPC) (Allgöwer et al., 1999, Mayne, 2000 and Rao and Rawlings, 2000) has increasingly been applied to the control of processes with interacting dynamics. The basic idea of MPC is to employ a plant model that predicts the reaction of the plant to the past and future inputs, and to optimize a number of future inputs such that the predicted outputs follow the desired trajectory over a certain period of time (called the prediction horizon). This process is iterated, only the next input is applied to the plant, and new inputs are computed for a prediction horizon that is shifted one step into the future, taking new measurements into account. Thus, the behavior of the real process and disturbances are taken into account.

In the classical concept of predictive control, the trajectory (or set-point) of the process is assumed to be known. Control is implemented in a discrete-time fashion with a fixed sampling rate, i.e. measurements are assumed to be available at a certain frequency and the control inputs are changed accordingly. The inputs are piecewise constant over the sampling intervals. The prediction horizon H_p represents the number of time intervals over which the future process behavior will be predicted using the model and the assumed future inputs, and over which the performance of the process is optimized (Fig. 9.1). Only those inputs located in the control horizon H_r are considered as optimization variables, whereas the remaining variables between H_{r+1} and H_p are set equal to the input variables in the time interval H_r. The result of the optimization step is a sequence of input vectors. The first input vector is applied immediately to the plant. The control and the prediction horizon are then shifted one interval forward in time and the optimization run is repeated, taking into account new data on the process state and, eventually, newly estimated process parameters. The full process state is usually not measurable, so state estimation techniques must be used. Most model-predictive controllers employed in industry use input–output models of the process rather than a state-based approach.

The linear MPC technique, introduced in the 1960s, is now well-established in the petrochemical industry. A major advantage of this approach is the ability to include constraints in the optimization, thus exploiting the full range of available actuators (pumps, valves) and respecting operating limits of the equipment as well as hard constraints on the outputs (e.g. product purities). In current industrial practice,

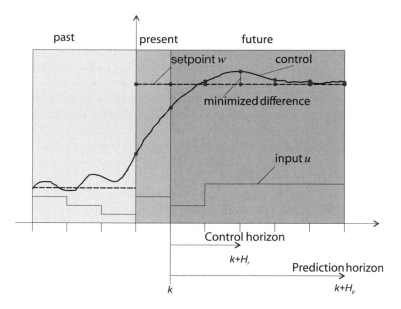

past present future

setpoint w control

minimized difference

input u

Control horizon

$k+H_r$

Prediction horizon

k $k+H_p$

Figure 9.1   Principle of model predictive control.

model-predictive control is almost exclusively used based upon linear process models that are only valid around a certain operating point of the plant, while the operating point itself is determined off-line by model-based optimization. This approach can guarantee to meet the constraints for those process outputs that are under feedback control, but it will, in general, not lead to an optimal operation unless all degrees of freedom are used as actuated variables to control at least as many measured outputs.

For linear plant models and a quadratic cost function, optimized solutions arising from MPC formulations and solved online computationally are feasible for slow industrial processes. Qin and Badgwell (2003) reported over 4500 industrial implementations of linear MPC by leading MPC software providers.

Linear process models do not give sufficiently accurate predictions for processes operated either over a wide operating window, such as polymerization reactors, or which exhibit strongly nonlinear behavior, such as SMB processes with nonlinear isotherms. For such processes, nonlinear process models have to be used. If a nonlinear process model is employed within a model predictive scheme, the resulting control strategy is termed nonlinear MPC (NMPC). NMPC requires the online solution of nonlinear optimization problems, with the associated problems of long computation times and convergence problems. If constraints are present, it must be made sure that in each sampling interval a solution is found (not necessarily exactly at the optimum) that satisfies the constraints. NMPC has been a key area of academic research in process control in the last 10 years, but according to Qin and Badgwell (2003) there have been relatively few successful applications. Most of these applications are based on empirical (black box) models. Rigorous first-principle models often lead to complex optimization problems that are hard to solve in real-time.

A major advantage of this approach is that not only regulatory control can be performed (i.e. to keep some variables at their specified values) but also the degrees of freedom can be used to maximize or minimize a cost functional online while meeting all relevant constraints. Of course, this nonlinear optimization requires sufficient computing power to solve the complex nonlinear optimization problems, and sufficiently accurate process models. As chromatographic SMB processes are relatively slow, so that a modern PC suffices to solve the optimization problem within a switching period, and accurate rigorous models are available and can be adapted to the real process online using the available measurements, nonlinear model-predictive control of SMB processes is, as we show in this chapter, feasible, and solves the problem of operating the process at its optimum while meeting the product specifications (possibly with some safety margin) and respecting limitations of the equipment.

9.3
State-of-the-art in Control of SMB Processes

In industry, the control of SMB processes is widely understood as the way in which pumps are appropriately adjusted or how the pressure of the SMB loop is stabilized (Holt, 1995 and Hotier, 1998). We refer to such controllers as basic controllers.

Automatic control of the flow rates or the switching time to meet purity specifications is difficult due to the extremely long time delays and complex dynamics described by nonlinear distributed parameter models, and mixed discrete and continuous dynamics, leading to small operating windows and a strongly nonlinear response to input variations.

To reach the desired product purities, the flow rates are usually varied manually. Modification of the operating parameters is based either on heuristic rules or relies on operator expertise. Antia (2003) proposed the following practical scheme:

- Start with low feed concentrations to achieve linear separation conditions.
- Increase $\dot{V}_1$ to a large value and decrease $\dot{V}_4$ to a low value so that the design criteria for the sections 1 and 4 are satisfied by a large margin. Attention is then focused on the appropriate choice of flow rates in the central sections 2 and 3.
- Increase the concentration of the feed in steps. Determine which outlet is polluted and correct the flow rates according to predefined rules. This can be repeated until the feed concentrations reach their upper-limits.
- Once the flow rates in the central sections are chosen appropriately, increase $\dot{V}_4$ and decrease $\dot{V}_1$. This ensures that a minimal flow of eluent is used and thus near-optimal process performance is reached.

See Chapter 7.3 for more details on the background of these rules.

In practice, SMB processes are controlled using similar manual schemes (Küsters et al., 1995, Juza, 1999 and Miller et al., 2003). Antia (2003) suggested that these heuristic rules are included in a fuzzy controller to achieve full automatic control of SMB processes, but no applications have been described so far. Cox et al. (2003) recently reported a successful control and monitoring system for the separation of an enantiomer mixture based on the concentration profiles in the recycle loop.

Automatic purity control has been reported for the separation of aromatic hydrocarbons where online Raman spectroscopy can be used to measure the concentrations of the compounds at the outlet of the chromatographic columns (Marteau et al., 1994). This approach, and the geometric nonlinear control concept described by Kloppenburg and Gilles (1999b), is based on a model that describes the corresponding true moving bed (TMB) process, so that cyclic port switching is neglected. For SMB processes with few columns (8 or less) the TMB process does not approximate the SMB process well, and so the applicability of this control scheme to plants with few columns is questionable.

Schramm et al. (2001) have presented a model-based control approach for direct control of the product purities of SMB processes. Based on wave theory, relationships between the front movements and the flow rates of the equivalent TMB process were derived. Using these relationships, a simple control concept with two PI controllers was proposed. This concept is very easy to implement; however, it does not address the issue of optimizing the operating regime in the presence of disturbances or model mismatch.

Natarajan and Lee (2000) have investigated the application of a repetitive model predictive control (RMPC) technique to SMB processes. RMPC is a model-based control technique that results from incorporating the basic concept of repetitive control into the MPC framework. The switching period of the process is assumed to be constant. This is limiting, since the switching time can also be manipulated to control the process. The rigorous model is linearized along the optimal trajectory. In recent work a periodic Kalman filter, which reconstructs the process state, was included in the RMPC controller (Erdem et al., 2004 and Abel et al., 2004b). The applicability of this scheme in the presence of strong nonlinearities as they occur in enantiomer separations and in the reactive case is an open question.

Klatt et al. (2002) have proposed a two-layer control architecture where, on the upper level, the optimal operating regime is calculated at a low sampling rate by dynamic optimization based on a rigorous process model, and applied a similar approach to batch chromatography. Model parameters are adapted based on online measurements. The low-level control task is to keep the process on the optimal trajectory despite disturbances and plant/model mismatch. This controller is based on identified models gained from simulation data of the rigorous process model along the optimal trajectory. For the linear (linear adsorption isotherm) case, linear autoregressive exogenous (ARX) models are sufficient, as shown by Klatt et al. (2002), whereas in the nonlinear case neural networks (NN) were applied (Klatt et al., 2000 and Wang et al., 2003). A disadvantage of this two-layer concept is that the stabilized front positions do not guarantee product purities if plant/model mismatch occurs. Thus, an additional purity controller is required (Hanisch, 2003).

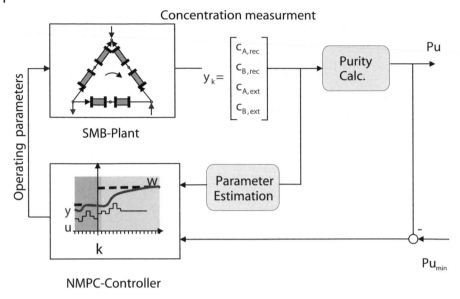

Concentration measurment

$$y_k = \begin{bmatrix} c_{A,rec} \\ c_{B,rec} \\ c_{A,ext} \\ c_{B,ext} \end{bmatrix}$$

SMB-Plant

Purity Calc.

Pu

Parameter Estimation

NMPC-Controller

Figure 9.2 Online optimizing control structure.

Toumi and Engell (2004b) have recently proposed a control scheme, where nonlinear model-based optimization is performed online, and have applied it successfully to the control of a three-section reactive SMB process for glucose isomerization. The key feature of this approach is that the production cost is minimized online while product purities are considered as constraints. Figure 9.2 shows the control structure used here. Product purity is measured on-line in order to correct the actual operating point. The NMPC controller uses a rigorous general rate type process model, the parameters of which are re-estimated online during plant operation, which reduces plant–model mismatch and enables one to compensate for a drift or sudden changes in plant parameters. The next section describes this concept and its laboratory-scale plant implementation in detail.

9.4
Online Optimizing Control of a Reactive Simulated Moving Bed Process

9.4.1
Process Description

The reactive simulated moving bed process considered here is the isomerization of glucose to fructose. The plant consists of six reactive chromatographic fixed beds that are interconnected to form a closed-loop arrangement (Fricke and Schmidt-Traub, 2002). As shown in Fig. 9.3, a pure glucose solution is injected to the system at the feed line. At the extract line, a mixture of glucose and fructose, called high fructose

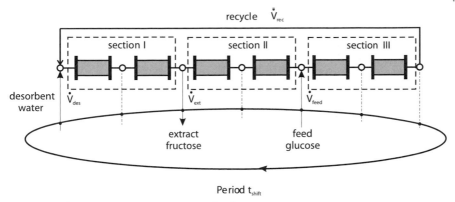

Figure 9.3 Three-section reactive SMB process for glucose isomerization.

corn syrup (HFCS), is withdrawn. Water is used as solvent and is fed continuously to the system at the desorbent line. Between the positions of the ports, three sections are established that have different roles (Chapter 8):

- Section I between the desorbent and the extract line: regeneration of the solid
- Section II: reaction/separation section
- Section III between the feed and the eluent line: reaction/separation section and recycling of solvent.

In this special SMB process, no attempt is made to completely separate glucose and fructose since the most common type of fructose syrup, usually called high-fructose syrup, is mainly produced either as HFCS42 (42 % fructose) or as HFCS55 (55 % fructose). For some purposes, syrup with more than 55 % fructose, called a higher-fructose syrup, is desirable. In any case, the objective is to transform a feed containing pure glucose into a stream where the glucose is partially converted into fructose.

9.4.2
Formulation of the Online Optimizing Controller

As discussed above, the task of the controller is to optimize the performance of the process over a certain horizon in the future, the prediction horizon. Specifications of product purities, equipment limitations and the dynamic process model (a full hybrid model of the process, including the switching of the ports and a general rate model of all columns) appear as constraints. The control algorithm solves the following nonlinear optimization problem online:

$$\min_{[\beta_k, \beta_{k+1}, \ldots, \beta_{k+H_r}]} \quad \Gamma = \sum_{i=k}^{k+H_p} (C_{\text{spec}}(i) + \Delta\beta_i^T R_i \Delta\beta_i)$$

$$\text{subject to} \quad 0 = f(\dot{x}_i, x_i, \beta_i, p_i),$$

$$\frac{1}{H_r} \sum_{j=k}^{k+H_r} \text{Pu}_{\text{ext},j} + \Delta\text{Pu}_{\text{ext},i} \geq \text{Pu}_{\text{ext,min},i},$$

$$\frac{1}{H_p} \sum_{j=k}^{k+H_p} \text{Pu}_{\text{ext},j} + \Delta\text{Pu}_{\text{ext},i} \geq \text{Pu}_{\text{ext,min},i}, \tag{9.1}$$

$$\dot{V}_1^i \leq \dot{V}_{\max},$$

$$g(\beta_i) \geq 0,$$

$$i = k, (k+1), \ldots, (k+H_p)$$

The natural degrees of freedom of the process are the flow rates of desorbent, $\dot{V}_{\text{des}}$, $\dot{V}_{\text{feed}}$, and recycle, $\dot{V}_{\text{IV}}$, and the switching period, t_{shift}. However, this results in an ill-conditioned optimization problem. Numerical tractability is improved by introducing the so-called β-factors via a nonlinear transformation of the natural degrees of freedom:

$$\dot{V}_{\text{solid}} = \frac{1-\varepsilon}{V_s t_{\text{shift}}}$$

$$\beta_1 = \frac{1}{H_A} \left(\frac{\dot{V}_{\text{I}}}{\dot{V}_{\text{solid}}} - \frac{1-\varepsilon}{\varepsilon} \right)$$

$$\beta_2 = \frac{1}{H_B} \left(\frac{\dot{V}_{\text{II}}}{\dot{V}_{\text{solid}}} - \frac{1-\varepsilon}{\varepsilon} \right) \tag{9.2}$$

$$\beta_3 = \frac{1}{H_A} \left(\frac{\dot{V}_{\text{III}}}{\dot{V}_{\text{solid}}} - \frac{1-\varepsilon}{\varepsilon} \right)$$

Here $\dot{V}_{\text{solid}}$ is the apparent solid flow rate, H_A and H_B describe the slopes of the adsorption isotherm, which are calculated in the nonlinear case by linearization of the adsorption isotherm for the feed concentration $c_{\text{feed},i}$. The transformation reflects the fact that, in a counter-current process, it is not the net flow rates that are important but rather their values relative to the apparent solid movement. For this reason, Morbidelli et al. introduced the m factors in their graphical design (known as the triangle theory) (Biressi et al., 2000 and Mazzotti et al., 1997c).

The *prediction horizon* is discretized in cycles, where a cycle is a switching time t_{shift} multiplied by the total number of columns. Equation 9.1 constitutes a dynamic optimization problem with the transient behavior of the process as a constraint; f describes the continuous dynamics of the columns based on the general rate model (GRM) as well as the discrete switching from period to period. To solve the PDE models of columns, a Galerkin method on finite elements is used for the liquid phase and orthogonal collocation for the solid phase. The switching of the node equations is considered explicitly, i.e. a full hybrid plant model is used. The objective function Γ is the sum of the costs incurred for each cycle (e.g. the desorbent consumption) and a regularizing term that is added to smooth the input sequence in order to avoid high fluctuations of the inputs from cycle to cycle. The first equality constraint represents the plant model evaluated over the finite prediction horizon H_p. Since the maximal attainable pressure drop by the pumps must not be exceeded, constraints are imposed on the flow rates in section I. Further inequality constraints $g(\beta)$ are added to avoid negative flow rates during optimization.

The objective of meeting the product specifications is reflected by the purity constraint over the control horizon H_r, which is corrected by a bias term δPu_{ext} resulting from the difference between the last simulated and the last measured process output to compensate un-modeled effects:

$$\Delta Pu_{ext,i} = Pu_{ext,(i-1),exp} - Pu_{ext,(i-1)}. \tag{9.3}$$

The second purity constraint over the whole prediction horizon acts as a terminal (stability) constraint, forcing the process to converge towards the optimal cyclic steady state. The goal of feedback control in a standard control approach (i.e. to fulfill the extract purity) is introduced as a constraint here. A feasible path SQP algorithm is used for the optimization (Zhou et al., 1997), which generates a feasible point before it starts to minimize the objective function.

Concentration profiles in the recycle line are measured and collected during a cycle. Since this measurement point is fixed in the closed-loop arrangement, the sampled signal includes information from all three sections. During the start-up phase, an online estimation of the actual model parameters is started in every cycle. The quadratic cost functional

$$J_{est}(p) = \sum_{i=1}^{n_{sp}} \left\{ \int_0^{N_{col}} \left[c_{i,exp}(t) - c_{i,rec}(t) \right]^2 \, dt \right\} \tag{9.4}$$

is minimized with respect to the parameters p. For this purpose, the least-squares solver E04UNF from the NAG library is used (NAG, 1991). The model parameters are equal for all columns.

To predict the evolution of the process, the actual state (i.e. the concentration profiles over the columns) is needed. It is computed by simulation of the process model using the measurements in the recycle line as input functions. As the model is adapted online, this is sufficiently accurate.

9.4.3
Simulation Study

Figure 9.4 shows a simulation scenario where the desired extract purity was set to 70% at the start of the experiment. The desired extract purity was then changed to 60% at period 60. At period 120, the desired extract purity was increased to 65%.

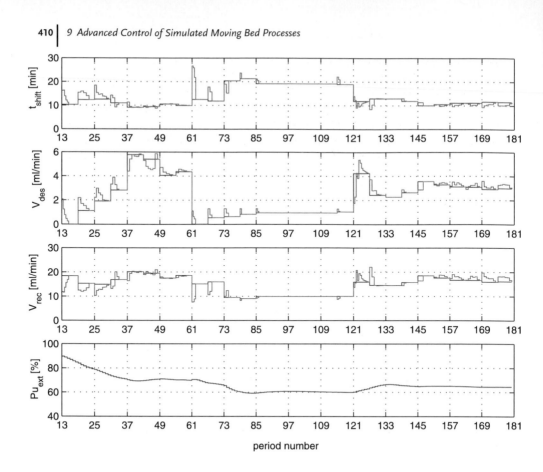

Figure 9.4 Simulation of the optimizing controller. $H_r = 2$, $H_p = 10$.

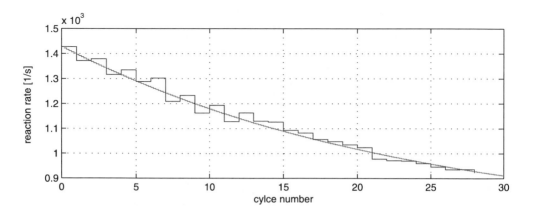

Figure 9.5 Estimation of reaction rate.

Enzyme activity is assumed to decay exponentially over the experiment. Fast controller response in both directions can be observed. Compared with the uncontrolled case, the controller controls the product purity and compensates the drift in the enzyme activity. The evolution of the results of the optimization algorithm during each cycle is plotted as a dashed line, shifted by one cycle to the right in order to vitalize the convergence. This shows that a feasible solution is found rapidly and that the controller can be implemented under real-time conditions. In this example, the control horizon was set to two cycles and the prediction horizon was set to ten cycles. A diagonal matrix $R_j = 0.02\ I\ (3,3)$ was chosen for regularization.

Figure 9.5 shows the results of the parameter estimation. A good fit is achieved and the estimated reaction rate accurately follows the drift of the reaction rate in the model.

9.4.4
Experimental Study

Figure 9.6 shows a photograph of the pilot plant used for experimental investigations. It consists of a LicoSEP 12-50 plant (NovaSEP, France) that can be used with up to 12 preparative chromatographic columns with a maximum diameter of 50 mm. Columns are packed homogeneously with the ion-exchange resin (Amberlite CR-13Na) and the immobilized enzyme, Sweetzyme IT (supplied by Novo Nordisk Bio-industrial). The complete process is controlled by a modern programmable logic controller (PLC) PC-S7 of the Siemens S7-400 series (CPU S7-414-2DP). A human-machine interface was implemented in the Windows Control Center (WinCC) envi-

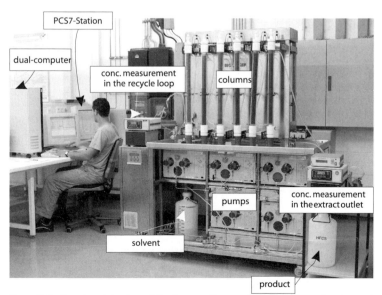

Figure 9.6 Experimental set-up of the three-section reactive SMB process for glucose isomerization.

ronment, permitting the control, visualization and storage of all process variables. Since the algorithms implemented to control the process require considerable computing power, they cannot be implemented on the CPU of the PLC. The control algorithm runs on a separate PC (Dual-Athlon, two 1.5 GHz CPUs), which communicates via an industrial Ethernet with the WinCC tool using the C-script interface Global Script.

The SMBR plant is equipped with several sensors: the temperature of the columns and the incoming streams are kept at around 60 °C by a closed heating water circuit that is controlled by a thermostat. Concentrations in the product line and in the recycling loop are measured online using a combination of a density measurement unit and a polarimeter (Jupke et al., 2002). The pressure in the recycling loop (i.e. at the inlet of the recycling pump) is maintained constant using a P-controller that varies the extract flow rate.

Figure 9.7 shows the effect of a temperature decrease on the conversion of glucose into fructose. The first six cycles show the start-up period of the SMBR until cyclic steady state. At the end of the sixth cycle, the water temperature was reduced by

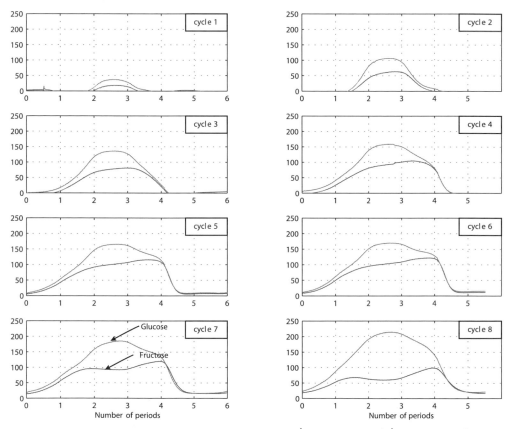

Figure 9.7 Effect of column temperature (t_{shift} = 16.7 min, $\dot{V}_{\text{rec}}$ = 10.47 ml min^{-1}, $\dot{V}_{\text{feed}}$ = 1.3 ml min^{-1}, $\dot{V}_{\text{des}}$ = 1.53 ml min^{-1}).

10 °C. Figure 9.7 clearly shows a considerable decrease in the amount of fructose produced in subsequent cycles.

An analysis based upon the Fisher information matrix showed that the process model is highly sensitive to the Henry coefficients, mass transfer resistances and reaction rate. These parameters are, therefore, re-estimated online at every cycle. Figure 9.8 compares the concentration profiles collected in the recycling line with the simulated ones. The parameters were initialized with the values given in Table 9.1. At the end of the experiment, all system parameters have converged towards stationary values (Fig. 9.9). Thus, the mathematical model describes the behavior of the SMBR process sufficiently well.

The formulation of the optimization problem of the NMPC controller was slightly modified for the experimental investigation. The sampling time of the controller was reduced to one switching period instead of one cycle so that the controller reacts faster. The switching time was still used as a controlled variable, but modified only from cycle to cycle. This is due to the asymmetry of the SMBR process that results from the dead volume of the recycling pump in the closed loop. As this would disturb the overall performance of the process, it is corrected by adding a delay for the switching of the inlet/outlet line passing the recycling pump. Therefore, the shift of the valves is not synchronous to compensate for the technical imperfection of the

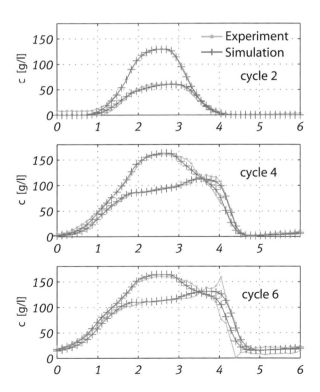

Figure 9.8 Comparison of simulated and measured concentrations in the recycle line.

Table 9.1 Model parameters of the 3-sections reactive SMB process.

Process parameters

Column length	57.5 cm	Column diameter	2.6 cm
Particle porosity	0.01 (–)	Column porosity	0.4 (–)
Particle radius	8.125 om	Diffusion coefficient	$1.0 \times 10^{-3} \, cm^2 \, s^{-1}$
Mass transfer coeff. (A)	$3.8 \times 10^{-5} \, s^{-1}$	Mass transfer coeff. (B)	$2.05 \times 10^{-5} \, s^{-1}$
Parabolic isotherm	$q_i = H_i c_i + k_i c_i^2 + k_{ij} c_i c_j$	Henry coeff. H_A	0.2545 (–)
Henry coeff. H_B	0.1958 (–)	Parabolic term k_A	$1.46 \times 10^{-1} \, cm^3 \, g^{-1}$
Parabolic term k_B	$1.33 \times 10^{-1} \, cm^3 \, g^{-1}$	Parabolic term k_{AA}	$2.90 \times 10^{-1} \, cm^3 \, g^{-1}$
Parabolic term k_{BB}	$9.30 \times 10^{-2} \, cm^3 \, g^{-1}$	Particle diffusion D_p	$1.0 \times 10^{-3} \, cm^2 \, s^{-1}$
Reaction kinetic	$r_i = \nu_i k_m \left(c_{b,i} - \dfrac{c_{b,j}}{k_{eq}} \right)$	Reaction rate k_m	$4.70 \times 10^{-3} \, s^{-1}$
Catalyst amount	10%	Stoichiometric coeff.	[+1, -1]
Equilibrium constant K_{eq}	1.0798 (–)		

Operating parameters (typical values)

Max. flow rate	20.0 ml min^{-1}	Feed flow rate	1.3 ml min^{-1}
Column configuration	[2 2 2]	Feed concentrations	[0, 300] (g l^{-1})
Eluent flow rate	3.77 ml min^{-1}	Recycle flow rate	19.84 ml min^{-1}
Raffinate flow rate	0.0 ml min^{-1}	Extract flow rate	5.07 ml min^{-1}
Switching period	10.41 min	β-factors	[1.1,1.0,1.4,0.81]
Extract purity	51.73%		

Numerical settings (axial discretization)

Number of finite elements (liquid phase)	12	Number of collocation points (solid phase)	1
Integration tolerance	1.0×10^{-7}	Steady state tolerance	3.0×10^{-4}
Max. mass balance error	0.25%		

A: fructose, B: glucose.

real system and to get closer to the ideal symmetrical SMB system. A patent provides a detailed description of this method (Hotier, 1998).

To avoid port overlapping, the switching time must be held constant during a cycle. In the real process, the enzyme concentration changes from column to column. In addition, the geometrical lengths of the columns differ slightly. Moreover, the temperature is not constant over the columns due to the inevitable gradient of the closed heating circuit. These problems cause a fluctuation of the concentration profiles at the product outlet. Even at the cyclic steady state, the product purities change form period to period. Using the bias term as given by Eq. 9.3 would cause large variations of the controlled inputs from period to period. In the experimental study, this was dampened by using the minimal value of the extract purity over the last cycle as the measured purity.

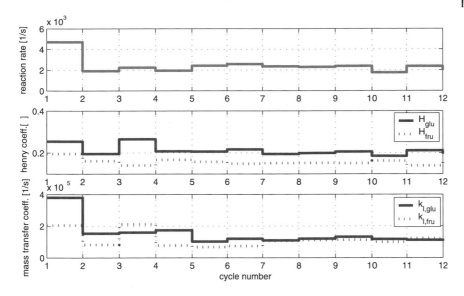

Figure 9.9   Evolution of estimated parameters.

The desired purity for the experiment reported below was set to 55.0% and the controller was started at the 60th period. As in the simulation study, a diagonal matrix $R_j = 0.02\ I\ (3,3)$ was chosen for regularization. The control horizon was set to $H_r = 1$ and the prediction horizon was $H_p = 60$ periods. Figure 9.10 shows the evolution of both the product purity and the controlled variables. In the open-loop mode, where the operating point was calculated based on the initial model, the product purity constraint was violated at periods 48 and 54. After one cycle, the controller drove the purity above 55.0% and kept it there. The controller first reduces the desorbent consumption. This action seems to contradict the intuitive idea that more desorbent injection should enhance separation. However, in the presence of a reaction, this is not true, as shown by this experiment. The controlled variables converge towards a steady state, but they still change from period to period, due to the non-ideality of the plant.

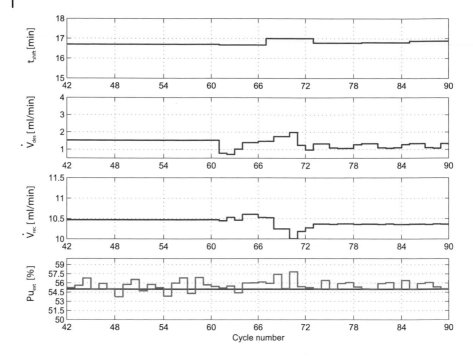

Figure 9.10 Experimental control result.

9.5
Conclusions

In this chapter we have advocated the use of online model-based optimization for the automatic control of SMB plants. This approach has the advantage that the process is automatically operated at its economic optimum while meeting all relevant constraints on purities and flow rates. Application to a pilot-plant-scale reactive SMB process for glucose isomerization showed that implementation at a real plant is feasible – the requirements for additional hardware are moderate (a high-level PC and online concentration measurements in the recycle line). The experiments confirmed the excellent properties of the proposed control scheme. The scheme is extremely versatile, and the cost function and constraints can easily be adapted to any specific separation task.

Outlook

After several hundred pages of text, pictures, graphs and equations on the 'how to's in separation technologies we want to end our review of past and present of preparative chromatography by trying to glimpse future needs and the resulting tasks for research and development. Where do we go from here and what will be the drivers for the future in preparative chromatography? Comments on these questions are not predictions but descriptions of issues that should be addressed. Keys to a future expanded use of chromatography will continue to be better understanding, process integration and interdisciplinary cooperation.

From the process point of view there are three main areas for future research and development in preparative chromatography: separation of small molecules corresponding to the focus of this book; separation of macromolecules, and integration of chromatography in the overall design and optimization of down stream processes. A different view identifies the following topics: development of tailored stationary phases, thermodynamics of phase equilibria and adsorption kinetics, process design, simulation and optimization of chromatography as well as downstream processes and last but not least process control.

Chemistry and material sciences are key disciplines for the development of advanced and more specific adsorbents. The stability and reproducibility of chromatographic columns was significantly increased by the introduction of spherical instead of irregular stationary phases. Recently, another step forward was made by the development of high efficiency monolithic columns with a rather low pressure drop. Future, further improvements, which include surface activation and internal pore structures of stationary phases, should help to tailor stationary phases for certain applications. But, besides the need for more specific and efficient solid phases, their cost is often a major problem for the widespread application of preparative chromatography.

As pointed out in Chapter 4, thermodynamics are of major concern in selecting a chromatographic system as well as in process optimization, where academic research is still at the initial stages. Today, the phase equilibrium has to be measured for each individual application. It is desirable, and a big challenge, to reach the same

Preparative Chromatography. Edited by H. Schmidt-Traub
Copyright © 2005 WILEY-VCH Verlag GmbH & Co. KGaA, Weinheim
ISBN: 3-527-30643-9

theoretical standard for phase equilibria as has been established for gas-liquid systems. However, adsorption is far more complex so that these methods can hardly be copied or transformed. Therefore, indirect methods, which combine modeling and advanced mathematical evaluation procedures, need to be used to derive thermodynamic properties and model parameters from a few chromatograms. Additionally, a better understanding and quantification of local transport mechanisms will help to describe adsorption processes better and to develop more goal-oriented stationary phases. The necessity for such thermodynamic research increases still further when macromolecules are involved. Here complex and variable 3D configurations of, e.g. proteins, play an important part. There is also strong evidence that both adsorption kinetics and a correlation between pore size and molecule structure have to be taken into account. With this in mind molecular modeling might also help, for instance, to simulate the movement of a macromolecule within pores.

Proven models and simulation tools for isocratic chromatography are available. They describe with very good accuracy the separation of small molecules. Extension of these methods to macromolecules is another challenge for future research. Here new models, e.g. for ion-exchange, have to be developed and evaluated. However these efforts will be very closely linked to progress in thermodynamics and molecular modeling. Better understanding of transport phenomena on a microscale (mentioned above) will also be a key factor for sustainable improvements in process design.

Another most important issue is solubility. To date, no reliable methods are available to calculate and forecast the solubility of solids in liquids.

Not all problems and open questions can be solved by basic research. As pointed out, time pressure is an important and increasing constraint for laboratory work that very often leads to „quick and dirty" solutions. Another severe constraint is the availability of only very small amounts of samples for the determination of model parameters. Therefore, more efficient experimental methods and faster evaluation tools would help to improve small-scale separations.

Process integration offers significant potential for future improvement. Process optimization has to focus on the entire process and look for a global optimum instead of the optimum of each process step. Flowsheeting tools that are common in chemical industries do not yet exist for downstream processing. Such systems should also include all steps of chemical synthesis because the consequences of a certain type of reaction and the respective intermediate products on mass separation has to be taken into account for the overall process design. Heuristic rules based on a better understanding of process integration may help to place chromatography in a process sequence where it can made the largest contribution to creating value.

Expansion into new applications in bioseparations will mean, in particular, the use of modern process design and development concepts, i.e. in downstream processing of biological products. While typically, in bioseparations a whole sequence of chromatographic steps is involved in the recovery of low concentration of target molecules from large volumes of fermentation broth at significant costs, there has not been much improvement in process design or reengineering in this field for the last 30 years. The resistance, or rather the inability, to change is mainly because, unlike with small molecules, the target molecules in bioprocesses cannot be comprehensively characterized by analytical methods. Therefore, the process, by which the product is made, has become a part of the product specification, which makes a process change of an approved product very difficult.

Significantly enhanced analytics, on one hand, and increasing cost pressures on biopharmaceutical products, on the other hand, may help to change this situation in the coming years, allowing a transfer of modern separation concepts to bioprocessing. This will include various aspects of SMB technology, and also new material design for enhanced productivity.

Practical application of advanced process control for chromatographic processes is still at the beginning. Optimal operation of large-scale production should also include the adaptation and optimization of thermo- and fluid-dynamic properties, which might vary during extended operation. Progress in this field is very closely linked to improved detectors and other measurement and control devices.

A boost for new developments may stem from the „Process Analytical Technology" (PAT) program of the US-FDA (2003), which is a drastic paradigm change for quality control and certification. PAT has been developed based on the realization that it is more and more difficult to cope with escalating documentations and the present solidified regulations cause a loss of innovation and increase costs. The basic idea of PAT is that quality cannot be tested into products but should be designed and built in the process. The change from product- to process-oriented work flow is characterized by the following elements. PAT Tools: new tools are applied, e.g. for pharmaceutical development, analytical chemistry, process analysis, simulation as well as control and knowledge management. Process Understanding: instead of ensuring product quality by organizational measures, particularly prescribed quality tests, the backbone of quality assurance will be detailed process understanding in order to install quality into the process and to allow continuous improvements. Risk-based Approach: establishment of quality systems based on process understanding. Integrated Systems Approach: system thinking brings closer together development, manufacturing, quality assurance and management. Real Time Release: finally, PAT leads to a quicker product release for distribution and should improve confidence in product quality. PAT will initiate intensive research and development over the next decade and offers, besides assurance of product quality, great potential to improve process economics.

Interdisciplinary cooperation and integration needs to be further enhanced between the small and large molecule world, the extremely small and the very large process scale, as well as the process engineer, development chemist and biochemist, and even the molecular biologist.

It does not take much foresight to predict that chromatographic separation technologies will remain a key to modern technologies and be indispensable in life science applications.

References

Abel, S., Mazzotti, M., Morbidelli, M. Solvent gradient operation of simulated moving beds I: Linear isotherms, *J. Chromatogr. A*, **2002**, 944, 23–39.

Abel, S., Mazzotti, M., Morbidelli, M. Solvent gradient operation of simulated moving beds II: Langmuir isotherms, *J. Chromatogr. A*, **2004a**, 1026, 47–55.

Abel, S., Erdem, G., Mazzotti, M., Morari, M., Morbidelli, M. Optimizing control of simulated moving beds – linear isotherm, *J. Chromatogr. A*, **2004b**, 1033, 229–239.

Aced, G., Möckel, H. J. *Liquidchromatogaphie: Apparative, theoretische und Methodische Grundlagen der HPLC*, VCH, Weinheim, **1991**.

Adam P., Nicoud, R.-M., Bailly, M., Ludemann-Hombourger, O. Process and device for separation with variable length, US Patent 6.136.198, **2000**.

Adams, B. A., Holmers, E. L. Adsorbtive properties of synthetic resins, *J. Soc. Chem. Ind. (London)*, **1935**, 54, 1T.

Allgöwer, F., Badgwell, T., Qin, J., Rawlings, J., Wright, S. Nonlinear predictive control and moving horizon estimation – an introductory overview, in *Advances in Control*, Frank, P. (Ed.), **1999**, 391–449.

Altenhöner, U., Meurer, M., Strube, J., Schmidt-Traub, H. Parameter estimation for the simulation of liquid chromatography, *J. Chromatogr. A*, **1997**, 769, 59–69.

Altshuller, D. Design equations and transient behaviour of the counter current moving bed chromatographic reactor, *Chem. Eng. Commun.*, **1983**, 19, 363–375.

Antia, F. A simple approach to design and control of simulated moving bed chromatographs, *Chromatogr. Sci. Series*, **2003**, 173–202.

Antos, D., Kaczmarski, K., Wojciech, P., Seidel-Morgenstern, A. Concentration dependence of lumped mass transfer coefficients – Linear versus non-linear chromatography and isocratic versus gradient operation, *J. Chromatogr. A*, **2003**, 1006, 61–76.

Arangio, M. Chromatographische charakterisierung von umkehrphasen, Dissertation, Mathematisch-Naturwissenschaftliche Fakultät, Saarbrücken, **1998**.

Arnott, S. A., Fulmer, A., Scott, W. E., Dea, I. C. M., Moorehouse, R., Rees, D. A. Agarose double helix and its function in agarose gel structure, *J. Mol. Biol.*, **1974**, 90, 269–284.

Ashley, J.W., Reilley, C. N. De-Tailing and sharpening of response peaks in gas chromatography, *Anal. Chem.*, **1965**, 37(6), 626–630.

Atkins, P.W. *Physikalische Chemie*, VCH, Weinheim, **1990**.

Baerns, M., Hofmann, H., Renken, A. Chemische Reaktionstechnik in *Lehrbuch der Technischen Chemie Band 1*, Georg Thieme Verlag, Stuttgart, **1999**.

Bangs, L. B. Uniform latex particles, *Am. Biotechnol. Lab.*, **1987**, 5(3), 10, 12–16.

Barker, P.E., Ganetsos, G., Ajongwen, J., Akintoye, A. Bioreaction-separation on continuous chromatographic systems, *Chem. Eng. J.*, **1992**, 50, B23–B28.

Barker, P. E., Ganetsos, G. *Preparative and Production Scale Chromatography*, Chromatogr. Sci. Series 61, Marcel Dekker Inc., New York, **1993**.

Barns, J. W. *Statistical Analysis for Engineers and Scientists*, McGraw-Hill, **1994**.

Bassett, D. W., Habgood, H. W. A gas chromatographic study of the catalytic isomerisation of cyclopropane, *J. Phys. Chem.*, **1960**, 64, 769–773.

Preparative Chromatography. Edited by H. Schmidt-Traub
Copyright © 2005 WILEY-VCH Verlag GmbH & Co. KGaA, Weinheim
ISBN: 3-527-30643-9

Bayer, E., Müller, W., Illg, M., Albert, K. Abbildung der chromatographischen trennung mit NMR-bildgebung (NMR-imaging), *Angew. Chem.* **1989**, 101, 1033.

Bellot, J. C., Condoret J. S. Liquid chromatography modelling: a review, *Process Biochem.*, **1991**, 26, 363–376.

Bellot, J. C., Condoret, J. S. Modelling of liquid chromatography equilibria, *Process Biochem.*, **1993**, 28, 365–376.

Berninger, J. A., Whitley, R. D., Zhang, X., Wang, N.-H. Versatile model for simulation of reaction and nonequilibrium dynamics in multicomponent fixed-bed adsorption processes, *Comput. Chem. Eng.*, **1991**, 15(11), 749–768.

Beste, Y. A., Lisso, M., Wozny, G., Arlt, W. Optimization of simulated moving bed plants with low efficient stationary phases: separation of fructose and glucose, *J. Chromatogr. A*, **2000**, 868, 169–188.

Bidlingmeyer, B. A., Waren, F. V. Column efficient measurement, *Anal. Chem.*, **1984**, 56, 1588–1595.

Biressi, G., Ludemann-Homburger, O., Mazzotti, M., Nicoud, R.-M., Morbidelli, M. Design and optimisation of a simulated moving bed unit: role of deviation from equilibrium theory *J. Chromatogr. A*, **2000**, 976, 3–15.

Bluhm, L. H., Wang, Y., Li, T. An alternative procedure to screen mixture combinatorial libraries for selectors for chiral chromatography, *Anal. Chem.*, **2000**, 72, 5201–5205.

Borren, T., Schmidt-Traub, H. Vergleich chromatographischer reaktorkonzepte, *Chem. Ing. Tech.*, **2004**, 76(6), 805–814.

Boysen, H., Wozny, G., Laiblin, T., Arlt, W. CDF simulation of preparative chromatography columns considering adsorption isotherms, *Chem. Ing. Tech.*, **2002**, 74, 294–298.

Brandt, A., Kueppers, S. Practical aspects of preparative HPLC in pharmaceutical and development production, *LC GC Eur.*, **2002**, 15(3), 147–151.

Brauer, H. *Grundlagen der Einphasen- und Mehrphasenströmungen*, Verlag Sauerländer, Aarau, Frankfurt am Main, **1971**.

Broughton, D. B. Adsorption isotherms for binary gas mixtures, *Ind. Eng. Chem.*, **1948**, 40(8), 1506–1508.

Broughton, D. B., Gerhold, C. G. Continuous Sorption Process Employing Fixed Bed of Sorbent and Moving Inlets and Outlets, US Patent No. 2.985.589, **1961**.

Carr, R. W. Continuous reaction chromatography, in *Preparative and Production Scale Chromatography*, Ganetsos, G., Barker, P. E. (Eds.), Marcel Dekker Inc., New York, **1993**.

Carr, R. W., Dandekar, H. W. Adsorption with reaction, in *Reactive Separation Processes*, Kulprathipanja, S. (Ed.), Taylor & Francis, New York, **2002**.

Chang, J.-P., el Rassi, Z., Horvath, C. Polyethyle glycol-bounded phases for protein separation by high-performance hydrophobic interaction chromatography, *J. Chromatogr.*, **1985**, 319, 396.

Chang, J.-P., An, J.G. Silica-bound polyethylenglycol as stationary phase for separation of proteins by high-performance liguid chromatography, *Chromatographia*, **1988**, 25, 350.

Charton, F., Nicoud, R.-M. Complete design of a simulated moving bed, *J. Chromatogr. A*, **1995**, 702, 97–112.

Cherrak, D. E., Al-Bokari, M., Drumm, E. C., Guiochon, G. Behavior of packing materials in axially compressed chromatographic column, *J. Chromatogr. A*, **2002**, 943, 15–31.

Ching, C. B., Ruthven, D. M. Experimental study of a simulated counter-current adsorption system, *Ind. Eng. Sci.*, **1986**, 41, 3063–3071.

Ching, C. B., Ho, C., Ruthven, D. M. An improved adsorption process for the production of high-fructose syrup, *AIChE J.*, **1986**, 32, 11, 1876–1880.

Ching, C. B., Chu, K. H., Hidajat, K., Ruthven, D. M. Experimental study of a simulated counter-current adsorption system – vii effects of non-linear and interacting isotherms, *Chem. Eng. Sci.*, **1993**, 48(7), 1343–1351.

Cho, B. K., Aris, R., Carr, R. W. The mathematical theory of a countercurrent catalytic reactor, *Proc. R. Soc. London*, **1982**, A 383, 147–189.

Chu, K., Hashim, M. Simulated contercurrent absorption processes: A comparison of modelling strategies, *Chem. Eng. J.*, **1995**, 56, 59–65.

Chung, S. F., Wen, C. Y. Longitudinal dispersion of liquid flowing through fixed and fluidised beds, *AIChE J.*, **1968**, 14, 6, 857–866.

Clavier, J.-Y., Nicoud, R. M., Perrut, M. A new efficient fractionation process: The simulated moving bed with supercritical eluent in *High-Pressure Chemical Engineering*, von Rohr, P. R., Trepp, C. (Eds), Elsevier Science, London, **1996**.

Coca, J., Bravo, M., Abascal, E., Adrio, G. Dicyclopentadiene dissociation in a chromatographic reactor – effect of the liquid phase polarity on the reaction rate, *Chromatographia*, **1989**, 28, 300–302.

Coca, J., Adrio, G., Jeng, C. Y., Langer, S. H. Gas and liquid chromatographic reactors in *Preparative and Production Scale Chromatography*, Ganetsos, G., Barker, P.E. (Eds.), Marcel Dekker Inc., New York, **1993**.

Colin, H., Hilaireau, P., Martin, M. Flip-Flop elution concept in preparative liquid chromatography, *J. Chromatogr. A*, **1991**, 557, 137–153.

Cox, G. B., Khattabi, S., Dapremont, O. Real-time monitoring and control of a small-scale SMB unit from a polarimeter-derived internal profile, *16th International Symposium on Preparative Chromatography*, San Francisco, USA, **2003**, 41–42.

Cox, G. B., Amoss, C. W. Extending the range of solvents for chiral analysis using a new immobilized polysaccharide chiral stationary phase CHIRALPAK (R) IA, *LC GC North America*, **2004**.

Craig, L. C. Identification of small amounts of organic compounds by distribution studies II: Separation by counter-current distribution, *J. Biol. Chem.*, **1944**, 155, 519–534.

Danckwerts, P. V. Continuous flow systems – Distribution of residence times, *Chem. Eng. Sci.*, **1953**, 2(1), 1–13.

Davis, M. E. *Numerical Methods and Modeling for Chemical Engineers*, John Wiley & Sons, **1984**.

Denet, F., Hauck, W., Nicoud, R. M., Giovanni, O. D., Mazzotti, M., Jaubert, J. N., Morbidelli, M. Enantioseparation through supercritical fluid simulated moving bed (SF-SMB) chromatography, *Ind. Eng. Chem. Res.*, **2001**, 40, 4603–4609.

Deckert, P., Arlt, W. Pilotanlage zur simulierten Gegenstromchromatographie- Ergebnisse für die Trennung von Fructose und Glucose, *Chem. Ing. Tech.*, **1997**, 69, 115–119.

Depta, A., Giese, T., Johannsen, M., Brunner, G. Separation of stereoisomers in a simulated moving bed-supercritical fluid chromatography plant, *J. Chromatogr. A*, **1999**, 865, 175–186.

De Vault, D., *J. Am. Chem. Soc.*, **1943**, 65, 532.

Dingenen, J. Polysaccharide phases in enantioseparations, in *A Practical Approach to Chiral Separations by Liquid Chromatography*, Subramanian, G. (Ed.), Wiley-VCH, Weinheim, **1994**.

Dingenen, J., Kinkel, J. Preparative chromatographic resolution of racemates on chiral stationary phases on laboratory and production scales by closed-loop recycling chromatography, *J. Chromatogr. A*, **1994**, 666, 627–650.

Du Chateau, P., Zachmann, D. *Applied Partial Differential Equations*, Harper & Row, **1989**.

Dünnebier, G., Klatt, K.-U. Optimal operation of simulated moving bed chromatographic separation processes, *Comp. Chem. Eng.*, **1999**, 23, 189–192.

Dünnebier, G. Effektive Simulation und mathematische Optimierung chromatographischer Trennprozesse, Dissertation, Universität Dortmund, Shaker Verlag, **2000**.

Dünnebier, G., Fricke, J., Klatt, K.-U. Optimal design and operation of simulated moving bed chromatographic reactors, *Ind. Eng. Chem. Res.*, **2000a**, 39, 2292–2304.

Dünnebier, G., Jupke, A., Klatt, K.-U. Optimaler betrieb von smb-Chromatographieprozessen, *Chem. Ing. Tech.*, **2000b**, (6), 589–593.

Engelhardt, H., Grüner, R., Scherer, M. The polar selectivities of non-polar reversed phases, *Chromatographia*, **2001**, Suppl. 53, 154–161.

Epping, A., Modellierung, Auslegung und Optimierung chromatographischer Batch-Trennung, Shaker-Verlag, Aachen, **2005**.

Epton, R., *Hydrophobic, Ion Exchange and Affinity Methods*, in *Chromatography of Synthetic and Biological Polymers*, Vol. 2, E. Horwood, Chichester, **1978**, 1–9.

Erdem, G., Abel, S., Morari, M., Mazzotti, M., Morbidelli, M., Lee, J.H. Automatic control of simulated moving beds, *Ind. Eng. Chem. Res.*, **2004**, 43, 405–421.

Ettre, L. M.S. Tswett and the 1918 Nobel Prize in Chemistry, *Chromatographia*, **1996**, 42, 343–351.

Everett, D. H. Thermodynamics of adsorption from solutions, in *Fundamentals of Adsorption Processes*, Meyers, A. L., Belfort G. (Eds.), Eng. Found., New York, **1984**.

Falk, T., Seidel-Morgenstern, A. Analysis of a discontinuously operated chromatographic reactor, *Chem. Eng. Sci.*, **2002**, 57, 1599–1606.

Felinger, A., Guiochon, G., Optimization of the experimental conditions and the column design parameters in overloaded elution chromatography, *J. Chromatogr. A*, **1992a**, 591, 31–45.

Felinger, A., Guiochon, G., Optimization of the experimental conditions and the column design parameters in displacement chromatography, *J. Chromatogr. A*, **1992b**, 609, 35.

Felinger, A., Guiochon, G., Comparison of maximum production rates and optimum operating/design parameters in overloaded elution and displacement chromatography, *Biotechnol. Bioeng. A*, **1993**, 41, 134–147.

Felinger, A., Guiochon, G., Optimizing experimental conditions for minimum production cost in preparative chromatography, *AIChE J.*, **1994**, 40(4), 594–605.

Felinger, A., Guiochon, G., Optimizing preparative separations at high recovery yield, *J. Chromatogr. A*, **1996**, 752, 31–40.

Felinger, A., Guiochon, G. Comparing the optimum performance of the different modes of preparative liquid chromatography, *J. Chromatogr. A*, **1998**, 796, 59–74.

Ferziger, J. H. *Numerical Methods for Engineering Applications*, 2nd Edition, Wiley-Interscience, **1998**.

Figge, H., Deege, A., Köhler, J., Schomburg, G. Stationary phases for reversed-phase liquid chromatography – coating of silica by polymers of various polarities, *J. Chromatogr.*, **1986**, 351, 393.

Finlayson, B. *Numerical Analysis in Chemical Engineering*, McGraw Hill, New York, **1980**.

Fish, B. B., Carr, R. W. An experimental study of the countercurrent moving-bed chromatographic reactor, *Chem. Eng. Sci.*, **1989**, 44, 1773–1783.

Foley, J. P., Dorsey, J. G. Equations for calculation of chromatographic figures of merit for ideal and skewed peaks, *Anal. Chem.*, **1983**, 55, 730–737.

Foley, J. P., Dorsey, J. G. A review of the exponentially modified Gaussian (EMG) function: Evaluation and subsequent calculation of universal data, *J. Chromatogr. Sci.*, **1984**, 22, 40–46.

Francotte, E. Achiral Derivatization as a means of improving the chromatographic resolution of racemic alcohols on benzoylcellulose CSPs, *Chirality*, **1998**, 10, 492–498.

Francotte, E., Enantioselective chromatography as a powerful alternitive for the preparation of drug enantiomers, *J. Chromatogr. A*, **2001**, 906, 379–397.

Francotte, E., Huynh, D. Immobilized halogenophenylcarbamate derivatives of cellulose as novel stationary phases for enantioselective drug, *J.Pharm Biomed Anal.*, **2002**, 27, 421–424.

du Fresne von Hohenesche, C., Ehwald, V., Unger, K. K. Development of standard operation procedures for the manufacture of n-octadecyl bonded silicas as packing material in certified reference columns for reversed-phase liquid chromatography, *J. Chromatogr. A*, **2004**, 1025, 2, 177–187.

Fricke, J., Schmidt-Traub, H. Design of chromatographic SMB-reactors, Oral presentation at the International Symposium on Preparative and Industrial Chromatography and Allied Techniques, Heidelberg, **2002**.

Fricke, J., Schmidt-Traub, H. A new method supporting the design of simulated moving bed chromatographic reactors, *Chem. Eng. Proc.*, **2003**, 42, 237–248.

Fricke, J. *Entwicklung einer Auslegungsmethode für chromatographische SMB-Reaktoren*, VDI-Verlag, Düsseldorf, **2005**.

Furuya, F., Takeuchi, Y, Noll, K. E., Intraparticle diffusion of phenols within bidispersed macrorecticular resin particles, *J. Chem. Eng. Japan*, **1989**, 22(6), 670.

Galushko, S. V. Calculation of retention and selectivity in reversed-phase liquid chromatography, *J. Chromatogr.*, **1991**, 552, 91–102.

Gant, J.R., Dolan J.W., Snyder, L.R. Systematic approach to optimizing resolution in reversed phase liquid chromatography, with emphasis on the role of temperature, *J. Chromatogr.*, **1979**, 185, 153–177.

Geiss, F., Schlitt, H., Klose, A. Reproducibility in thin-layer chromatography: Influence of humidity, chamber form, and chamber atmosphere, *Zeitschrift für Anal. Chem.*, **1965**, 213, 5.

Gentilini, A., Migliorini, C., Mazzotti, M., Morbidelli, M. Optimal operation of simulated moving-bed units for non-linear chromatographic separations, II. Bi-langmuir isotherm, *J. Chromatogr. A*, **1998**, 805, 37–44.

Ghethie, V., Schell, H. D. Electrophoresis and immunoelectrophoresis of proteins on DEAE-agarose gels, *Rev. Roum. Biochim.*, **1967**, 4, 179–184.

Giovanni, O. D., Mazzotti, M., Morbidelli, M., Denet, F., Hauck, W., Nicoud, R. M. Supercritical fluid simulated moving bed chromatography. II Langmuir isotherm, *J. Chromatogr. A*, **2001**, 919.

Glueckauf, E., *J. Chem. Soc.*, **1947**, 1302 ff.

Glueckauf, E., Coates, J. I. Theory of chromatography, Part 4: The influence of incomplete equilibrium on the front boundary of chromatogram and on the effectiveness of separation, *J. Chem. Soc.*, **1947**, 1315–1321.

Glueckauf, E., *Disc. Faraday Soc.*, **1949**, 7, 12.

Golshan-Shirazi, S., Guiochon, G. Experimental characterization of the elution profiles of high concentration chromatographicbands using the analytical solution of the ideal model, *Anal. Chem.*, **1989**, 61, 462–467.

Golshan-Shirazi, S., Guiochon, G. Comparison of the various kinetic models of non-linear chromatography, *J. Chromatogr. A*, **1992**, 603, 1–11.

Graham, D. The characterization of physical adsorption systems. I. the equilibrium function and standard free energy of adsorption, *J. Phys. Chem.*, **1953**, 57, 665–669.

Grill, C. M., Miller, L. Separation of a racemic pharmaceutical intermediate using closed-loop steady state recycling, *J. Chromatogr. A*, **1998**, 827, 359–371.

Grill, C. M., Miller, L.; Yan, T. Q. Resolution of a racemic pharmaceutical intermediate – A comparison of preparative HPLC, steady state recycling, and simulated moving bed, *J. Chromatogr. A*, **2004**, 1026, 101–108.

Grushka, E. S., Snyder, L. R.; Knox, J. H. Advances in band spreading theories, *J. Chromatogr. Sci.*, **1975**, 13, 25–37.

Gu, T., Tsai, G.-J., Tsao, G. T. New approach to a general nonlinear multicomponent chromatography model, *AIChE J.*, **1990a**, 36(5), 784–788.

Gu, T., Tsai, G.-J., Tsao, G. T., Ladisch, M. R. Diplacement effect in multicomponent chromatography, *AIChE J.*, **1990b**, 36(8), 1156–1162.

Gu, T., Tsai, G.-J., Tsao, G. T. Multicomponent adsorption and chromatography with uneven saturation capacities, *AIChE J.*, **1991**, 37(9), 1333–1340.

Gu, T., Tsai, G.-J., Tsao, G. T. Modeling of nonlinear multicomponent chromatograpy *Advances in Biochemical Engineering/Biotechnology*, Fiechter, A. (Ed.), Vol. 49, pp. 45–71, Springer Verlag, New York, **1993**.

Gu, T. *Mathematical Modeling and Scale-up of Liquid Chromatography*, Springer Verlag, New York, **1995**.

Guiochon, G., Golshan-Shirazi, S. Theory of optimizing of the experimental conditions of preparative elution using the ideal model of liquid chromatography, *Anal. Chem.*, **1989a**, 61, 1276–1287.

Guiochon, G., Golshan-Shirazi, S. Theory of optimizing of the experimental conditions of preparative elution chromatography: optimizing the column efficiency, *Anal. Chem.*, **1989b**, 61, 1368–1382.

Guiochon, G., Czok, M. The physical sense of simulation models of liquid chromatography: propagation through a grid or solution of the mass balance equation, *Anal. Chem.*, **1990**, 62, 189–200.

Guiochon, G., Golshan-Shirazi, S. Solutions of the equilibrium and semi-equilibrium models of chromatography, *J. Chromatogr.*, **1990a**, 506, 495.

Guiochon, G., Golshan-Shirazi, S. Optimizing of the experimental conditions in preparative liquid chromatography with touching bands, *J. Chromatogr.*, **1990b**, 517, 229–256.

Guiochon, G., Golshan-Shirazi, S. Optimizing of the experimental conditions in preparative liquid chromatography – Trade-offs between recovery yield and production rate, *J. Chromatogr.*, **1991**, 536, 57–73.

Guiochon, G. Basic Principles of Chromatography in *Ullmann's Encyclopedia of Industrial Chemistry*, Vol. B5, VCH Verlagsgesellschaft, **1994**.

Guiochon, G., Charton, F., Bailly, M. Recycling in preparative liquid chromatography, *J. Chromatogr. A*, **1994a**, 687, 13–31.

Guiochon, G., Golshan-Shirazi, S., Katti, A. *Fundamentals of Preparative and Nonlinear Chromatography*, Academic Press, Boston, **1994b**.

Guiochon, G. Preparative liquid chromatography, *J. Chromatogr. A*, **2002**, 65, 129–161.

Guiochon, G., Lin, B. *Modeling For Preparative Chromatography*, Academic Press, London, **2003**.

Hallmann, M. Optimierung der packmethode zum füllen von präparativen kieselgelsäulen mit 50 mm innendurchmesser, Diploma thesis, Universität, Mainz, **1992**.

Hallmann, M. Untersuchungen an schüttungen, suspensionen und gepackten säulen von kieselgelen für die hochleistungsflüssig-chromatographie, Ph.D. thesis, Universität Mainz, **1995**.

Hanisch, F. Prozessführung präparativer Chromatographieverfahren, Dissertation, Universität Dortmund, **2003**.

Hansen, S. H., Helböe, P., Thomasen, M. Agar derivates for chromatography, electrophoresis and gel-bound enzymes – VII. Influence of apparent surface pH of silica compared with the effects in straight-phase chromatography, *J. Chromatogr.*, **1986**, 368, 39–47.

Hanson, M., Unger, K.K., Schomburg, G. Nonporos polybutadiene-coated silicas as stationary phases in reversed-phase chromatography, *J. Chromatogr.*, **1990**, 517, 269.

Hashimoto, K., Adachi, S., Noujima, H., Maruyama, H. Models for separation of glucose-fructose mixtures using a simulated moving bed adsorber, *J. Chem. Eng. Japan.*, **1983a**, 16(4), 400–406.

Hashimoto, K., Adachi, S., Noujima, H., Ueda, Y. A new process combining adsorption and enzyme reaction for producing higher-fructose syrup, *Biotechnol. & Bioeng.*, **1983b**, 25, 2371–2393.

Hashimoto, K., Adachi, S., Shirai, Y. Operation and design of simulated moving-bed adsorbers in *Preparative and Production Scale Chromatography*, Ganetsos, G., Barker, P. E. (Eds.), Marcel Dekker Inc., New York, **1993a**.

Hashimoto, K., Adachi, S., Shirai, Y. Development of a simulated moving-bed adorber in *Preparative and Production Scale Chromatography*, 395–419, Ganetsos, G., Barker, P. E. (Eds.) Marcel Dekker Inc., New York, **1993b**.

Heitz, W. Gel chromatography, *Angew. Chem. Int. Ed.*, **1970**, 9, 689–702.

Helfferich, F. G., Klein, G. *Multicomponent Chromatography – Theory of Interference*, Marcel Dekker Inc., New York, **1970**.

Helfferich, F. G., Whitley, R. D. Non-linear waves in chromatography II. Wave interference and coherence in multicomponent systems, *J. Chromatogr. A*, **1996**, 734, 7–47.

Henke, H., Ruelke, K., Präparative Trennbeispiele mit Lobar®-LiChroprep-Fertigsäulen, *Swiss Chem.*, **1987**, 9, 23–30.

Herbsthofer, R., Bart, H. J., Brozio, J. Einfluss der reaktionskinetik auf die funktion eines chromatographischen reaktors, *Chem. Ing. Tech.*, **2002**, 74, 1006–1011.

Heuer, C., Küsters, E., Plattner, T., Seidel-Morgenstern, A. Design of the simulated moving bed process based on adsorption isotherm measurements using a perturbation method, *J. Chromatogr. A*, **1998**, 827, 175–191.

Hill, T. L. *An Introduction to Statistical Thermodynamics*, Addison-Wesley, Reading (MA), **1960**.

Hjerten, S. Preparation of agarose spheres for chromatography of molecules and particles, *Biochim. Biophys. Acta*, **1964**, 79, 393–398.

Hjerten, S. *Protides of Biological Fluids*, Peeters, H. (Ed.), Pergamon Press, Oxford, **1983**, Vol. 30, 9–17.

Holt, G. Control process for simulated moving bed separations, US Patent 5.457.260, **1995**.

Hotier, G., Nicoud, R.M. Separation by simulated moving bed chromatography with dead volume correction by desynchronization of periods, Europäisches Patent, EP 688589 A1, **1995**.

Hotier, G. Process for simulated moving bed separation with a constant recycle rate, US Patent 5.762.806, **1998**.

Houwing, J., Billiet, H. A. H., van der Wielen, L. A. M. Optimization of azeotropic protein separation in gradient and isocratic ion-exchange simulated moving bed chromatography, *J. Chromatogr. A*, **2002a**, 944, 189–201.

Houwing, J., van Hateren, S. H., Billiet, H. A. H., van der Wielen, L. A. M. Effect of salt gradients on the separation of dilute mixtures of proteins by ion-exchange in simulated moving beds, *J. Chromatogr. A*, **2002b**, 952, 85–98.

Ishikawa, A., Shibata, T. Cellulosic chiral stationary phase under reversed-phase condition, *J. Liq. Chromatogr.*, **1993**, 16, 4, 61–68.

IUPAC Manual of Symbols and Terminology, Appendix 2, Pt. 1, Colloid and Surface Chemistry, *Pure Appl. Chem.*, **1972**, 31, 578.

Jandera, P., Guiochon, G. Effect of the sample solvent on band profiles in preparative liquid chromatography using non-aqueous reversed-phase high-performance liquid chromatography, *J. Chromatogr.*, **1991**, 588, 1–14.

Janowski, F., Heyer, W., *Poröse Gläser*, VEB Deutscher Verlag für Grundstoffindustrie, Leipzig, **1982**.

Janson, J.-C., *Chromatographia*, **1937**, 23, 361.

Jeansonne, M. S.; Foley, J. P. Review of the exponentially modified Gaussian chromatographic peak model since 1983, *J. Chromatogr. Sci.*, **1991**, 29, 258–266.

Jeansonne, M. S., Foley, J. P. Improved Equations for Calculation of Chromatographic Figures of Merit for Ideal and Skewed Peaks, *J. Chromatogr.*, **1992**, 594, 1–8.

Jeng, C. Y., Langer, S. H. Hydroquinone oxidation for the detection of catalytic activity in liquid chromatographic columns, *J. Chromatogr. Sci.*, **1989**, 27, 549–552.

Jiping, M., Yafend, G., Lingxin, C., *J. Sep. Sci.*, **2003**, 26, 307–312.

Jupke, A. Experimentelle Modellvalidierung und modellbasierte Auslegung von Simulated Moving Bed (SMB) Chromatographieverfahren, *Fortschrittbericht VDI: Reihe 3 Nr. 807*, VDI Verlag GmbH, Düsseldorf, **2004**.

Jupke, A., Epping, A., Schmidt-Traub, H. Optimal Design of Batch and SMB Chromatographic Separation Processes, *J. Chromatogr. A*, **2002**, 944, 93–117.

Juza, M. Development of a high-performance liquid chromatographic simulated moving bed separation form an industrial perspective, *J. Chromatogr. A*, **1999**, 865, 35–49.

Kaczmarski, K., Antos, D. Modified Rouchon and Rouchon-like algorithms for solving different models of multicomponent preparative chromatography, *J. Chromatogr. A*, **1996**, 756, 73–87.

Kaczmarski, K., Mazzotti, M., Storti, G., Morbidelli, M. Modeling fixed-bed adsorption columns through orthogonal collocation on moving finite elements, *Comp. Chem. Eng.*, **1997**, 21, 641–660.

Kaiser, R. E., Oelrich, E. *Optimisation in HPLC*, Dr. Alfred Hüthig Verlag, Heidelberg, **1981**

Kalbé, J., Höcker, H., Berndt, H. Design of enzyme reactors as chromatographic columns for racemic resolution of amino acid esters, *Chromatographia*, **1989**, 28, 193–196.

Kärger, J., Pfeifer, H., Heink, W. Principles and application of self-diffusion measurements by nuclear magnetic resonance, *Adv. Magn. Resonan.*, **1988**, 12, 1.

Katti, A. M., Jagland, P. Development and optimisation of industrial scale chromatography for use in manufacturing, *Anal. Mag.*, **1998**, 26(7), 38–46.

Kawase, M., Suzuki,T.B., Inoue, K., Yoshimoto, K., Hashimoto, K. Increased esterification conversion by application of simulated moving-bed reactor, *Chem. Eng. Sci.*, **1996**, 51, 11, 2971–2976.

Kawase, M., Inoue, Y., Araki, T., Hashimoto, K. The simulated moving-bed reactor for production of bisphenol A, *Catal. Today*, **1999**, 48, 199–209.

Kawase, M., Pilgrim, A., Araki, T., Hashimoto, K. Lactosucrose production using a simulated moving bed reactor, *Chem. Eng. Sci.*, **2001**, 56, 453–458.

Kearney, M. Control of fluid dynamics with engineered fractals – Adsorption process applications, *Chem. Eng. Comm.*, **1999**, 173, 43–52.

Kearney, M. Engineered fractals enhance process applications, *Chem. Eng. Prog.*, **2000**, 96, 12, 61–68.

Kemball, C., Rideal, E. K., Guggenheim, E. A. *Trans. Faraday Soc.*, **1948**, 44, 948.

Klatt, K. U. Modellierung und effektive numerische Simulation von chromatographischen Trennprozessen im SMB-Betrieb, *Chem. Ing. Tech.*, **1999**, 71, 6.

Klatt, K.-U., Hanisch, F., Dünnebier, G., Engell, S. Model-based optimization and control of chromatographic processes, *Comp. Chem. Eng.*, **2000**, 198–203.

Klatt, K.-U., Hanisch, F., Dünnebier, G. Model-based control of a simulated moving bed chromatographic process for the separation of fructose and glucose, *J. Proc. Control*, **2002**, 12, 203–219.

Kloppenburg, E., Gilles, E. D. A new concept for operating simulated moving-bed processes, *Chem. Eng. Technol.*, **1999a**, 22, 10, 813–817.

Kloppenburg, E., Gilles, E. D. Automatic control of the simulated moving bed process for C8 aromatics separation using asymptotically exact input/output linearization, *J. Proc. Control*, **1999b**, 9, 41–50.

Kniep, H., Blümel, C., Seidel-Morgenstern, A. Efficient design of the SMB process based on a perturbation method to measure adsorption isotherms and on a rapid solution of the dispersion model, Oral presentation at *SPICA 98*, Strasbourg, August 23–25th, **1998**.

Kniep, H., Mann, G., Vogel, C., Seidel-Morgenstern, A. Enantiomerentrennung mittels Simulated Moving Bed-Chromatographie, *Chem. Ing. Tech.*, **1999**, 71, 708–713.

Knox, J. H. Practical Aspects of LC Theory, *J. Chromatogr.*, **1977**, 15, 352–364.

Korns, G. A. *Mathematical Handbook for Scientists and Engineers*, Dover Publications Inc., **2000**.

Kraak, J., Ostervink, R., Poppe, H., Esser, U., Unger, K. K. Hydrodynamic chromatography of macromolecules on 2 μm non-porous spered silica gel packings, *Chromatographia*, **1989**, 27, 585.

Kubin, M. Beitrag zur theorie der chromatographie, *Collect. Czech. Chem. Commun.*, **1965**, 30, 1104–1118.

Kucera, E. Contribution to the theory of chromatography/linear non-equilibrium elution, *J. Chromatogr.*, **1965**, 19, 2, 237–248.

Kulprathipanja, S. (Ed.) *Reactive Separation Processes*, Taylor & Francis, New York, **2002**.

Kummer, M., Palme, H. J. Resolution of enantiomeric steroids by high-performance liquid chromatography on chiral stationary phases, *J. Chromatogr. A*, **1996**, 749, 1–2, 61–68.

Kümmel, R., Worch, E. *Adsorption aus Wäßrigen Lösungen*, VEB Deutscher Verlag für Grundstoffindustrie, Leipzig, **1990**.

Kurganov, A., Trüdinger, U., Isaeva, T., Unger, K.K. Native and modified alumina, titania and zirkonia in normal and reversed-phase high-performance liquid chromatography, *Chromatographia*, **1996**, 42, 217–222.

Küsters, E., Gerber, G., Antia, F. Enantioseparation of a chiral epoxide by simulated moving bed processes, *AIChE J.*, **1995**, 42, 154–160.

Laas, T. Agar derivates for chromatography, electrophoresis and gel-bound enzymes – IV. Benzylated dibromopropanol cross-linked sepharose as an amphophilic gel for hydrophobic salting-out chromatography of enzymes with special emphasis on denaturing risks, *J. Chromatogr.*, **1975**, 111, 373–387.

Lapidus, L., Amundson, N.R. A descriptive theory of leaching: Mathematics of adsorption beds, *J. Phys. Chem.*, **1952**, 56, 984–988.

Lapidus, L. *Digital Computation for Chemical Engineers*, McGraw-Hill, New York, **1962**.

Lauer, K. Technische Herstellung von Fructose, *Starch/Stärke*, **1980**, 32, 11–13.

Lea, D. J., Sehon, A. H. Preparation of synthetic gels for chromatography of macromolecules, *Can. J. Chem.*, **1962**, 40, 159–160.

Leonard, B. A stable and accurate convective modelling procedure based on quadratic upstream procedure, *Com. Meth. Appl. Mech. Eng.*, **1979**, 19, 59–98.

Levenspiel, O. *Chemical Reaction Engineering*, 3rd Edition, John Wiley & Sons, New York, **1999**.

Levenspiel, O., Bischoff, K. B. Patterns of flow in chemical process vessels, *Adv. Chem. Eng.*, **1963**, 4, 95 ff.

Liapis, A. I., Rippin, D. W. T. The simulation of binary adsorption in continuous counter-current operation and a comparison with other operation modes, *AIChE J.*, **1979**, 25(3), 455–460.

Lin, B., Ma, Z., Golshan-Shirazi, S. Guiochon, G., *J. Chromatogr. A*, **1989**, 475.

Lisso, M., Wozny, G., Arlt, W., Beste, Y.A. Optimierung der HETP präparativer HPLC-Säulen, *Chem. Ing. Tech.*, **2000**, 72, 494–502.

Lode, F., Houmard, M., Migliorini, C., Mazzotti, M., Morbidelli, M. Continuous reactive chromatography, *Chem. Eng. Sci.*, **2001**, 56, 269–291.

Lode, F. A simulated moving bed reactor (SMBR) for esterifications, Dissertation, ETH Zürich, **2002**.

Lode, F., Mazzotti, M., Morbidelli, M. Comparing true countercurrent and simulated moving-bed chromatographic reactors, *AIChE J.*, **2003a**, 49, 4, 977–990.

Lode, F., Francesconi, G., Mazzotti, M., Morbidelli, M. Synthesis of Methylacetate in a simulated moving bed reactor: experiments and modelling, *AIChE J.*, **2003b**, 49, 6, 1516–1524.

Lu, Z. Ching, C. B. Dynamics of simulated moving-bed adsorption separation processes, *Sep. Sci. Technol.*, **1997**, 32, 1118–1137.

Ludemann-Hombourger, O., Bailly, M., Nicoud, R.-M. Design of a Simulated Moving Bed: Optimal Size of the Stationary Phase, *Sep. Sci. Tech.*, **2000a**, 35(9), 1285–1305.

Ludemann-Hombourger, O., Bailly, M., Nicoud, R.-M. The VARICOL-Process: A new multicolumn continuous chromatographic process, *Sep. Sci. Technol.*, **2000b**, 35(12), 1829.

Ludemann-Hombourger, O., Pigorini, G., Nicoud, R.M., Ross, D., Terfloth G. Application of the VARICOL process to the separation of the isomers of the SB-553261 racemate, *J. Chromatogr. A*, **2002**, 947, 59–68.

Ma, J., Guan, Y., Chen, L. Study on conical columns for semi-preparative liquid chromatography, *J. Sep. Sci.*, **2003**, 26, 307–312.

Ma, Z., Whitley, R. D., Wang, N.-H. Pore and Surface Diffusion in Multicomponent Adsorption and Liquid Chromatography Systems, *AIChE J.*, **1996**, 42(5), 1244–1262.

Ma, Z., Wang, N.-H. Standing wave analysis of SMB-chromatography: Linear systems, *AIChE J.*, **1997**, 43, 2488–2508.

Maa, Y.-F., Horvath, C. Rapid analysis of proteins and peptides by reversed-phase chromatography with polymeric microcellular sorbent, *J. Chromatogr.*, **1988**, 445, 71–86.

Mackie, J. S., Meares, P. The diffusion of electrolytes in a cation-exchange resin membrane, *Proc. Roy. Soc. London, Ser. A*, **1955**, 267, 498–506.

Majors, R. E. The cleaning and regeneration of reversed-phase HPLC columns, *LC-GC Eur.*, **2003a**, 16, 7, 404–409.

Majors, R. E. A review of HPLC Column Packing Technology, *Am. Lab.* (Shelton, CT, United States), **2003b**, 35, 20, 46–54.

Mallmann, T., Burris, B., Ma, Z., Wang, N.-H. Standing wave design for nonlinear SMB systems for fructose purification, *AIChE J.*, **1998**, 44, 2628–2646.

Mannschreck, A. Chiroptical detection during liquid chromatography, *Chirality*, **1992**, 4, 163–169.

Marco, V., Bombi, G. Mathematical functions for representation of chromatographic peaks, *J. Chromatogr. A*, **2001**, 931, 1–30.

Marme, St. Untersuchung der sorption von gelösten organischen verbindungen an zeolithischen molekularsieben silikalit-1, ZSM-5, zeolith beta, zeolith Y, VPI 5 und dem mesoprösen Materials MCM-41, Diploma thesis, Universität Mainz, **1991**.

Martin, A. J. P., Synge, R. L. M. *Biochem. J.*, **1941**, 35, 1358 ff..

Martin, A. J. P. Summarizing paper, *Discuss. Faraday Soc.*, **1949**, 7, 332.

Marteau, P., Hotier, G., Zanier-Szydlowski, N., Aoufi, A., Cansell, F. Advanced control of C_8 aromatics separation process with real-time multiport on-line raman spectroscopy, *Process and Quality*, **1994**, 6, 133–140.

Matsen, J.M., Harding, J.W., Magee, E.M. Chemical reactions in chromatographic columns, *J. Phys. Chem.*, **1965**, 69, 522–527.

Mayne, D. Nonlinear model predictive control: challenges and opportunities, in *Progr. Systems Control Theor.*, Allgöwer, F., Zheng, A. (Eds), **2000**, 26, 23–44.

Mazzotti, M., Storti, G., Morbidelli, M. Robust design of countercurrent adsorption separation processes: 2. Multicomponent systems, *AIChE J.*, **1994**, 40, 1825–1842.

Mazzotti, M., Kruglov, A., Neri, B., Gelosa, D, Morbidelli, M. A Continuous chromatographic reactor: SMBR, *Chem. Eng. Sci.*, **1996a**, 51, 10, 1827–1836.

Mazzotti, M., Storti, G., Morbidelli, M. Robust design of countercurrent adsorption separation processes: 3. Nonstoichiometric systems, *AIChE J.*, **1996b**, 42, 2784–2796.

Mazzotti, M., Storti, G., Morbidelli, M. Supercritical fluid simulated moving bed chromatography, *J. Chromatogr. A*, **1997a**, 786, 309–320.

Mazzotti, M., Storti, G., Morbidelli, M. Robust design of countercurrent adsorption separation processes: 4. Desorbent in the feed, *AIChE J.*, **1997b**, 43, 64–72.

Mazzotti M., Storti G., Morbidelli M. Optimal operation of simulated moving bed units under nonlinear chromatographic separation. *J. Chromatogr. A*, **1997c**, 769, 3–24.

McCarthy, J. P. Direct Enantiomeric Separation of the 4 stereoisomers of nadolol using normal-phase and reversed-phase high performance liquid chromatography with Chiralpak AD, *J. Chromatogr. A*, **1994**, 685, 2, 349–355.

Meurer, M., Altenhöner, U. Strube, J., Schmidt-Traub, H. Dynamic simulation of simulated moving bed chromatographic reactors, *J. Chromatogr. A*, **1997**, 769, 71–79.

Meyer, V. R. Fallstricke und Fehlerquellen der HPLC in Bildern, Wiley VCH, Heidelberg, **1999**.

Meyers, J.J., Liapis, A.I. Network modeling of the intraparticle convection and diffusion of molecules in porous particles packed in a chromatographic column, *J. Chromatogr. A*, **1998**, 827, 197–213.

Meyers, J.J., Liapis, A.I. Network modeling of the convective flow and diffusion of molecules adsorbing in monoliths and in porous particles packed in a chromatographic column, *J. Chromatogr. A*, **1999**, 852, 3–23.

Michel, M., Schmidt-Traub, H., Ditz, R., Schulte, M. Kinkel, J. Stark, W., Küpper, M., Vorbrodt, M. Development of an integrated process for electrochemical reaction and chromatographic SMB-separation, *J. Appl. Electrochem.*, **2003**, 33, 939–949.

Migliorini, C., Mazzotti, M., Morbidelli, M. Simulated moving-bed units with extra-column dead volume, *AIChE J.*, **1999**, 45, 1411–1421.

Migliorini, C., Mazzotti, M., Morbidelli, M. Robust design of countercurrent adsorption separation processes: 5. Nonconstant Selectivity, *AIChE J.*, **2000**, 46, 1384–1399.

Migliorini, C., Wendlinger, M., Mazotti, M., Morbidelli, M. Temperature gradient operation of a simulated moving bed unit, *Ind. Eng.Chem. Res.*, **2001**, 40, 2606–2617.

Migliorini, C., Mazzotti, M., Gianmarco, Z., Morbidelli, M. Shortcut experimental method for designing chiral SMB separations, *AIChE J.*, **2002**, 48, 1.

Mihlbachler, K., Fricke, J., Yun, T., Seidel-Morgenstern, A., Schmidt-Traub, H., Guiochon, G. Effect of the homogeneity of the column set on the performance of a simulated moving bed unit, part I: Theory, *J. Chromatogr. A*, **2001**, 908, 49–70.

Mihlbachler, K., Jupke, A., Seidel-Morgenstern, A., Schmidt-Traub, H., Guiochon, G. Effect of the homogeneity of the column set on the performance of a simulated moving bed unit,

part II: Experimental study, *J. Chromatogr. A*, **2002**, 944, 3–22.

Mikeš, O., Štrop, P., Zbrožek, J., Čoupek, J. Chromatography of biopolymers and their framents on ion-exchange derivates of the hydrophilic macroporous synthetic gel spheron, *J. Chromatogr.*, **1976**, 119, 339–354.

Mikeš, O. HPLC of Biopolymers and Biooligomers, *J. Chromatogr. Library*, 41A, Elsevier, Amsterdam, **1988**, A142–A146.

Miller, J. B. Techniques for NMR imaging of solids, *TrAC*, **1991**, 10, 9.

Miller, L., Grill, C., Yan, T., Dapremont, O., Huthmann, E., Juza, M. Batch and simulated moving bed chromatographic resolution of a pharmaceutical racemate, *J. Chromatogr. A*, **2003**, 1006, 267–280.

Moore, J.C. Gel permeation chromatography. I. A new method for molecular weight distribution of high polymers, *J. Polymer. Sci.*, Part A, **1964**, 2, 835.

Müller, W. New phase supports for liquid–liquid partition chromatography of biopolymers in aqueous poly(ethyleneglycol)-dextran systems. Synthesis and application for the fractionation of DNA restriction fragments, *Eur. J. Biochem.*, **1986**, 155, 213.

Müller, W. New ion exchangers fort he chromatography of biopolymers, *J. Chromatogr.*, **1990**, 510, 133.

Müller, H. D. J., Delp, A., Spurk, J. H. Chromatographic column, *Eur. Pat. Appl.*, **1997**, EP 754482 A2 19970122.

Murer, P., Lewandowski, K., Svec, F., Frechet, J. M. On-bead combinatorial approach to the design of chiral stationary phases for HPLC, *Anal. Chem.*, **1999**, 71, 1278–1284.

Myers, A. L., Prausnitz, J. M. Thermodynamics of mixed-gas adsorption, *AIChE J.*, **1965**, 11, 1, 121–127.

NAG. The NAG Fortran Library Mark 15, Numerical Algorithms Group Ltd., Oxford, **1991**

Natarajan, S., Lee, J. H. Repetitive model predictive control applied to a simulated moving bed chromatography system, *Comp. Chem. Eng.*, **2000**, 24, 1127–1133.

Nawrocki, J., Rigney, M. P, Mc Cormick, A., Carr, P. W. Chemistry of zirconia and its use in chromatography, *J. Chromatogr. A*, **1993**, 657, 229–282.

Neue, U. D., Mazza, C. B., Cavanaugh, J. Y., Lu, Z., Wheat, T. E. At-column dilution for improved loading in preparative chromatography, *Chromatographia*, **2003**, 57, 121–127.

Nicoud, R. M. The simulated moving bed: A powerful chromatographic process, *Mag. Liquid Gas Chromatogr.*, **1992**, 5, 43–47.

Nicoud, R. M. Simulated Moving Bed (SMB): Some possible applications for biotechnology in *Bioseparation and Bioprocessing: Volume I*, Subramanian, G. (Ed.), Wiley-VCH Verlag GmbH, Weinheim, **1998**.

Ning, J. G. Direct chiral separation with Chiralpak AD converted into the reversed-phase mode, *J. Chromatogr. A*, **1998**, 805(1–2), 309–314.

Nyiredy, S., Meier, B., Erdelmeier, C. A. J., Sticher, O. PRISMA: a geometrical design for solvent optimization in HPLC, *J. High Res. Chrom. & Chrom. Comm.*, **1985**, 8, 186–188.

Nyiredy, S., Fater, Z. Automatic mobile phase optimization, using the "PRISMA" model for the TLC separation of apolar compounds, *Jpc-J. Planar Chromatography-Mod. TLC*, **1995**, 8, 5, 341–345.

Nyiredy, S. Planar chromatographic method development using the PRISMA optimization system and flow charts, *J. Chromatogr. Sci.*, **2002**, 40, 10, 553–563.

Okamoto, M., Nakazawa, H., Reversal of elution order during direct enantiomeric separation of pyriproxyfen on a cellulose-based chiral stationary phase, *J. Chromatogr.*, **1991**, 588, 177–180.

Pai, V. B., Gainer, J. L., Carta, G. Simulated moving bed chromatographic reactors in *Fundamentals of Adsorption*, LeVan, M. D. (Ed.), Kluwer Academic Publishers, Boston, **1996**.

Pais, L., Loureiro, J., Rodrigues, A. Separation of 1,1-bi-2-naphthol enantiomers by continuous chromatography in simulated moving bed, *Chem. Eng. Sci.*, **1997a**, 52, 245–257.

Pais, L., Loureiro, J., Rodrigues, A. Modeling, simulation and operation of a simulated moving bed for continuous chromatographic separation of 1,1-bi-2-naphthol enantiomers, *J. Chromatogr. A*, **1997b**, 769, 25–35.

Pais, L., Loureiro, J., Rodrigues, A. Modelling strategies for enantiomeres separation by SMB chromatography, *AIChE J.*, **1998a**, 44(3), 561–569.

Pallaske, U. Kritik an zwei bekannten Verweilzeitverteilungsformeln für das Diffusionsströmungsrohr, *Chem. Ing. Tech.*, **1984**, 56, 1, 46–47.

Pandey, R. C. Isolation and purification of paclitaxel from organic matter containing paclitaxel, cephalomannine and other related taxanes, US patent, **1997**.

Pandey, R. C., Yankov, L. K., Poulev, A. Nair, R., Caccamese, S. Synthesis and separation of potential anticancer active dihalocephalomannine diastereomers from extracts of Taxus yunnanensis, *J. Nat. Prod.*, **1998**, 61, 1, 57–63.

Pelz, P. F., Spurk, J. H., Müller, H. D. J. Reducing mixing at the outlet bore of a cylinder, *Arch. Appl. Mechan.*, **1998**, 68, 395–406.

Pirkle, W. H., Welch, C. J., Lamm, B. Design, synthesis and evaluation of an improved enantioselective naproxen selector, *J. Org. Chem.*, **1992**, 57, 3854–3860.

Pirkle, W. H., Däppen, R. Reciprocity in chiral recognition – Comparison of several chiral stationary phases, *J. Chromatogr.*, **1987**, 404, 107–115.

Poole, C. F. *The Essence of Chromatography*, Elsevier, Amsterdam, **2003**.

Porath, J., Flodin, P. Gel filtration: a method for desalting and group separation, *Nature*, **1959**, 183, 1657.

Porath, J., Janson, J. C., Laas, T. Agar derivates for chromatography, electrophoresis and gel-bound enzymes – I. Desulphatet and reduced cross-linked agar and agarose in spherical bead form, *J. Chromatogr.*, **1971**, 60, 167–177.

Porath, J., Laas, T., Janson, J.-Ch. Agar derivates for chromatography, electrophoresis and gel-bound enzymes – III. Rigid agarose gels cross-linked with divinyl sulphone (DVS), *J. Chromatogr.*, **1975**, 103, 49–62.

Press, W. H., Flannery, B. P., Teukolsky, S. A., Vetterling, W. T. *Numerical Recipes in C++: The Art of Scientific Computing*, 2nd Edition, Cambridge University Press, Cambridge, **2002**.

Pröll, T., Küsters, E. Optimization strategy for simulated moving bed systems, *J. Chromatogr. A*, **1998**, 800, 135–150.

Qin, S., Badgwell, T., A survey of industrial model predictive control technology, *Control Eng. Prac.*, **2003**, 11, 733–764.

Rabel, F. Use and maintenance of microparticle high performance liquid chromatography columns, *J. Chromatogr. Sci.*, **1980**, 18, 394.

Radke, C. J., Prausnitz, J. M. Thermodynamics of multi-solute adsorption from liquid solutions, *AIChE J.*, **1972**, 18, 1, 761–768.

Rao, C., Rawlings, J. Nonlinear moving horizon state estimation, in *Progress in Systems and Control Theory*, Allgöwer, F., Zheng, A. (Eds), **2000**, 26, 45–69.

Regnier, F. E. HPLC of biological macromolecules: The first decade, *Chromatographia*, **1987**, 24, 241–251.

Rhee, H.-K., Aris, R., Amundson, N. R. On the theory of multicomponent chromatography, *Phil. Trans. Roy. Soc. London A*, **1970**, 267, 419–455.

Rhee, H.-K., Aris, R., Amundson, N. R. *First-order Partial Differential Equations, Vol. 1*, Prentice-Hall, Englewood Cliffs, NJ, **1986**.

Rhee, H.-K., Aris, R., Amundson, N. R. First-order Partial Differential Equations, Vol. 2, Prentice-Hall, Englewood Cliffs, NJ, 1989.

Reilley, C. N., Hildebrand, G. P., Ashley Jr., J. W. Gas chromatographic response as a function of sample input profile, Anal. Chem., 1962, 34, 1198–1213.

Roumeliotis, P., Chatziathanassiou, M., Unger, K. K. Study on the efficiency of assembled packed microbore columns in HPLC, Chromatographia, 1984, 19, 145–150.

Roussel, C., Piras, P., Heitmann, I. An approach to discriminating 25 commercial chiral stationary phases from structural data sets extracted from a molecular database, Biomed. Chromatogr., 1997, 11, 311–316.

Ruthven, D. M. Principles of Adsorption and Adsorption Processes, John Wiley & Sons, New York, 1984.

Ruthven, D. M., Ching, C. B. Review article no. 31: Counter-current and simulated counter-current adsorption separation processes, Chem. Eng. Sci., 1989, 44(5), 1011–1038.

Ruthven D.M. (Ed.) Encyclopedia of Separation Technology, Vol. 1, John Wiley and Sons, New York, 1997.

Sardin, M., Schweich, D., Villermaux, J. Preparative fixed-bed chromatographic reactor in Preparative and Production Scale Chromatography, Ganetsos, G., Barker, P.E. (Eds.), Marcel Dekker Inc., New York, 1993.

Sarmidi, M. R., Barker, P. E. Saccharification of modified starch to maltose in a continuous rotating annular chromatograph, J. Chem. Tech. Biotechnol., 1993a, 57, 229–235.

Sarmidi, M.R., Barker, P. E. Simultaneous biochemical reaction and separation in a rotating annular chromatograph, Chem. Eng. Sci., 1993b, 48, 2615–2623.

Saska, M., Clarke, S. J., Wu, M. D., Iqbal, K. Applications of continuous chromatographic separation in the sugar industry, Part I. Glucose/fructose equilibria on Dowex Monosphere 99 Ca resin at high sugar concentrations, Int. Sugar J., 1991, 93, 1115.

Saska, M., Clarke, S. J., Iqbal K. Continuous Separation of Sugarcane Molasses with a Simulated-Moving-Bed Adsorber: Absorption Equilibria, Kinetics and Applications, Sep. Sci. Tech., 1992, 27, 1711–1732.

Sattler, K. Thermische Trennverfahren – Grundlagen, Auslegung, Apparate, VCH, Weinheim, 1995.

Schenk, F. W. Glucose and Glucose-Containing Syrup, in Ullmann's Encyclopedia of Industrial Chemistry, 7th Edition, Wiley-VCH, Weinheim, 2003.

Schmidt, S., Wu, P., Konstantinov, K., Kaiser, K., Kauling, J., Henzler, H.-J., Vogel, J. H. Kontinuierliche Isolierung von Pharmawirkstoffen mittels annularer Chromatographie, Chem. Ing. Tech., 2003, 75, 3, 302–305.

Schomburg, G. Stationary phases in high performance liquid chromatography: Chemical modification by polymer coating, LC GC, 1988, 6, 1, 37.

Schramm, H., Grüner, S., Kienle, A., Gilles, E. D. Control of moving bed chromatographic processes, Proceedings of the European Control Conference, 2001, 2528–2533.

Schulte, M. Chiral derivatization chromatography, in Chiral Separation Techniques, Subramanian, G. (Ed.), Wiley-VCH, Weinheim, 2001

Schramm, H., Kienle, A., Kaspereit, M., Seidel-Morgenstern, A. Improved operation of simulated moving bed processes through cyclic modulation of feed flow and feed concentration, Chem. Eng. Sci., 2003, 58, 5217–5227.

Scott, R. P. W. Liquid Chromatography Detectors, Amsterdam, Elsevier Science Publishers, 1986.

Seidl, J., Malinsky, J., Dušek, D., Heitz, W. Macroporöse Styrol-Divinylbenzol-Copolymere und ihre Verwendung in der Chromatographie und zur Darstellung in Ionenaustauschern, Adv. Polymer Sci., 1967, 5, 113–213.

Seidel-Morgenstern, A. Mathematische Modellierung der präparativen Flüssigchromatographie, Deutscher Universitätsverlag, Wiesbaden, 1995.

Seidel-Morgenstern, A., Heuer, C., Hugo, P. Experimental Investigation and Modelling of Closed-Loop Recycling in Preparative Chromatography, Chem. Eng. Sci., 1995, 50(7), 1115–1127.

Seidel-Morgenstern, A., Nicoud, R.-M., Adsorption isotherms: experimental determination and application to preparative chromatography, Isolation & Purific., 1996, 2, 165–200.

Seidel-Morgenstern, A., Heuer, C., Kniep, H., Falk, T. Review: comparison of various process engineering concepts of preparative chromatography, Chem. Eng. Tech., 1998, 21, 469–477.

Seidel-Morgenstern, A., Heuer, C., Hugo, P. Experimental and theoretical study of recycling in preparative chromatography, Sep. Sci. Tech., 1999, 34, 173–199.

Seidel-Morgenstern, A. Experimental determination of single solute and competitive adsorption isotherms, J. Chromatogr. A, 2004, 1037, 1–2, 255–272.

Shieh, M. T., Barker, P. E. Combined bioreaction and separation in a simulated counter-current chromatographic bioreactor-separator for the hydrolysis of lactose, *J. Chem. Tech. Biotechnol.*, **1996**, 66, 265–278.

Shih, T. M. Numerical Heat Transfer, *Series in Computational Methods in Mechanics and Thermal Sciences*, Springer-Verlag, New York, **1984**.

Snyder, L. R. *Principles of Adsorption Chromatography*, Marcel Dekker, New York, **1968**.

Snyder, L. R., Kirkland, J. J. *Introduction to Modern Liquid Chromatography*, Wiley, New York, **1979**.

Snyder, L. R., Kirkland, J. J., Glajch, J. L. *Practical HPLC Method Development*, Wiley, New York, **1997**.

Snyder, L. R., Dolan, J. W. The linear-solvent-strength model of gradient elution, *Adv. Chromatogr.*, **1998**, 38, 115–187.

Soczewinski, E. Solvent composition effects in thin-layer chromatography systems of the type silica gel-electron donor solvent, *Anal. Chem.*, **1969**, 41, 1, 179–182.

Solms, J. Kontinuierliche papierchromatographie, *Helv.Chim. Acta*, **1955**, 38 , 1127–1133.

Stahl, E. *Dünnschichtchromatographie – Ein Laboratoriumshandbuch*, Springer-Verlag, Berlin **1967**, 90.

Storti, G., Masi, M., Morbidelli, M. Optimal design of SMB adsorption separation units through detailed modelling and equilibrium theory, *Proceedings of NATO Meeting 1988 in Vimeiro (Portugal)*, Research and Technology Agency, Neuilly sur Seine (France), **1988**.

Storti, G., Masi, M., Morbidelli, M. Modeling of Countercurrent Adsorption Processes in *Preparative and Production Scale Chromatography*, Ganetsos, G., Barker, P. E. (Eds.), Marcel Dekker Inc., New York, **1993a**.

Storti, G., Mazzotti, M., Morbidelli, M., Carrà, S. Robust design of binary countercurrent adsorption separation processes, *AIChE J.*, **1993b**, 39, 471–492.

Storti, G., Baciocchi, R., Mazzotti, M., Morbidelli, M. Design of optimal conditions of simulated moving bed adsorptive separation units, *Ind. Eng. Res.*, **1995**, 34, 288–301.

Strube, J. Simulation und Optimierung kontinuierlicher Simulated-Moving-Bed (SMB)-Chromatographie-Prozesse, Dissertation, Universität Dortmund, **1996**.

Strube, J., Altenhöner, U., Meurer, M., Schmidt-Traub, H. Optimierung kontinuierlicher Simulated-Moving-Bed-Chromatographie-Prozesse durch dynamische Simulation, *Chem. Ing. Tech.*, **1997a**, 69(3), 328–331.

Strube, J., Altenhöner, U., Meurer, M., Schmidt-Traub, H., Schulte, M. Dynamic simulation of simulated moving-bed chromatographic processes for the optimisation of chiral separations, *J. Chromatogr. A*, **1997b**, 769, 328–331.

Strube, J., Schmidt-Traub, H., Schulte, M. Auslegung, Betrieb und ökonomische Betrachtung chromatographischer Trennprozesse, *Chem. Ing. Tech.*, **1998a**, 70(10), 1271–1279.

Strube, J., Haumreisser, S., Schmidt-Traub, H., Schulte, M., Ditz, R. *Comparison of Batch Elution and Continuous Simulated Moving Bed Chromatography*, Org. Proc. Res. Dev., **1998b** 2, 5, 305–319.

Strube, J., Jupke, A., Epping, A., Schmidt-Traub, H., Schulte, M., Devant, R. Design, optimization and operation of smb chromatography in the production of enantiomerically pure pharmaceuticals, *Chirality*, **1999**, 11, 440–450.

Strube, J., Technische Chromatographie: Auslegung, Optimierung, Betrieb und Wirtschaftlichkeit, Dissertation, Universität Dortmund, Shaker-Verlag, **2000**.

Sun, T., Wang, G., Feng, L., Liu, B., Ma, Y., Jiang, L., Zhu, D. Reversible Switching between Superhydrophilicity and Superhydrophobicity, *Angew. Chem.*, **2004**, 116, 361–364.

Suzuki, M. *Adsorption Engineering*, Elsevier, Amsterdam, **1990**.

Svec, F., Wulff, D., Fréchet, J. M. Combinatorial Approaches to recognition of chirality: preparation and use of materials for the separation of enantiomers, in *Chiral Separation Techniques*, Subramanian, G. (Ed.), Wiley-VCH, Weinheim, **2001**.

Szanya, T., Argyelan, J., Kovats, S., Hanak, L. Separation of steroid compounds by overloaded preparative chromatography with precipitation in the fluid phase, *J. Chromatogr. A*, **2001**, 908(1–2), 265–272.

Takeuchi, K., Miyauchi, T. and Uraguchi, Y. Computational studies of a chromatographic moving-bed reactor for consecutive and reversible reactions, *J. Chem. Eng. Jpn.*, **1978**, 11, 216–220.

Tallarek, U., Bayer, E., Guiochon, G. Study of dispersion in packed chromatographic columns by pulsed field gradient nuclear magnetic resonance, *J. Am. Chem. Soc.*, **1998**, 120, 7, 1494–1505.

Tanaka, N., Hashizume, K. and Araki, M. Comparison of polymer-based stationary phases with silica-based stationary phases in reversed-phase liquid chromatography, *J. Chromatogr.*, **1987**, 400, 33–45.

Thomas, H. Heterogeneous ion exchange in a flowing system, *J. Am. Chem. Soc.*, **1944**, 66, 1664–1666.

Toth, J., Acta Chim. Acad. Sci. Hung., **1971**, 69, 311.

Toumi, A., Engell, S., Ludemann-Hombourger, O., Nicoud, R. M., Bailly, M. Optimization of simulated moving bed and VARICOL processes, *J. Chromatogr. A*, **2003**, 1006, 15–31.

Toumi, A., Engell, S. Optimierungsbasierte regelung eines integrierten reaktiven chromatographischen trennprozesses, *Automatisierungstechnik*, **2004a**, 52, 13–22.

Toumi, A., Engell, S. Optimal operation and control of a reactive simulated moving bed process, oral presentation at 7th *International Symposium on Advanced Control of Chemical Processes*, Hong Kong, **2004b**.

Trapp, O., Schurig, V. Novel direct access to enantiomerization barriers from peak profiles in enantioselective dynamic chromatography: enantiomerization of dialkyl-1,3-allene-dicarboxylates, *Chirality*, **2002**, 14, 465–470.

Tsotsas, E., Über die Wärme- und Stoffübertragung in durchströmten Festbetten, *VDI Fortschrittsberichte*, 3rd Series, No. 223, VDI-Verlag, Düsseldorf, **1987**.

Tweeten, K. A., Tweeten, T. N. Reversed-phase chromatography of proteins on resin-based wide-pore packings, *J. Chromatogr.*, **1986**, 359, 111.

Ugelstad, J., Mørk, P. C., Kaggernd, K. H., Ellingsen, T., Berge, A. Swelling of oligomer-polymer particles. New methods of preparation of emulsions and polymer dispersions, *Adv. Colloid Interface Sci.*, **1980**, 13, 101–140.

Unger K.K., Janzen R., Jilge G. Packings and stationary phases for biopolymer separations by HPLC, *Chromatographia*, **1987**, 24, 144–154.

Unger, K.K., *Packings and Stationary Phases in Chromatographic Techniques*, Marcel Dekker, New York, **1990**.

Unger, K. K. (Ed.) *Handbuch der HPLC, Teil 2: Präparative Säulenflüssig-Chromatographie*, GIT Verlag, Darmstadt, **1994**.

Unger, K. K., Weber, E. *A Guide to Practical HPLC*, GIT Verlag, **1999**.

Unger K. K., Bidlingmaier, B., du Fresne von Hohenesche, C., Lubda, D. Evaluation and comparison of the pore structure and related properties of particulate and monolithic silicas for liquid phase separation processes, COPS VI Proceedings in *Stud. Surf. Sci. Catal.*, **2002**, 144, 115–122.

US-FDA, *Draft Guidance for Industry: PAT – A Framework for Innovative Pharmaceutical Manufacturing and Quality Assurance*, **2003**.

Van Deemter, J. J., Zuiderweg, F. J., Klinkenberg, A. Longitudinal Diffusion and resistance to mass transfer as causes of nonideality in chromatography, *Chem. Eng. Sci.*, **1956**, 5, 271–289.

Villadsen, J., Stewart, W. E., *Chem. Eng. Sci.*, **1967**, 22, 1483 ff..

Villermaux, J. The Chromatographic Reactor in *Percolation Processes: Theory and Applications*, Rodrigues, A. E., Tondeur, D. (Eds.), Sijthoff & Noordhoff International Publishers B. V., Alphen aan den Rijn, **1981**.

Voigt, U., Hemple, R., Kinkel, J.N., Deutsches Patent, DE 196 11 094 A1.

Wang, C., Klatt, K., Dünnebier, G., Engell, S., Hanisch, F. Neural network based identification of SMB chromatographic processes, *Control Eng. Prac.*, **2003**, 11, 949–959.

Wang, Y., Bluhm, L. H., Li, T. Identification of chiral selectors from a 200-member parallel combinatorial library, *Anal.Chem.*, **2000**, 72, 5459–5465.

Weber, S. G., Carr, P. W. The Theory of the Dynamics of Liquid Chromatography in *High Performance Liquid Chromatography*, Brown, P. R., Hartwick, R. A. (Eds.), Wiley, New York, **1989**.

Wekenborg, K.; Susanto, A.; Schmidt-Traub, H.; Nicht-isokratische SMB-Trennung von Proteinen mittels Ionenaustauschchromatographie, *Chem. Ing. Tech.*, **2004**, 76, 6, 815–819.

Wen, C. Y., Fan, L. T. *Models for Flow Systems and Chemical Reactors*, Marcel Dekker, New York, **1975**.

Welsh, C. J., Bhat, G., Protopopova, M. N. Selection of an optimised adsorbent for preparative chromatographic enantioseparations by microscale screening of a second-generation chiral stationary phase library. *J. Com. Chem.*, **1999**, 1, 364–367.

Whitley, R. D., Van Cott, K. E., Wang, N.-H. Analysis of nonequilibrium adsorption/desorption kinetics and implications for analytical and preparative chromatography, *Ind. Eng. Chem. Res.*, **1993**, 32., 149–159.

Wicke, E., *Kolloid Z.*, **1939**, 86, 285 ff.

Wilson, E.J. Geankoplis, C. J. Liquid mass transfer at very low reynolds numbers in packed beds, *Ind. Eng. Chem. Fundam.*, **1966**, 5, 9.

Wolfgang, J., Prior, A. Modern advances in chromatography, *Adv. Biochem. Eng./Biotech.*, **2002**, 76, 233–255.

Wu, Y., Wang, Y., Yang, A., Li, T. Screening of mixture combinatorial libraries for chiral selectors: A reciprocal chromatographic approach using enantiomeric libraries, *Anal. Chem.* **1999**, 71, 1688–1691.

Yun, T., Zhong, G., Guiochon, G. Simulated moving bed under linear conditions: experimental vs. calculated results, *AIChE J.*, **1997a**, 43(4), 935–945.

Yun, T., Zhong, G., Guiochon, G. Experimental study of the influence of the flow rates in SMB chromatography, *AIChE J.*, **1997b**, 43(11), 2970–2983.

Zafar, I., Barker, P. E. An experimental and computational study of a biochemical polymerisation reaction in a chromatographic reactor separator, *Chem. Eng. Sci.*, **1988**, 43, 2369–2375.

Zang, Y., Wankat, P. C. SMB operation strategy – partial feed, *Ind. Eng. Chem. Res.*, **2002**, 41, 2504–2511.

Zhang, Z., Mazzotti, M., Morbidelli, M. Power-Feed operation of simulated moving bed units: changing flow-rates during the switching interval, *J. Chromatogr. A*, **2003**, 1006, 87–99.

Zhong, G., Guiochon, G. Analytical solution for the linear ideal model of simulated moving bed chromatography, *Chem. Eng. Sci.*, **1996**, 51(18), 4307–4319.

Zhou, J., Tits, A., Lawrence, C. *Users's Guide for FFSQP Version 3.7: A FORTRAN code for solving constrained nonlinear (minimax) optimization problems, generating iterates satisfying all inequality and linear constraints*, University of Maryland, **1997**.

Appendix A

A1 Scale-up Factors Between Different Columns

Table A.1 allows a quick determination of the volumetric scale factor between two different columns. For scale up, select the diameter and length of the original column from the horizontal line and of the new column from the vertical line. The scale up factor is given at the intersection. For scale down, the procedure is reversed. The scale factor should be used for all process parameters that depend on column volume.

A1.1 Example Determination of the Scale-up Factor

The separation with an analytical column (d_c = 4.6 mm ; L_c = 125 mm) should be transferred to the separation on a preparative column (d_c =100 mm ; L_c = 200 mm). The volumetric scale-up factor can be calculated by Eq. 3.10:

$$F = \frac{d_{c,j}^2}{d_{c,i}^2} \cdot \frac{L_j}{L_i} = \frac{(100 \text{ mm})^2}{(4.6 \text{ mm})^2} \cdot \frac{200 \text{ mm}}{125 \text{ mm}} = 756.14$$

The scale-up factor can also be found with the help of Tab. A.1. In this case the column has to be scaled for constant length or cross section. The scale factor is given by the ratio of the cross section to the respective column length.

A.2 Standard Operating Procedures for Column Packing

Note:

Not all details for the different packing procedures are given here. Please refer to the column manufacturers' instructions for details. To achieve optimal packed columns, some experiments are at least necessary for optimization of the packing technology.

Please also make sure of your standard laboratory safety regulations and operating safety instructions for hydraulic equipment are followed when using the column packing equipment.

Preparative Chromatography. Edited by H. Schmidt-Traub
Copyright © 2005 WILEY-VCH Verlag GmbH & Co. KGaA, Weinheim
ISBN: 3-527-30643-9

Table A.1 (a) Volumetric scale factors between analytical and preparative columns.

d_c	L_c	4		4.6		10		21.2		25		50	
		125	250	125	250	125	250	125	250	125	250	125	250
4	125												
	250	2.00											
4.6	125	1.32	0.66										
	250	2.65	1.32	2.00									
10	125	6.25	3.13	4.73	2.36								
	250	12.50	6.25	9.45	4.73	2.00							
21.2	125	28.09	14.05	21.24	10.62	4.49	2.25						
	250	56.18	28.09	42.48	21.24	8.99	4.49	2.00					
25	125	39.06	19.53	29.54	14.77	6.25	3.13	1.39	0.70				
	250	78.13	39.06	59.07	29.54	12.50	6.25	2.78	1.39	2.00			
50	125	156.25	78.13	118.15	59.07	25	12.50	5.56	2.78	4.0	2.0		
	250	312.50	156.25	236.29	118.15	50	25	11.12	5.56	8.0	4.0	2.0	
100	200	1000	500	756.14	378.07	160	80	35.60	17.80	25.6	12.8	6.4	3.2
	300	1500	750	1134.22	567.11	240	120	53.40	26.70	38.4	19.2	9.6	4.8
200	100	2000	1000	1512.29	756.14	320	160	71.20	35.60	51.2	25.6	12.8	6.4
	200	4000	2000	3024.57	1512.29	640	320	142.40	71.20	102.4	51.2	25.6	12.8
	300	6000	3000	4536.9	2268.43	960	480	213.60	106.80	153.6	76.8	38.4	19.2
300	100	4500	2250	3402.6	1701.3	720	360	160.20	80.10	115.2	57.6	28.8	14.4
	200	9000	4500	6805.3	3402.6	1440	720	320.40	160.20	230.4	115.2	57.6	28.8
	300	13500	6750	10207.9	5104.0	2160	1080	480.60	240.30	345.6	172.8	86.4	43.2
450	100	10125	5062.5	7656	3828.0	1620	810	360.45	180.22	259.2	129.6	64.8	32.4
	200	20250	10125	15311.9	7656.0	3240	1620	720.90	360.4	518.4	259.2	129.6	64.8
	300	30375	15187.5	22967.9	11483.9	4860	2430	1081.3	540.7	777.6	388.8	194.4	97.2
600	100	18000	9000	13610.6	6805.3	2880	1440	640.8	320.4	460.8	230.4	115.2	57.6
	200	36000	18000	27221.2	13610.6	5760	2880	1281.6	640.8	921.6	460.8	230.4	115.2
	300	54000	27000	40831.8	20415.9	8640	4320	1922.4	961.2	1382.4	691.2	345.6	172.8
800	100	32000	16000	24196.6	12098.3	5120	2560	1139.2	569.6	819.2	409.6	204.8	102.4
	200	64000	32000	48393.2	24196.6	10240	5120	2278.4	1139.2	1638.4	819.2	409.6	204.8
	300	96000	48000	72589.8	36294.9	15360	7680	3417.6	1708.8	2457.6	1228.8	614.4	307.2
1000	100	50000	25000	37807.2	18903.6	8000	4000	1780.0	890.0	1280.0	640.0	320.0	160.0
	200	100000	50000	75614.4	37807.2	16000	8000	3560.0	1780.0	2560.0	1280.0	640.0	320.0
	300	150000	75000	113421.6	56710.8	24000	12000	5340.0	2670.0	3840.0	1920.0	960.0	480.0

(b) Volumetric scale factors between different preparative columns.

d_c	L_c	100/200	100/300	200/100	200/200	200/300	300/100	300/200	300/300	450/100	450/200	450/300	600/100	600/200	600/300	800/100	800/200	800/300	1000/100	1000/200
100	250																			
100	200																			
100	300	1.50																		
200	100	2.00	1.33																	
200	200	4.00	2.67	2.00																
200	300	6.00	4.00	3.00	1.50															
300	100	4.50	3.00	2.25	1.13	0.75														
300	200	9.00	6.00	4.50	2.25	1.50	2.00													
300	300	13.50	9.00	6.75	3.38	2.25	3.00	1.50												
450	100	10.13	6.75	5.06	2.53	1.69	2.25	1.13	0.75											
450	200	20.25	13.50	10.13	5.06	3.38	4.50	2.25	1.50	2.00										
450	300	30.38	20.25	15.19	7.59	5.06	6.75	3.38	2.25	3.00	1.50									
600	100	18.0	12.00	9.0	4.50	3.00	4.00	2.00	1.33	1.78	0.89	0.59								
600	200	36.0	24.00	18.0	9.00	6.00	8.00	4.00	2.67	3.56	1.78	1.19	2.00							
600	300	54.0	36.00	27.0	13.50	9.00	12.00	6.00	4.00	5.33	2.67	1.78	3.00	1.50						
800	100	32.0	21.33	16.0	8.00	5.33	7.11	3.56	2.37	3.16	1.58	1.05	1.78	0.89	0.59					
800	200	64.0	42.67	32.0	16.00	10.67	14.22	7.11	4.74	6.32	3.16	2.11	3.56	1.78	1.19	2.00				
800	300	96.0	64.00	48.0	24.00	16.00	21.33	10.67	7.11	9.48	4.74	3.16	5.33	2.67	1.78	3.00	1.50			
1000	100	50.0	33.33	25.0	12.50	8.33	11.11	5.56	3.70	4.94	2.47	1.65	2.78	1.39	0.93	1.56	0.78	0.52		
1000	200	100.0	66.67	50.0	25.00	16.67	22.22	11.11	7.41	9.88	4.94	3.29	5.56	2.78	1.85	3.13	1.56	1.04	2.00	
1000	300	150.0	100.0	75.0	37.50	25.00	33.33	16.67	11.11	14.81	7.41	4.94	8.33	4.17	2.78	4.69	2.34	1.56	3.00	1.50

Column packing – slurry preparation

1. Weigh the appropriate amount of packing material into a container that can hold at least twice the amount of final slurry volume. Table A.2 gives some general remarks on the amount of packing material per ml of column volume.

Table A.2 Amount of packing material per L of bed volume.

Type of adsorbents	Adsorbent amount per ml bed volume
Silica, normal and reversed phase	approx. 0.4–$0.6 \, kg \, L^{-1} \, V_c$
Polymeric adsorbents	approx. 0.3–$0.4 \, kg \, L^{-1} \, V_c$

2. Add the amount of solvent in several portions, making sure that the adsorbent is wetted by the solvent and does not form any aggregates. For the choice of solvent there are many different recommendations, not all of them using solvents, which can be used in preparative chromatography according to availability or toxicity. Table A.3 gives some general suggestions. Nevertheless, other solvents might be used that are either the eluent of the initial step or at least chemically close to that eluent to avoid other solvents and the resulting tedious conditioning steps.

Table A.3 Suggestions for slurry solvents.

Normal phase	Methanol (vacuum)
	Toluene–isopropanol (DAC)
	Isopropanol (circle suspension flow packing)
Reversed phase	Acetone, ethanol
	Isopropanol (CSF-packing)
Polymeric adsorbents	Ethanol, isopropanol
Chiral stationary phases	Isopropanol

Table A.4 lists the ratio of total slurry volume to packed bed volume for the different packing technologies. Slurry volumes might have to be adjusted to the given column geometry.

Table A.4 Slurry volumes for different packing technologies.

Packing technology	Volume slurry : volume packed bed
Dynamic axial compression	3:1
Vacuum packing	2:1
Circle suspension flow packing	9:1

3. Homogenize and wet the adsorbent intensively. For small slurry volumes and stable adsorbents this might be done by ultrasonification for 5 min. For larger slurry volumes this might be carried out in a mixing device or the suspension has to be stirred for at least 15 min with a paddle stirrer. *Never use a magnetic stirrer in any slurry preparation and make sure that with a paddle stirrer no sediment is built up underneath the stirrer, as this might damage the adsorbent particles.*

4. During ultrasonification the adsorbent tends to form a sediment on the ground of the container, which has to be carefully re-homogenized. After this final homogenisation the slurry should be immediately transferred to the prepared column and the packing procedure started.

Column preparation

1. Clean the frits carefully and inspect them for obvious damage.
2. Mount, clean and examine the column for proper sealing according to the manufacturer's operation manual.
3. Prior to packing, check and adjust the vertical orientation of the column using a spirit level. Move the piston to its maximum bed length position.
4. Flush the empty column with the slurry solvent prior to use to ensure that the column, column frits, and plumbing are clean and wetted. Empty the column before introducing the slurry suspension.

Column packing – dynamic axial compression technology

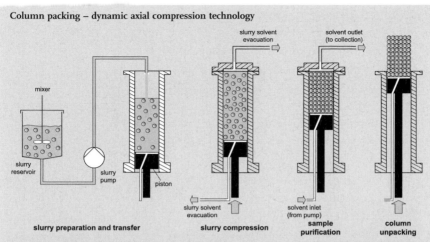

Figure A.1 Scheme of the axial dynamic compression technique. (Courtesy of H. Colin, formerly Promochrom, Champigneulles, France.)

1. Transfer the slurry to the column. If some slurry has been accidentally spilt at the top of the column, rinse and clean the column carefully.
2. Install the outlet flange and put the clamp in place. Tighten the nut of the clamp by hand and give one additional turn with a wrench. Connect the solvent line to the flange and secure the other ends of both solvent lines in a container of sufficient volume to accommodate the slurry solvent that will be expelled during the packing process. Select a line with a diameter as large as possible to reduce the pressure drop in the line.
3. Make sure the hydraulic valve is in the NEUTRAL position and adjust the (air) pressure to get the expected piston pressure (= packing pressure). Set the hydraulic valve to the COMPRESSION position. At this time the packing procedure starts. After a short while, the solvent starts flowing through both outlet flanges. As the bed forms, the oil pressure increases. After a certain time, the oil pressure stabilizes at the value selected. When the bed is almost completely formed, the flow of solvent decreases and finally stops almost completely. Depending on the packing conditions (pressure and packing material), it may be necessary to wait (5 to 30 min) before the flow of solvent has completely stopped.
4. The hydraulic valve should be left in the COMPRESSION position if rigid materials are used. In this case the **dynamic** axial compression mode is used, which ensures an immediate re-compression of the bed if some bed settlement takes place. If less stable adsorbents are packed, especially organic polymers, the hydraulic valve can be moved back to the NEUTRAL position, thus fixing the bed compression. If, after a while, a re-compression is necessary, move the hydraulic valve again to the COMPRESSION position for a short time.

Column packing – vacuum technology

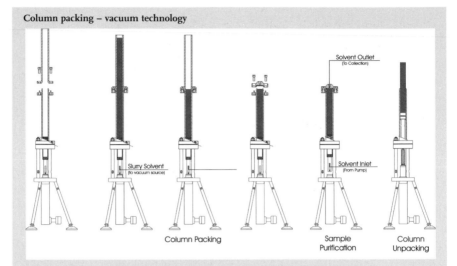

Figure A.2 Vacuum packing technology.

1. Transfer the slurry to the column, equipped with the packing reservoir. *After* introducing the slurry apply the vacuum by connecting the column inlet line with the vacuum source through a safety bottle. The vacuum should be approximately 100 mbar. Check the quality of the vacuum before introducing the slurry by closing the top of the packing reservoir.
2. Maintain the vacuum constant over the suction time (5 to 45 min, depending on the particle size and the adsorbent quality). Immediately after the top of the column bed falls to dryness disconnect the vacuum line.
3. Remove the packing reservoir, close the column with the top flange and connect the outlet line to a solvent container. Compress the column with the hydraulic to the appropriate compression and secure the piston from moving back by means of the safety stop. The compression varies from 10 bar for polymeric materials up to 80 bar for very rigid silica adsorbents.
4. Place the column in the system; do not connect it to any subsequent unit yet to avoid damaging it by particles that may come out of the column.
5. Start your preparative system with a mobile phase of known viscosity, e.g. methanol–water (80:20%) for RP systems, or isopropanol for normal phase systems. Use a linear flow-rate at least as high as the flow-rate later used in operation of the column. The pressure drop should be slightly higher than the operational pressure drop to be used. Let the system run for a short time
6. Stop the pump, recompress the column to the compression pressure and readjust the safety stop.
7. Repeat steps 5 and 6 up to three to four times. During the last repetition there should be no further movement of the safety stop when compressed.
8. Test the column according to the test procedure.

Column packing – (circle suspension) flow packing technology

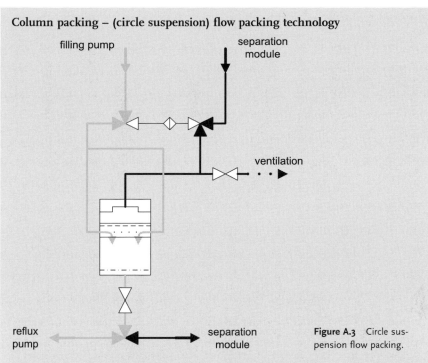

Figure A.3 Circle suspension flow packing.

1. Close the column outlet line after wetting the bottom flange and leave 1–2 cm of solvent on top of the bottom frit.
2. Pour the slurry slowly down the inside of the column to prevent air entrapment. Allow the suspension to stand until there is a 2–5 cm headspace above the settling resin. Rinse any resin particles from the inner wall of the column into the slurry and add fresh solvent to fill up the column. Place the column adapter on top and tighten to create a seal. With the inlet tube open, lower the top adapter until it is below the surface of the liquid. **Make sure that the column tube and entire length of the inlet tubing is full of liquid and free of air.**
3. Place the end of the suction tube of the pump in a large container of packing solvent and the outlet tube to waste. Simultaneously open the outlet tube and turn on the pump and pack the column at a starting flow rate of approx. $1/3$ of the final packing velocity.
4. Slowly ramp up to the target flow rate. This prevents hydraulic shock to the forming bed and therefore prevents uneven packing of the column. Adjust the pump speed to maintain a constant pressure drop over the resin bed. The operating pressure must never exceed the pressure limit of the column. Continue to flow pack at constant pressure until the resin bed stabilizes.
5. Turn off the pump 5 min after a stable bed has been achieved. Close the bottom valve and switch the inlet valve to waste. Expect the resin bed to rebound higher than the packed bed height as soon as the flow is stopped. Lower the adapter gradually to eliminate the headspace and compress the resin to the final bed height.
6. Switch the inlet valve to pump, open the bottom valve, and allow the column to run in down flow mode for 30 min at the previous packing pressure.

Differences for circle suspension flow packing

The difference between flow packing and circle suspension flow packing is the introduction of the slurry suspension, which is done in the case of circle suspension flow packing by lateral openings in the upper part of the otherwise closed column. The slurry suspension is pumped from a stirred tank into the column, which is filled with slurry solvent by means of a slurry pump and the column bed is built up over the outlet frit. The slurry solvent is recovered in the slurry suspension tank and circulated for a certain time. After introducing all adsorbent continue following the normal flow packing modes.

Appendix B: Data of Test Systems

B.1 EMD53986

EMD53986 [5-(1,2,3,4-tetra-hydroquinolin-6-yl)-6-methyl-3,6-dihydro-1,3,4-thiadiazin-2-one, Fig. B.1] is a chiral precursor for a pharmaceutical reagent. After chemical synthesis, it is present as a racemic mixture of the R- and S-enantiomers. The target component for the separation is the R-enantiomer.

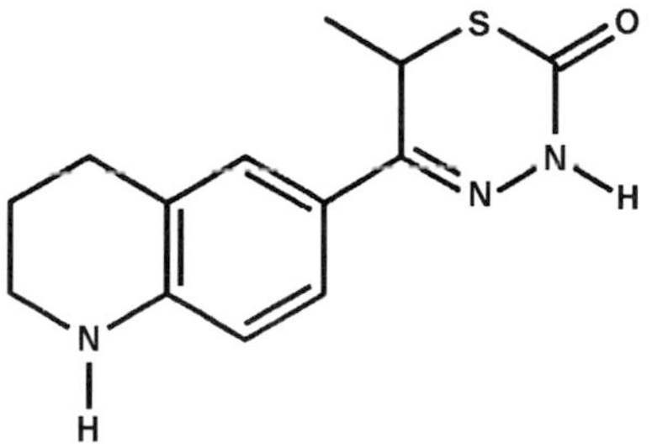

Figure B.1 Chemical structure of EMD53986.

Given below are the separation conditions (Table B.1) and typical model parameters (Tables B.2 and B.3) used in this book (Jupke, 2004 and Epping 2005).

Eluent: ethanol (gradient grade), LiChrosolv®, 99–100%, Merck (Darmstadt)
Adsorbent: Chiralpak AD, Daicel (Japan)
                 (Amylose-tris[3,5-dimethylphenylcarbamet]-phase)
Temperature: 25 °C

Preparative Chromatography. Edited by H. Schmidt-Traub
Copyright © 2005 WILEY-VCH Verlag GmbH & Co. KGaA, Weinheim
ISBN: 3-527-30643-9

Table B.1 Typical design and model parameters for EMD53986.

Design parameter	Value
Particle diameter	$20\,\mu m$
Column diameter (d_c)	2.5 cm
Column length (L_c)	11.5 cm
Division (for SMB)	2/2/2/2

Model parameter	Value
Void fraction (ε)	0.355
Total porosity (ε_t)	0.72
Dispersion coefficient (D_{ax})	Eq. 6.171 (Chapter 6.5.6.2)
Viscosity (η_l)	$0.0119\,\mathrm{g\,cm^{-1}\,s^{-1}}$
Density (ϱ_l)	$0.799\,\mathrm{g\,cm^{-3}}$
Effective transfer coefficient (k_{eff}) R-enantiomer	$1.50 \times 10^{-4}\,\mathrm{cm\,s^{-1}}$
Effective transfer coefficient (k_{eff}) S-enantiomer	$2.00 \times 10^{-5}\,\mathrm{cm\,s^{-1}}$

Table B.2 Additional parameters for model validation batch column (EMD53986).

Typical plant parameters (batch experiments)	Value
Pipe volume (V_{pipe})	$12.3\,\mathrm{cm^3}$
Dispersion coefficient pipe $(D_{ax,pipe})$	approx. $3000\,\mathrm{cm^2\,s^{-1}}$
Pipe diameter (d_{pipe})	0.05 cm
Tank volume (V_{tank})	$3.1\,\mathrm{cm^3}$

Table B.3 Parameters for model validation SMB (EMD53986).

SMB experiments (Fig. 6.42)	Unit	Value
Shifting time (t_{shift})	s	431
Asynchronus shift time	s	40.4
Feed flow rate	$\mathrm{ml\,min^{-1}}$	5.54
Desorbent flow rate	$\mathrm{ml\,min^{-1}}$	57.5
Extract flow rate	$\mathrm{ml\,min^{-1}}$	47.97
Raffinate flow rate	$\mathrm{ml\,min^{-1}}$	15.06
Recycle flow rate	$\mathrm{ml\,min^{-1}}$	72.0
m_I	–	30.14
m_{II}	–	8.34
m_{III}	–	10.86
m_{IV}	–	4.02

HETP correlation:

R-enantiomer: $\text{HETP}_R = 5.03 \times 10^{-3} + 1.9u_{\text{int}}$ (B.1)

S-enantiomer: $\text{HETP}_S = 1.26 \times 10^{-2} + 1.63u_{\text{int}}$ (B.2)

Isotherms for 25 °C (Chapter 6.5.7.5)

R-enantiomer: $q_R = 2.054c_R + \dfrac{5.847c_R}{1 + 0.129c_R + 0.472c_S}$

S-enantiomer: $q_S = 2.054c_S + \dfrac{19.902c_S}{1 + 0.129c_R + 0.472c_S}$ (B.3)

Alternatively, the isotherm can also be expressed by a multi-Langmuir equation:

R-enantiomer: $q_R = \dfrac{7.859c_R}{1 + 0.29543c_R + 0.08293c_S}$

S-enantiomer: $q_S = \dfrac{19.75c_S}{1 + 0.29543c_R + 0.08293c_S}$ (B.4)

Pressure drop correlations:

$$\Delta p = 1233.4 \frac{u_{\text{int}}\, \eta_l\, L_c}{d_{\text{p}}^2}$$ (B.5)

B.2 Tröger's Base

Tröger's base (Fig. B.2) was the first example of a chromatographic separation of an enantiomeric mixture. Since then it has been the subject of several publications.

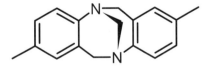

Figure B.2   Chemical structure of Tröger's base.

Given below are the separation conditions and typical model parameters used in this book (Tables B.4 and B.5) (Jupke, 2004, Mihlbachler et al. 2001 and 2002).

Eluent: isopropanol (HPLC grade), Fisher Scientific (Pittsburgh, USA)
Adsorbent: Chiralpak AD, Daicel (Japan)
                 amylose-tris[3,5-dimethylphenylcarbamet]-phase
Solute: Tröger's base, Aldrich (Milwaukee, USA)
Temperature: 25 °C

Isotherms from breakthrough experiments (Chapter 6.5.7.8):

$$R\text{-enantiomer: } q_R = \frac{54c_R(0.035 + 0.0046c_S)}{1 + 0.035c_R + 0.062c_S + 0.0046c_Rc_S + 0.0052c_S^2}$$

$$S\text{-enantiomer: } q_S = \frac{54c_S(0.062 + 0.0046c_R + 20.0052c_S)}{1 + 0.035c_R + 0.062c_S + 0.0046c_Rc_S + 0.0052c_S^2}$$

(B.6)

Isotherms from perturbation experiments (Chapter 6.5.7.8) are

$$R\text{-enantiomer: } q_R = \frac{0.0311\,c_R(54 + 0.732\,c_S)}{1 + 0.0311\,c_R} + \frac{0.732 \cdot 0.0365\,c_R\,c_S}{1 - 0.0365\,c_R}$$

$$S\text{-enantiomer: } q_S = \frac{27\,c_S(0.1269 + 2 \cdot 0.0153\,c_S)}{1 + 0.1269\,c_S + 0.0153\,c_S^2}$$

(B.7)

Table B.4 Typical design and model parameters for Tröger's base.

Design parameter	Value
Particle diameter	20 µm
Column diameter (d_c)	1 cm
Column length (L_c)	10 cm
Division (for SMB)	2/2/2/2
Model parameter	**Value**
Void fraction (ε)	0.355
Total porosity (ε_t)	0.648
Dispersion coefficient (D_{ax})	Eq. 6.171 (Chapter 6.5.6.2)
Viscosity (η_l)	0.012 g cm^{-1} s^{-1}
Density (ϱ_l)	0.785 g cm^{-3}
Effective transfer coefficient (k_{eff}) R-enantiomer	5 × 10^{-4} cm s^{-1}
Effective transfer coefficient (k_{eff}) S-enantiomer	4.5 × 10^{-5} cm s^{-1}

Table B.5 Parameters for model validation SMB (Tröger's base).

SMB experiment (Fig. 6.44)	Unit	Value
Shifting time (t_{shift})	s	962
Feed flow rate	ml min^{-1}	0.30
Desorbent flow rate	ml min^{-1}	0.46
Extract flow rate	ml min^{-1}	0.33
Raffinate flow rate	ml min^{-1}	0.43
Recycle flow rate	ml min^{-1}	1.05
m_I	–	4.24
m_{II}	–	2.50
m_{III}	–	3.43
m_{IV}	–	1.70

B.3 Glucose and Fructose

Glucose (Fig. B.3) and fructose (Fig. B.4) are monosaccharides. Separation of their isomeric mixture is of industrial importance to produce sugar syrups or for further synthesis of sorbitol, gluconic acid and vitamin C. Due to higher sweetness, fructose is often used as an alternative sweetener for sucrose.

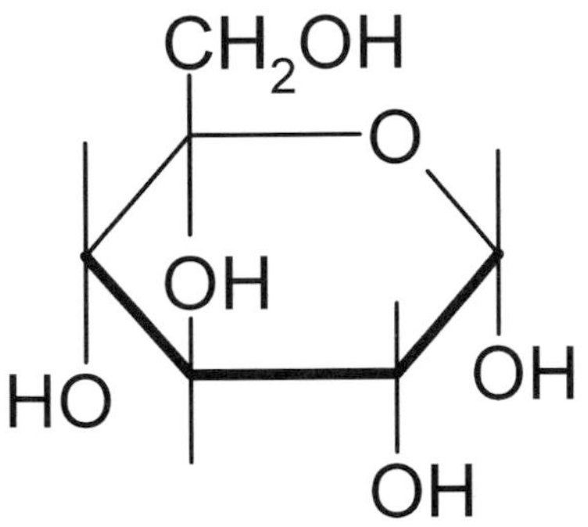

Figure B.3   Chemical structure of glucose.

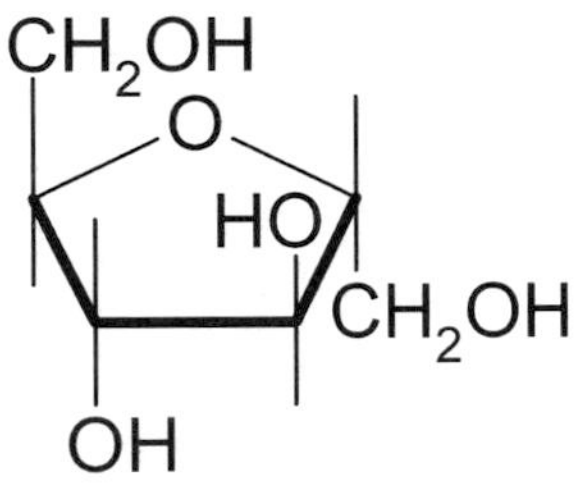

Figure B.4   Chemical structure of fructose.

Separation conditions and typical model parameter used in this book (Tables B.6 and B.7) (Jupke, 2004) are given below.

Eluent:	deionised and microfiltered water
Adsorbent:	ion exchange resin Amberlite CR 1320 Ca (Rohm u. Haas, Frankfurt)
Solute:	glucose, fructose (pure, Merck, Darmstadt)
Temperature:	60 °C

For chromatographic reactors additionally:

Catalyst:	immobilised glucose invertase Sweetzyme IT (Novozymes, Mainz)

Table B.6 Typical design and model parameters for fructose–glucose.

Design parameter	Value
Particle diameter	325 μm
Column diameter (d_c)	2.6 cm
Column length (L_c)	approx. 56.5 cm [a]
Division (for SMB)	2/2/2/2

Model parameter	Value
Void fraction (ε)	approx. 0.36
Total porosity (ε_t)	approx. 0.36
Dispersion coefficient (D_{ax})	Eq. 6.171 (Chapter 6.5.6.2)
Viscosity (η_l)	$4.7 \times 10^{-3}\,\mathrm{g\,cm^{-1}\,s^{-1}}$
Density (ϱ_l)	$0.983\,\mathrm{g\,cm^{-3}}$
Effective transfer coefficient (k_{eff}) glucose	$7 \times 10^{-4}\,\mathrm{cm\,s^{-1}}$ $5.7 \times 10^{-5}\,\mathrm{cm\,s^{-1}}$
Effective transfer coefficient (k_{eff}) fructose	$6.5 \times 10^{-5}\,\mathrm{cm\,s^{-1}}$ $8.9 \times 10^{-5}\,\mathrm{cm\,s^{-1}}$
Equilibrium constant (K_{eq})	1.079
Reaction rate constant (k_{reac})	$1.427 \times 10^{-3}\,\mathrm{s^{-1}}$

[a] Due to slightly different adsorbent amounts and packing compressions for different experiments.

Table B.7 Parameters for model validation SMB (fructose–glucose).

	Unit	Value	
SMB experiment		Fig. 6.45	Fig. 6.46
Shifting time (t_{shift})	s	1590	1590
Asynchronus shift time	s	170.6	171.8
Feed flow rate	ml min^{-1}	0.56	0.56
Desorbent flow rate	ml min^{-1}	4.24	4.24
Extract flow rate	ml min^{-1}	3.25	3.35
Raffinate flow rate	ml min^{-1}	1.55	1.45
Recycle flow rate	ml min^{-1}	10.0	10.0
m_I	–	0.82	0.82
m_{II}	–	0.37	0.36
m_{III}	–	0.45	0.43
m_{IV}	–	0.23	0.23

HETP correlations for 60 °C

glucose: $\mathrm{HETP}_{glu} = 0.1188 + 8.997 u_{int}$ (B.8)

fructose: $\mathrm{HETP}_{fru} = 0.1163 + 25.155 u_{int}$ (B.9)

Isotherms for 60 °C: (Section 6.5.7.5):

$$q_{glu} = 0.27 c_{glu} + 0.000122 c_{glu}^2 + 0.103 c_{glu} c_{fru}$$
$$q_{fru} = 0.47 c_{fru} + 0.000119 c_{fru}^2 + 0.248 c_{glu} c_{fru}$$

(B.10)

B.4 β-Phenethyl Acetate

β-Phenethyl acetate (Fig. B.5) is used for the production of scents and perfumes.

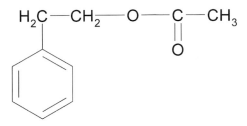

Figure B.5 Chemical structure of β-phenethyl acetate.

Separation conditions and model parameter used in this book (Table B.8) are given below.

Eluent: 1,4-dioxan
Adsorbent and catalyst: ion-exchange resin Amberlite 15 (Rohm u. Haas)
Solute: acetic acid and β-phenethyl alcohol
Temperature: 85 °C

Isotherms for 85 °C:

$$q_{\text{alcohol}} = \frac{0.884 c_{\text{alcohol}}}{1 + 24.16 c_{\text{alcohol}}}$$

$$q_{\text{acid}} = 0.687 c_{\text{acid}}$$

$$q_{\text{ester}} = 0.479 c_{\text{ester}}$$

$$q_{\text{water}} = \frac{11.126 c_{\text{water}}}{1 + 568.54 c_{\text{water}}}$$

Table B.8 Typical design and model parameters for β-phenethyl acetate.

Design parameter	Value
Particle diameter	590 μm
Column diameter (d_c)	1 cm
Column length (L_c)	30 cm

Model parameter	Value
Total porosity (ε_t)	0.36
Dispersion coefficient (D_{ax})	Eq. 6.172 (Chapter 6.5.6.2)
Viscosity (η_l)	$5.8 \times 10^{-3}\,\text{g cm}^{-1}\,\text{s}^{-1}$
Density (ϱ_l)	$1.03\,\text{g cm}^{-3}$
Effective transfer coefficient (k_{eff}) alcohol	$4.39 \times 10^{-4}\,\text{cm s}^{-1}$
Effective transfer coefficient (k_{eff}) acid	$4.82 \times 10^{-4}\,\text{cm s}^{-1}$
Effective transfer coefficient (k_{eff}) ester	$4.17 \times 10^{-4}\,\text{cm s}^{-1}$
Effective transfer coefficient (k_{eff}) water	$1.11 \times 10^{-3}\,\text{cm s}^{-1}$
Equilibrium constant (K_{eq})	3.18
Reaction rate constant (k_{reac})	$0.165326\,\text{cm}^3\,\text{s}^{-1}\,\text{g}^{-1}$

Index

Preparative Chromatography. Edited by H. Schmidt-Traub
Copyright © 2005 WILEY-VCH Verlag GmbH & Co. KGaA, Weinheim
ISBN: 3-527-30643-9